T0293175

Geographic Information, Geospatial Technologies and Spatial Data Science for Health

Geographic information, spatial analysis and geospatial technologies play an important role in understanding changes in planetary health and in defining the drivers contributing to different health outcomes both locally and globally. Patterns influencing health outcomes and disease in the environment are complex and require an understanding of the ecology of the disease and how these interact in space and time. Knowing where and when diseases are prevalent, who is affected and what may be driving these outcomes is important for determining how to respond. In reality, we all would like to be healthy and live in healthy places.

In this book, epidemiology and public health are integrated with spatial data science to examine health issues in dynamically changing environments. This is too broad a field to be completely covered in one book, and so, it has been necessary to be selective with the topics, methods and examples used to avoid overwhelming introductory readers while at the same time providing sufficient depth for geospatial experts interested in health and for health professionals interested in integrating geospatial elements for conducting health analysis. A variety of geographic information (some novel, some volunteered, some authoritative, some big and messy) is used with a mix of methods consisting of spatial analysis, data science and spatial statistics to better understand health risks and disease outcomes.

Key Features:

- Makes spatial data science accessible to health professionals.
- Integrates epidemiology and disease ecology with spatial data science.
- Integrates theoretical geographic information science concepts.
- Provides practical and applied approaches for examining and exploring health and disease risk.
- Provides spatial data science skill development ranging from map making to spatial modelling.

Justine Blanford is a professor of GeoHealth at ITC, University of Twente. She addresses a variety of local and global health challenges across different spatial and temporal scales. Her work is centred around three main facets that include (i) risk: understanding where and when health risks are, the mechanisms driving risk (why) and who may be affected; (ii) prevention: what response and actions are needed and where; and (iii) communication: what to communicate. She earned a PhD in Biology from Imperial College, UK; an MPhil from the University of Leicester, UK; and a BAH from Queen's University, Canada. She learned her GIS skills at the Centre of Geographic Sciences (COGS), Canada.

Geographic Information, Geospatial Technologies and Spatial Data Science for Health

Justine Blanford

CRC Press
Taylor & Francis Group
Boca Raton London New York

CRC Press is an imprint of the
Taylor & Francis Group, an **informa** business

A CHAPMAN & HALL BOOK

Front cover image: © Justine Blanford

First edition published 2025
by CRC Press
2385 NW Executive Center Drive, Suite 320, Boca Raton FL 33431

and by CRC Press
4 Park Square, Milton Park, Abingdon, Oxon, OX14 4RN

CRC Press is an imprint of Taylor & Francis Group, LLC

Library of Congress Cataloging-in-Publication Data
Names: Blanford, Justine, author.
Title: Geographic information, geospatial technologies and spatial data science for health / Justine Blanford.
Description: First edition. | Boca Raton, FL : CRC Press, 2024. |
Includes bibliographical references and index.
Identifiers: LCCN 2023057703 (print) | LCCN 2023057704 (ebook) |
ISBN 9781032563565 (hbk) | ISBN 9781032563572 (pbk) |
ISBN 9781003435082 (ebk)
Subjects: LCSH: Medical geography–Data processing. | Medical
informatics–Geographic information systems. | Geospatial data–Health aspects.
Classification: LCC RA792 .B53 2024 (print) | LCC RA792 (ebook) |
DDC 610.285–dc23/eng/20240524
LC record available at https://lccn.loc.gov/2023057703
LC ebook record available at https://lccn.loc.gov/2023057704

ISBN: 978-1-032-56356-5 (hbk)
ISBN: 978-1-032-56357-2 (pbk)
ISBN: 978-1-003-43508-2 (ebk)

DOI: 10.1201/9781003435082

Typeset in Palatino
by codeMantra

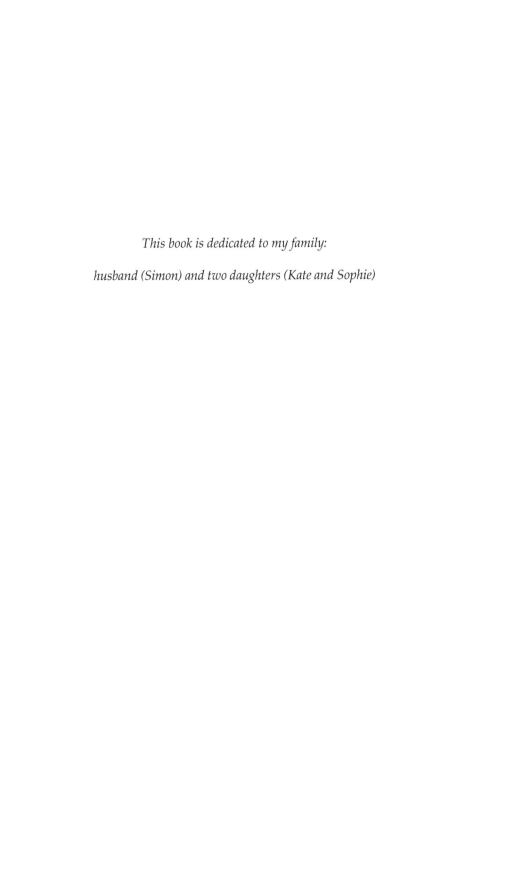

This book is dedicated to my family:

husband (Simon) and two daughters (Kate and Sophie)

Contents

Foreword

My journey into GeoHealth started at an early age with my own experience of being exposed to a variety of diseases (dengue and malaria; schistosomiasis and trypanosomiasis; mumps, cholera and tuberculosis). To stay healthy in an environment where one is constantly exposed to different diseases required vaccinations (smallpox, cholera, childhood vaccinations – pertussis, measles, mumps), regular deworming and the use of mosquito nets and anti-malaria medications, which at the time was still predominantly quinine. However, it wasn't until I started working on a project about plant diseases, in particular geminiviruses transmitted by whiteflies to economically important agricultural crops such as tomatoes while at CIAT (International Center for Tropical Agriculture, Colombia), that my interest in GeoHealth began. For my PhD, I moved onto another system that was also related to food security. This time with an emphasis on capturing host-pathogen-environment dynamics so as to better determine when and where to use a biopesticide against agricultural pests such as locusts and grasshoppers. This subsequently led to human vector-borne disease systems that included mosquito-vectored diseases such as malaria, dengue and West Nile virus, similar to biopesticides, the emphasis is on the disease ecology and capturing the host-pathogen-environment dynamics.

It is these experiences, along with my subsequent training and work, that have shaped my interest in health, GIS and the GISciences. Discovering the utility of GIS has married my formative experiences to an academic discipline ideally suited to answering complex questions about the world we live in and interact with. And now, if research was an early motivation, then teaching (and education) has become an equally important aspect of my work. I think you can all agree that having a good education and access to good educational materials opens us to new ideas and new possibilities, which is the premise of this book. It is necessary to enhance the use of geospatial information, methods and technologies for enabling a sustainable and healthy liveable planet so that we can contribute to achieving the sustainable development and planetary health.

And so in this book, I bring together multiple disciplines that include a biology perspective, geography, epidemiology and public health with the geospatial and spatial data sciences. To be able to address the many global challenges we face due to a loss of biodiversity, changing climates and unprecedented population growth, it is vital that we look at the challenges from many different perspectives, across different scales and within different contexts through the integration of different geographies. Geographic Information Systems (GIS) enables us to do this while also interacting with the data, making it possible to drill down into the information. It is quite surprising that although John Snow in 1854 used geovisualizations to better understand the cholera outbreak in London and identify the source of infection, GIS and spatial analysis are still largely absent from many public health and epidemiology curricula. Not only was he able to identify the source of infection, a novelty at the time but he used this information to take action and respond to the problem and reduce further infections.

Acknowledgements

This book is really a work that is the result of many collaborations, discussions, and insights gained from current and past colleagues, family, new and old friends, and feedback from the many current and past students who have passed through my classroom. In particular, I would like to thank my mum (Barbara), Hildi, my dad (George), sister (Sandra), David, and Patricia. I cannot say enough about the many places where I have lived and worked and the role these experiences have played in developing the insights that have enabled this work to be developed. It is through living in these places that I have gained a deeper understanding of the world around us from different perspectives.

Thank you, Bata, for enabling me to see, live, and experience the world. Thank you to the many institutes for providing a supportive and inspiring environment and the many colleagues for enriching discussions - CIAT (International Center for Tropical Agriculture), SEPA (Scottish Environment Protection Agency), Penn State University (Department of Geography, Dutton e-Education Institute and the GeoVISTA Center), and now ITC (Faculty of Geo-Information Science and Earth Observation, University of Twente). I am grateful to CRC Press for their support during the different phases of the book (inception to the final product) without which this would not have been possible. Lastly, a very big thanks to my husband for his continued support and for providing a listening ear.

1

Geographic Information and Geospatial Technologies: Applicability for Health and Disease

Overview

Are you at risk? We are all interested in staying healthy and minimizing our risk of getting sick. To avoid getting ill, we want to know not only *how* to avoid getting ill but also *where* a disease may be located and *when* it is present, if there is a temporal component to its prevalence, and *what* precautions we can take to stop getting sick. Knowing where, when, what, how and who is important for minimizing public health risks and developing intervention strategies.

The use of geography and maps for health-related studies is not new. In fact, maps have played a role in mapping where diseases are and are used for better understanding the drivers of disease and the spread of disease (Figure 1.1). In this first chapter, you will be introduced to a variety of maps that demonstrate how the integration of geographic information with health sciences are useful for improving our understanding of health and diseases risks:

> Geography and health are intrinsically linked....the air we breathe, the food we eat, [the water we drink] the viruses we are exposed to and the health services we can access.
>
> *(Dummer 2008)*

> Spatial location...plays a major role in shaping environmental risks as well as many other health effects
>
> *(Tunstall, Shaw and Dorling 2004)*

Where we live and the environment in which we live play a fundamental role in shaping our health (Chetty et al., 2016), be it through direct contact with our environment (e.g. altitude, climate, greenness, hazards and pollutants), our

DOI: 10.1201/9781003435082-1

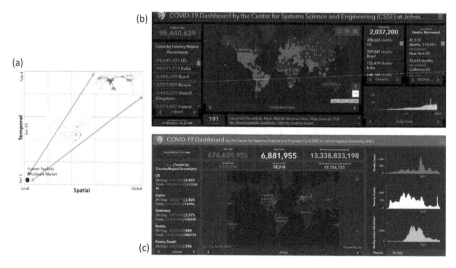

FIGURE 1.1

Distribution of COVID-19. (a) In 2020 COVID-19 went global in a matter of weeks during the winter of 2020 (Blanford and Jolly, 2021). This is a screen capture of the dashboard by the Johns Hopkins University (JHU, 2021) using software created by ESRI captured at (b) the start of the pandemic (2020) and (c) at the end of the pandemic (2023). Users were able to view the current situation of the pandemic globally or filter by country.

behaviour (e.g. social networks, risk behaviours and lifestyles) or economic conditions (e.g. access to healthcare and healthy environments, e.g., quality of food, physical activity spaces and quality of housing) (CSDGS_NRC_Ch6, 2010). Knowing how different factors affect our health and the health of our environment is important for determining how to respond and what interventions are needed (e.g. lockdowns, education, development of policies, infrastructure and service needs such as healthcare, electricity and/or access to clean water and sanitation). How to respond was well illustrated by the actions of John Snow when he removed the pump handle to reduce the cholera outbreak. But it was through the use of visualizations (in the form of a map) that provided the context (Figure 1.2).

John Snow, with his cholera map of 1854, demonstrated not only how geography and health are intrinsically linked but also the role of geovisualizations and the importance of integrating different types of information to provide context. It is this ability to integrate different types of information that is useful for enhancing our understanding of the risk factors that may be contributing to different health and disease outcomes. Since the early cholera map of 1854 along with improvements in digital technologies, disease mapping and spatial epidemiology have been greatly enhanced, making it easier to map and communicate disease prevalence, explore relationships and predict outcomes. Geographic information and geospatial technologies play a vital role in understanding the complex interactions between a host, the environment and a pathogen.

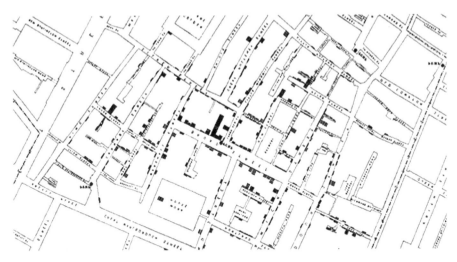

FIGURE 1.2
Map by John Snow of the cholera outbreak in Soho, London England in 1854 (Snow, 1855). In 1854, John Snow mapped the distribution of deaths by cholera to understand the aetiology of the disease and identify the potential source of infection, a water pump on Broad Street. Although geovisualizations are useful for understanding the spatial patterns of disease, it's ironic that geovisualization methods are still often missing from public health and epidemiological studies today. The image was obtained from Wikipedia Commons.

GeoHealth: Disease Mapping and Spatial Epidemiology

What Is GeoHealth?

Geospatial methods and information are not new for studying health and disease across time and place. You may already know GeoHealth by terms such as: *spatial epidemiology, landscape epidemiology, geographical epidemiology, health geographics, medical or health geography, geospatial health, digital health* and *planetary health*. Although how each of these is defined vary (see Table A.1 for an overview), in essence, they are about understanding the spatial variation of disease prevalence and the factors driving these patterns.

GeoHealth is about understanding *where* and *when* diseases/health risks occur in space and time, *why* they may be prevalent (**risk factors**), *what* the relationships are between different risk factors, *who* may be affected (*population* and *host(s)*) and *how* to respond (**intervention**). By bringing together different geographies (e.g. natural environment, population and settlements and economic development), we can add context which can then help improve our understanding of the factors influencing different health outcomes. Broadly, geohealth is the integration of data, specifically geographic information, the use of geospatial technologies and spatial methods with the ecology of disease, to study different health and disease risks and outcomes so that we can improve our understanding of the risk factors contributing to these outcomes. *GeoHealth is an integrative spatial data science approach to health. It is the integration of geographic information, technologies and spatial concepts with epidemiology.*

Being a GIScientist, I take a GIScience approach where I integrate the ecology of the disease to understand patterns of disease, some of which are approached from a theoretical perspective and others from a data-driven perspective, depending on what information and knowledge about the disease system is available. This will become clearer over the course of this book but particular in the chapter on vector-borne diseases. If we are going to tackle the health challenges of today and into the future, we need to take an integrated approach where we consider not only the prevalence of the disease but also what we can do about it (Figure 1.3) – how to respond and communicate. For response, I consider – (i) *diagnosis and treatment* to enable recovery from a disease or the management of long-term disease effects to (ii) *control & prevention* where policies and planning are designed to eliminate or reduce health risks. Lastly, communication involves informing the public and other health authorities about the current state of affairs about a disease.

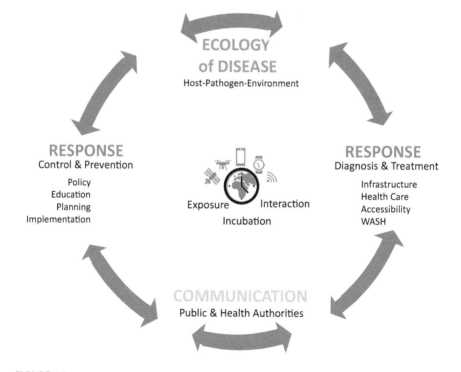

FIGURE 1.3

Integrated Health Approach. Ecology of Disease through monitoring, surveillance and analysis to improve our understanding of the ecological factors associated with the spatial distribution of a disease; recovery (provision and availability of public health services for timely diagnosis and treatment of the disease); and response (control and prevention strategies to minimize disease risk).

Geographic Information Systems, Spatial Analysis, Spatial Data Science Methods and Geospatial Technologies for Health and Disease

Geospatial technologies provide a rich suite of tools and analytical methods that enable us to understand the *where, when, why, who, what* and *how*.

A Geographic Information System (GIS), provide a system in which a variety of geographic (***where things are***) and non-geographic (***descriptive information about a person, place or thing***) information can be integrated WHILE ALSO providing the user with the ability to interact, query and analyse the data. A variety of GIS software are available today that is either proprietary software (e.g. ESRI ArcGIS Pro) or open source (e.g. QGIS).

Geospatial Competencies and Skills

Spatial data science is comprised of many parts that range from data acquisition, creation, and transformation to modelling and prediction to development of technology and applications. As such it requires a variety of skillsets (Figure 1.4) and involves the use of different methods, platforms, software and technological environments (e.g. desktop, virtual or cloud computing environments; see geospatial ecosystems (GeoSpatialSense, 2024)). These are well captured in the Geospatial Competency Model (DiBiase et al., 2010) and the Body of Knowledge (BOK) (UCGIS, 2023; eo4geo, 2024). The skills needed by geospatial professionals include:

- Data and Data Engineering: Data acquisition, management and integration
- Data Visualization and Exploration
- Spatial Analysis
- Ethics and Governance

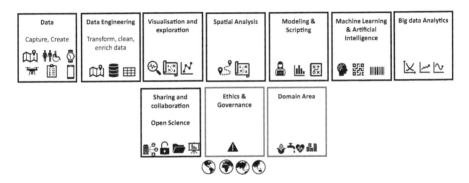

FIGURE 1.4
Overview of the skills needed by geospatial professionals. *Source*: Blanford (2023).

- Modelling and Scripting
- Application and Algorithm Development (mobile and web apps; new methods and tools)
- Machine Learning and Artificial Intelligence
- Big Data Analytics and Processing
- Application of GIScience to different domain areas such as health
- Open Science: Sharing and Collaboration

The evolution of GIS, the range of competencies and applications are well captured in studies such as Dangermond and Goodchild (2020) and Guo et al. (2020). However, a key part of GIS and Spatial Data Science is that spatial analysis can be used:

- to better understand mechanisms influencing health and well-being at different scales across space and time;
- to leverage diverse data sources to understand risk and changing risks;
- to determine how to effectively integrate the findings into a framework that enhances decision-making processes and provides sustainable interventions that minimizes future risks.

Spatial Analysis

Spatial analysis refers to the *"general ability to manipulate spatial data into different forms and extract additional meaning as a result"* (Bailey, 1994, p. 15) using a variety of methods or techniques *"requiring access to both the locations and the attributes of objects"* (Goodchild, 1992, p. 409).

A variety of spatial analysis functions (Cromley and McLafferty, 2012; de Smith et al., 2022) are useful for visualizing, exploring, analysing and predicting different health and disease outcomes. *"As such, spatial analysis covers a broad range of numerical methods"* (Cromley and McLafferty, 2012) that fall into six classes (Table 1.1). These range from spatial analysis methods centred around topological analysis, measurements, location and surface analysis to statistical and GeoAI methods. Statistical methods include spatial statistics and machine learning/artificial intelligence methodologies for analysing multiple dimensions (e.g. space-time; Franch-Pardo et al., 2020a; Purwanto et al., 2021), feature extraction (Mendes et al., 2021), and dealing with increasingly larger datasets of different velocities, volumes and validity (V's) (big data processing, analytics and uncertainties) (Li et al., 2016; Miller and Goodchild, 2015; Kwan, 2016).

In reality, one will likely conduct a variety of analyses using a range of different methods that may include statistics, spatial analysis, image analysis or data science methods (Figure 1.5) depending on the data that is available

TABLE 1.1

Spatial Analysis Functions Measurement, Topological Analysis, Network Analysis, Surface Analysis and Statistical Analysis

Function Class	Function	Description
Measurement	Distance, length, perimeter, area, centroid, buffering, volume, shape and measurement scale conversion	Used to calculate distance. This can be along straight-line distances between points, distances along paths, arcs or areas. Distance as a measure of separation in space is a key variable used in many kinds of spatial analysis and is often an important factor in interactions between people and places.
Topological analysis	Adjacency; polygon overlay; point-in-polygon, line-in-polygon, dissolve, merge, clip, erase, intersect, union, identity; spatial join and selection	Used to describe and analyse the spatial relationships among units of observation. Includes spatial database overlay and assessment of spatial relationships across databases including map comparison analysis. Topological analysis functions can identify features in the landscape that are adjacent or next to each other (contiguous). Topology is important in modelling connectivity and relationships in networks.
Network and location analysis	Connectivity, shortest path analysis, routing, service areas, location-allocation modelling and accessibility modelling	Used for investigating flows through a network. Network is modelled as a set of nodes and the links that connect the nodes. Links with graph theory and social network analyses.
Surface/Raster analysis	Slope, aspect, filtering, line-of-sight, viewsheds, contours, watersheds, surface overlays or multi-criteria decision analysis (MCDA)	Surface analysis is useful for analysing the real world through continuous surfaces and performs raster-based analyses. Often used to analyse terrain and other data that represent a continuous surface. Raster-based analyses can range from simple overlay analyses to more complex analyses involving complex mathematical operations that integrate and combine data layers (e.g. fuzzy logic, overlay and weighted overlay methods; dasymetric mapping and other geocomputational modelling). Filtering techniques include smoothing (remove noise from data to reveal broader trends) and edge enhancement (accentuate contrast and aids in the identification of features).

(Continued)

TABLE 1.1 *(Continued)*

Spatial Analysis Functions Measurement, Topological Analysis, Network Analysis, Surface Analysis and Statistical Analysis

Function Class	Function	Description
Statistical and mathematical analysis	Spatial sampling, spatial weights, exploratory data analysis, nearest neighbour analysis, global and local spatial autocorrelation, spatial interpolation, geostatistics and trend surface analysis	Spatial data analysis is influenced by spatial statistics and exploratory data analysis (Fotheringham and Rogerson, 1994; Bailey and Gatrell, 1995). These methods analyse information about the relationships being modelled based on attributes as well as their spatial relationship. Many of these methods take distance into account.
Machine learning and AI	Feature extraction Data mining and clustering of characteristics Big data processing and analytics	A set of data-driven algorithms and techniques used to automate predication, classification and clustering of data. Supervised ML methods (e.g, classifiers), unsupervised ML methods include clustering and semi-supervised methods such as label propagation (Safaei et al., 2021). Deep learning is a subset of machine learning techniques that uses artificial neural networks to learn from data. Machine learning and deep learning techniques are used for image classification to spatial pattern detection and multivariate predictions. Some uses may include: Feature extraction, object-based image analysis (OBIA) and natural language processing Pattern mining (include anomaly detection, outlier analysis, clustering and classification, trend analyses, change detection, missing data assessment and infilling) Big-data processing and analytics; agent-based modelling (e.g. identifying settlements; Mendes et al., 2021); determining Avian influenza disease reservoirs; Gulyaeva et al., 2020)

Sources: Cromley and McLafferty (2012); UCGIS (2023); ESRI (2023).

Data		
Structured Data / **Unstructured Data** / **Authoritative** / **Non-Authoritative**	**Spatial Data** • Image (UAV, Remote Sensing) • Polygon, Lines • Points (GPS, variety of sensors) • Digitised from maps • VGI/Citizen Science	**Non-Spatial Data** • Spreadsheet • Text documents • Video • Audio • Social Media • Citizen Science • Surveys (KAP)

Descriptive Analysis			
Statistics	**Spatial Analysis**	**Image Analysis**	**Data Science**
• Measures of Central Tendency (mean, median, mode); • Measures of variability: (standard deviation, minimum, maximum, skewness, kurtosis; • Data Distribution; • Standard Error of the mean	• Central Feature; • Mean Center (Weighted MC); • Standard Distance; • Standard Deviational Ellipse	• Spectral Band Assessment (Histogram; Scatter and density plots • class statistics (variance; separability: Euclidean Distance, Divergence, Transformed divergence, Jeffries- Matusita) • T-sne (t-distributed Stochastic Neighbor Embedding): dimensionality reduction technique • Standard PCA • Cloud cover • Filter noise	• Same as Statistics

Outliers

Visualisations of the Data (Interactive or Static)
(Maps and Graphs)

Questions, Hypothesis

Analysis			
Statistics	**Spatial Analysis**	**Image Analysis**	**Data Science**
• Difference between two means • Difference between two proportions • Contingency Analysis • Goodness of Fit Test • RMSE • Correlation Analysis • Regression Analysis • ANOVA • Kurskal-Wallis Test • Ratio • Summary Statistics	• Spatial Functions • Point Pattern Analysis • Spatial Autocorrelation • Cluster Analysis • Interpolations • Regression Analysis • Multi-criteria Decision Analysis CAUTION • MAUP (Modifiable Areal Unit Problem) • Edge Effects • Spatial Autocorrelation	• Classification, Enhancement, • Feature Detection, Feature Extraction • Segmentation, Fusion, Compression • Supervised Image Classification • Unsupervised Image Classification • Feature Extraction • Change Detection	• Classification • Clustering (supervised / unsupervised) • Regression • Natural Language Processing • Analytics • Machine Learning / AI • Anomaly Detection • Association Rule Learning • Outlier Analysis • Sequential Pattern Mining • Genetic Algorithms

Findings

Communication / Share Information	
• Cartographic Maps • Data visualisations	• Interactive & statistic visualisations and maps • Animations
Webmaps, dashboards	

FIGURE 1.5

Overview of data inputs and the different types of statistical, spatial, image and data science methods that may be needed to conduct an analysis. Adapted from GeoSpatialSense (2023).

and what you are trying to achieve. As data changes and new methods are developed, the methods that you will use will likely also change and expand. However, to help put these into perspective, I have provided an overview of the different types of analyses that you may need to think about. For example, the point pattern analysis methods are useful for analysing the cholera outbreak of 1854, and the Moran's I methods are useful for identifying spatial clusters of significance (potential hotspots) or spatial outliers.

Mapping Health and Disease: Past to Present

Geospatial information and spatial analysis methods have been used extensively for large variety of health and disease studies, as captured in Table 1.2.

TABLE 1.2

Summary of How Geospatial Data and Methods Have Been Used for Health and Disease

---

Create/Transform Geographic Information

Various	Create geographic data to enable for the visualization of disease risk. Various methods have been used that include conversion of data, transformation of data, geocoding, georeferencing, spatial join, aggregation of data or projection of data.
Collect data	UAV, sensors (stationary and moving), satellite, indoor mapping and apps.
	Accidental data collection – autonomous and surveillance – SLAM – simultaneous localisation and mapping (SLAM) capability. SLAM technology involves the use of cameras and/or sensors that capture 3D measurements and stitch the resulting data together to generate a map; feature extraction

Visualization

Cartographic maps	Disease maps provide a rapid visual summary of complex geographic information and may identify subtle patterns in the data that are missed in tabular presentations.
Web-based mapping	Use of the web-mapping tools and dashboards to map disease location and allow for interaction with the data and attributes

Explore Spatially Explicit Relationships and Evaluate and Analyse Spatial Relationships

Integration of geographic information and exploration of relationships	Examine where transmissions are taking place in relation to different geographies and information (see cartographic maps, web-based mapping and spatial methods)
Correlation studies	Examine variations in disease incidence/risk in relation to different geographies
Cluster analysis	• Evaluate whether features are clustered, dispersed or random. • Useful for searching for unusual patterns

(Continued)

TABLE 1.2 (*Continued*)

Summary of How Geospatial Data and Methods Have Been Used for Health and Disease

Connectivity	• Physical connectivity • Social networks • Phylogeography • Mobility
Neighbourhood structure and composition	Structure and composition of the landscape surrounding focal sites are important for understanding heterogeneity
Health infrastructure, planning and accessibility	Useful for planning of health infrastructure needs (e.g. vaccination programs, availability of health care and accessibility to healthcare)
Spatiotemporal dynamics of disease	Evaluate dynamically changing risk and/or spatial relationships
GeoAI: machine learning and deep learning	• Useful for sifting through large quantities of data both historically and in real time to identify patterns, assess similarity and correlation • Used for syndromic surveillance, analysis of symptoms, sentiment and perceptions. • Feature extraction • Data extraction • Natural language processing (including sentiment analysis)

Modelling: Simple to Advanced Geocomputational Methods

Suitability mapping	Determine suitability of environment for disease vectors or pests of disease. Useful when data is limited. Parameter estimations are subjective in nature. • Multi-criteria decision analysis (MCDA) or decision science • Niche modelling
Spatially explicit models	• Spatial interpolation and smoothing methods: interpolation and smoothing methods applied to spatial epidemiology, are useful for improving estimation of risk across a surface by creating a continuous surface from sampled data points • Mathematical models
Spatial regression	• Spatial autoregressive models • Bayesian regression models • Geographically weighted regression (GWR)
Agent-based modelling	• Diffusion of patterns of disease

Decision Support Systems & Mobile Apps

• Exploit multiple technologies (geographical information systems, statistical and mathematical models and decision-support modules), multiple data sources and permit widespread dissemination of epidemiological data.
• Spatial simulation and geocomputation
• Dashboards and geovisualizations
• Health apps for monitoring and managing health and data surveillance
• Early warning

Source: Adapted from Blanford and Jolly (2021) and compiled from a variety of sources: Elliott and Wartenberg (2004); Ostfeld et al. (2005); Dummer (2008); Clements and Pfeiffer (2009); Auchincloss et al. (2012); Cromley and McLafferty (2012); Kirby et al. (2017); Rushton (2003); Mandl et al. (2004). Polio (Mendes et al., 2021); COVID-19 (Franch-Pardo et al., 2020b).

These can range from visualization, risk assessment, source identification, hotspot detection, tracking and forecasting to early warning (Brownstein et al., 2023).

Some Examples of Disease Maps and Apps

Maps are useful for viewing where and when something is happening, providing information and updates and adding context. As technologies evolve so do mapping capabilities and the complexity of information that can be integrated and displayed. They can be as simple as dot maps to maps that are the result of complex modelled outputs or interactive geovisualizations with multiple layers of information. Over the next few pages, I have pulled together a variety of examples of different disease maps, ranging from static maps to different interactive maps.

Where a Disease Is Present

Cholera 160 years later. Cholera is still very much a health issue today, as illustrated by the map in Figure 1.6 created using data summarized in Table A.2. The map shows the number of cases reported by country between May and July 2023. GIS is being used not only to map cholera at a country

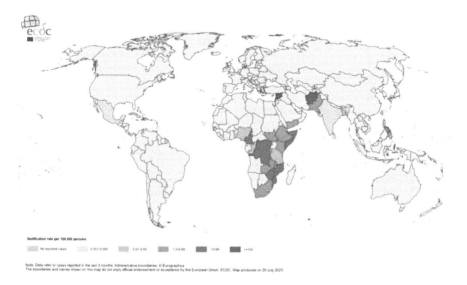

FIGURE 1.6
Geographical distribution of cholera cases reported worldwide from May to July 2023. Country level case numbers were extracted from the monthly report and summarized in Table A.2. *Source*: ECDC (2023).

level but also to assess where to deploy vaccinations in Kolkata (You et al., 2013) and track disease diffusion that may be the result of an accidental intro-duction (e.g. cholera introduced into Haiti in 2010 following the earthquake; Rubin et al., 2022; Keim et al., 2011; Chin et al., 2011).

> ECDC monitors cholera outbreaks globally through epidemic intelli-gence activities in order to identify significant changes in epidemiol-ogy and inform public health authorities. Reports are published on a monthly basis.
> - Since 21 June 2023 and as of 20 July 2023, 57,024 new cholera cases, including 399 new deaths, have been reported worldwide.
> - The five countries reporting the most cases are Afghanistan (23,298), the Democratic Republic of the Congo (8,469), Haiti (6,701), Ethiopia (5,974) and Cameroon (3,067).
> - The five countries reporting the most new deaths are Cameroon (95), the Democratic Republic of the Congo (74), Ethiopia (65), Haiti (42) and Zimbabwe (34). In addition, 62,615 new cases were reported or collected retrospectively from before 21 June 2023.

Vector-Borne Diseases: Malaria and Dengue

The Malaria Atlas Project Data Platform provides global maps of the distribu-tion of malaria from 2000 to 2020 (Figure 1.7). The platform provides a variety of tools and data so that you can explore and learn more about malaria and where it is distributed. Now that they have data spanning a number of years, you can also see how the distribution of malaria may be changing.

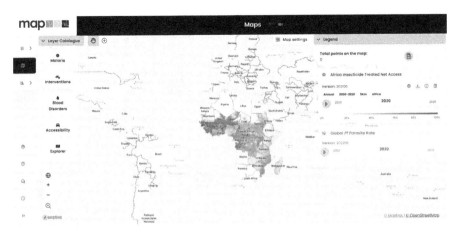

FIGURE 1.7
The Malaria Atlas Project maps the distribution of malaria around the world. *Source*: https://data.malariaatlas.org/maps.

ArboNET disease map was a National Arbovirus Surveillance System in the US that captured mosquito-vectored diseases present at a county level for the whole of the US (Figures 1.8 and 1.9). Unfortunately, this interactive map (Figure 1.8) is no longer available and was depreciated in 2023. The information is still available and is now presented differently (Figure 1.9).

FIGURE 1.8

Dengue cases recorded at the state and county level in the USA for 2023 (data are current as of September 13, 2023). *Source*: https://www.cdc.gov/fight-the-bite/at-risk/index.html; https://www.cdc.gov/dengue/statistics-maps/current-data.html. (a1 and a2) shows all reported cases; (b1 and b2) shows all locally acquired cases; and (c1 and c2) shows all travel-associated cases.

FIGURE 1.9
Mosquito-vectored diseases recorded at county level in the USA. Interactive map shows the now depreciated ArboNET. *Source*: https://www.cdc.gov/fight-the-bite/at-risk/index.html.

What I like about these maps is that they differentiate between locally acquired infections and imported or travel-associated infections. More and more geographic information on vector-borne diseases and their vectors are being captured and are being made available through interactive (e.g. https://atlas.ecdc.europa.eu/public/index.aspx) or static maps (e.g. https://www.ecdc.europa.eu/en/disease-vectors/surveillance-and-disease-data). More about vector-borne diseases in Chapter 6.

Where Diseases Are Circulating and the Role of Place in the Transmission and Diffusion of Diseases

The role of place and mobility in the transmission of some diseases may be important, as illustrated by the three examples in Figure 1.10. By mapping these data, it becomes clear what role 'place' has in the transmission of diseases. For example, where key transmission zones may be (Figure 1.10a), where the source of infection may be (Figure 1.10b) and where diseases may be circulating and diffusing (Figure 1.10c).

Pandemics and Outbreaks: Maps, Apps and Dashboards

COVID-19

In December 2019, a novel coronavirus, known as COVID-19, now named SARS-CoV-2 (Severe acute respiratory syndrome coronavirus 2), caused a series of acute atypical respiratory diseases in Wuhan, Hubei Province, China. This virus was highly transmittable between humans causing a global pandemic which resulted in substantial global morbidity and deaths.

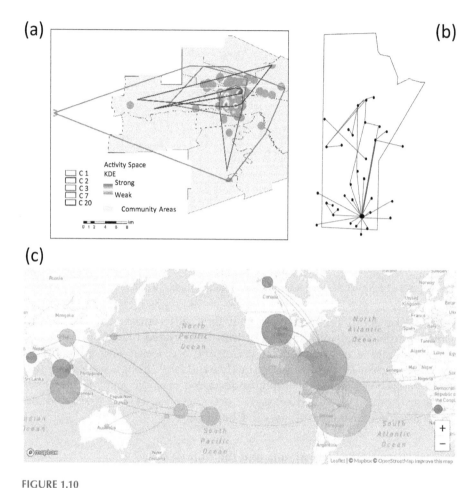

FIGURE 1.10

Examples of transmission and diffusion of diseases using different types of data and behaviour. (a) shows activity spaces of different social network groups and the geographic location of key "hang out" locations (Logan et al., 2016). (b) captures the diffusion of a gonorrhoea outbreak in Canada (simplified from Jolly and Wylie, 2013) and (c) shows the global distribution of Zika virus mapped by genetic variants. *Source*: Nextstrain https://nextstrain.org/zika, CC-BY-4.0 license (Hadfield et al., 2018).

The World Health Organization declared it a pandemic on March 11th, 2020 (see timeline of events in Blanford et al., 2022), leading governments to turn to non-pharmaceutical interventions to reduce the spread of the virus and lessen the burden on local facilities (e.g. health care systems (Miller et al., 2020) and deathcare management (Zavattaro, 2023)). In 2020 the COVID-19 pandemic affected populations around the world. By the end of 2021, more than 290 million people were infected, and 5.5 million people lost their lives. In response, several vaccines were developed in 2021 which led to many countries developing mass vaccination campaigns (He et al., 2022) using a variety of vaccination strategies (e.g. for the Netherlands see (Al-Huraibi et al., 2023)). During the COVID-19 pandemic of 2020–2023, a large number of dashboards were created that monitored the distribution of COVID-19 cases and mortality (Figures 1.11–1.15). I have pulled together a few examples of different dashboards that were regularly viewed in our house. These provided regular updates on the total number of people infected with COVID-19. These were created by various health organizations such as the World Health Organization (WHO), Rijksinstituut voor Volksgezondheid en Milieu (RIVM), European Centre for Disease Prevention and Control (ECDC) and Johns Hopkins University (Figures 1.11–1.15).

Many of the COVID-19 case maps were interactive and updated weekly (*Source*: https://www.rivm.nl/en/novel-coronavirus-covid-19/current-information). Over time we moved from clinical reporting of COVID-19 cases to

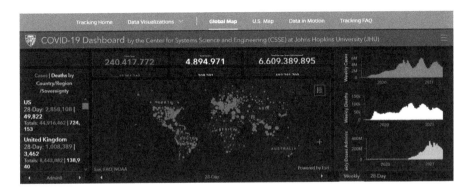

FIGURE 1.11
Johns Hopkins created an interactive coronavirus dashboard at the start of the pandemic showing the total number of infected people and the total number of deaths by COVID-19 around the world. Over time, this dashboard has continued to evolve to include the number of people who recovered and vaccination rates (see Figure 1.1). *Source*: Dong et al. (2020); JHU (2021). Map available at https://coronavirus.jhu.edu/map.html .

FIGURE 1.12
The World Health Organization has a variety of dashboards and continues to monitor the number of positive COVID-19 cases around the world. Summaries are provided at a country level. *Source*: https://covid19.who.int/.

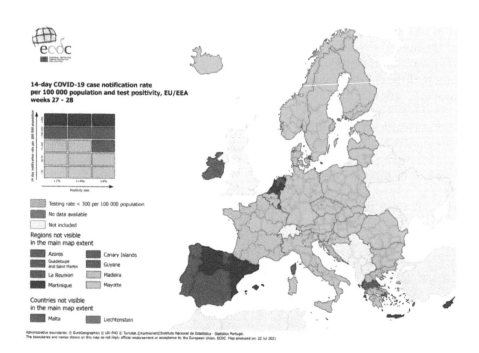

14-day COVID-19 case notification rate per 100 000 population and test positivity, EU/EEA weeks 27 - 28

Colour	Classification and How Defined	General Meaning of Classification
Green	• If the notification rate is less than 25 and the test positivity rate is less than 4% • If the 14-day notification rate is less than 50 and the test positivity rate is less than 4% • If the 14-day notification rate is less than 75 and the test positivity rate less than 1%	**No restrictions**, such as quarantine or testing, on travellers coming from 'green' regions provided have EUDCC certifications
Orange	• If the notification rate is less than 50 but the test positivity rate is 4% or more, or, if the notification rate ranges from 25 to 150 but the test positivity rate is less than 4% • If the 14-day notification rate is less than 50 and the test positivity rate is 4% or more • If the 14-day notification rate is 50 or more and less than 75 and the test positivity rate is 1% or more • If the 14-day notification rate is between 75 and 200 and the test positivity rate is less than 4%	Required to undergo quarantine/ self-quarantine; and/or take a test for COVID-19 infection before or after arrival. Up to member states to decide what measures to apply
Red	• If the notification rate is 50 or more and the test positivity rate is 4% or more, or if the notification rate is more than 150	**Discourage all** such travel to and from 'red' areas

(Continued)

Colour	Classification and How Defined	General Meaning of Classification
Red	• If the 14-day cumulative COVID-19 case notification rate ranges from 75 to 200 and the test positivity rate of tests for COVID-19 infection is 4% or more • If the 14-day cumulative COVID-19 case notification rate is more than 200 but less than 500	
Dark Red	• If the notification rate is 500 • If the 14-day cumulative COVID-19 case notification rate is 500 or more	Strongly discourage all non-essential travel to and from 'dark red' 'Dark red' areas, all member states should require persons travelling from such an area to do a pre-departure test and undergo quarantine/self-isolation. This should also apply to essential travellers provided it does not impact on the exercise function or need
Grey	• If there is insufficient information or if the testing rate is lower than 300 cases per 100,000	

FIGURE 1.13 (*Continued*)
The European Centre for Disease Prevetion and Control (ECDC) created regular risk maps for all European Countries (a) Map illustrating the traffic-light risk classification approach used for reporting weekly COVID-19 risk levels for each country in Europe and (b) the definition of each risk classification and suggested travel measure. *Source*: ECDC (2021); EuropeanCommission (2021).

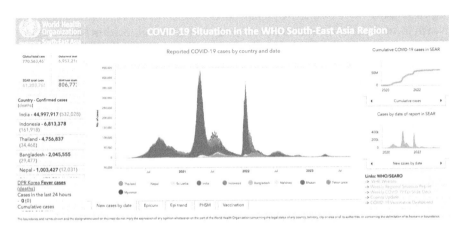

FIGURE 1.14
Coronavirus dashboard for South-East Asia provides an overview of COVID-19 cases by country. *Source*: https://experience.arcgis.com/experience/56d2642cb379485ebf78371e744b8c6a.

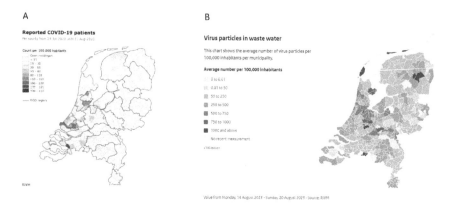

FIGURE 1.15
For the Netherlands, COVID-19 cases were monitored using (a) case data and (b) environmental data. (a) During the COVID-19 pandemic, the number of positively infected COVID-19 patients were reported at the municipality level using 2 weeks of information. *Source*: RIVM (2021). (b) Patient case maps were replaced by wastewater reports of the concentration of virus particles in waste water. Maps were updated weekly. *Source*: RIVM. https://www.rivm.nl/en/ coronavirus-covid-19/weekly-figures.

environmental monitoring of COVID-19. Wastewater information can be found at https://coronadashboard.government.nl/landelijk/rioolwater and the interactive map can be viewed at https://www.rivm.nl/en/coronavirus-covid-19/ weekly-figures.

Other Health and Disease Alerts

HealthMap (Brownstein et al., 2023; Freifeld et al., 2008) collects and integrates disease information from a variety of sources that range from news media (e.g. Google News), expert-curated messages and discussions (e.g. ProMED-mail) to validated official alerts. Since the information comes in a variety of formats and structures, text-processing algorithms are used to classify alerts by geographic location and disease. Processed information is integrated and displayed on an interactive geographic map. The accuracy of the information is evaluated, and if needed, human curation is used to correct misclassifications or inaccuracies.

Natural and Anthropogenic Hazards

Not only is it important to highlight where diseases are but also what is causing the occurrence of these diseases. These can be due to changes in the environment such as an event or a natural human-made change. Events may be hazards, such as droughts, floods, hurricanes or landslides, for example. Human-made changes can result from many things such as clearing of natural vegetation, building and expansion of settlements, the release of chemicals into the environment that may affect water quality, air quality and soil quality. These are but a few examples. Regardless of the change or the event,

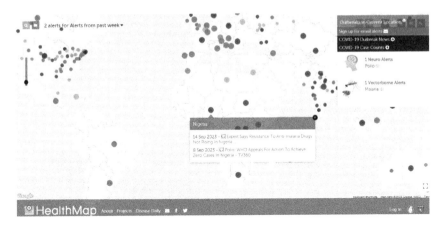

FIGURE 1.16

HealthMap captures current disease occurrences or alerts around the world. It is an interactive map that integrates information gathered from a variety of sources. *Source*: http://www. healthmap.org/en/.

each of these can affect the health and well-being of humans. More on hazards and the relationship of these to disease will be covered in the chapter on disease epidemiology, and more on pollution will be covered in the chapter on environment exposure. In the meantime, check out the Natural Hazards and Public Health Emergencies Geo Hub by the Pan-American Health Organization (PAHO) (Figure 1.17) and pollutants released into the environment through the air (e.g. Figure 1.18).

FIGURE 1.17

Health and emergencies dashboard by PAHO to monitor emergencies. *Source*: https://paho-health-emergencies-who.hub.arcgis.com/.

Concentrations of fine particulate matter (PM2.5)

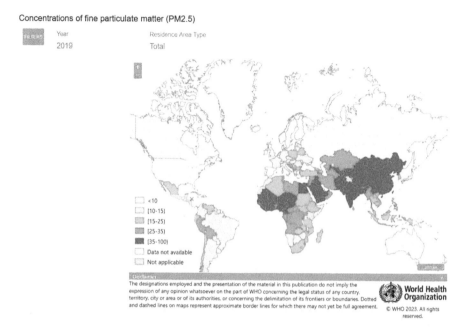

FIGURE 1.18
Air pollution concentrations around the world. *Source*: https://www.who.int/data/gho/data/themes/air-pollution/ambient-air-pollution.

All of the examples are really about communication and informing the public and health officials where something is happening and, in some cases, when something is happening. In 1854, John Snow used a map (Figure 1.2) to identify where the potential source of the cholera infections may have been, the Broad Street water pump. Using this information he took action to try to reduce further cholera infections by removing the water pump handle and access to the contaminated water source. Maps are not only useful for assessing where diseases are but can also be helpful for developing strategies for reducing further infections, eliminating or eradicating a disease? Next, I provide an example of how maps and other geospatial technologies are being used to eradicate polio globally.

Eliminating Polio – https://polioeradication.org/

The Global Polio Eradication Initiative is a public-private partnership led by national governments with six core partners – the World Health Organization (WHO), Rotary International, the US Centers for Disease Control and Prevention (CDC), the United Nations Children's Fund

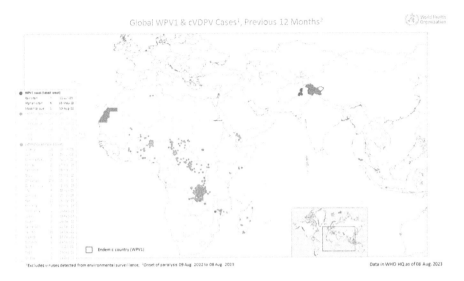

FIGURE 1.19
Distribution of Polio as of August 8, 2023. *Source*: https://polioeradication.org/polio-today/polio-now/.

(UNICEF), the Bill & Melinda Gates Foundation and Gavi, the Vaccine Alliance. The goal of this partnership is to eradicate polio worldwide.

As of 2023, polio remains endemic in two countries, Afghanistan and Pakistan (Figure 1.19). Until poliovirus transmission is interrupted in these countries, all countries remain at risk of the importation of polio, especially vulnerable countries with weak public health and immunization services and travel or trade links to endemic countries.

Getting to this point has taken an enormous amount of effort, which has required a digital and community-based approach to fill in data and knowledge gaps as well as track cases (e.g. along the Congo River; Kamadjeu, 2009) and vaccination uptake. A mixture of hand-drawn maps (Barau et al., 2014; Idris, 2020; Ajiri et al., 2021; el-Hamza, 2023) and digital technologies were used. In 2013, eHealth Africa and GIS technologies were implemented to locate settlements that were missed, improve vaccination coverage and guide local ward level focal persons on appropriate microplanning. An SMS-based mobile phone application was also implemented, AVADAR (Auto-Visual AFP Detection and Reporting System), to help identify cases (Idris, 2020). Vaccination teams were tracked during polio campaigns (Gammino et al., 2014) so as to ensure no one was missed (e.g. Northern Nigeria; Touray et al., 2016); mapathons and automated image feature extraction methods were used to fill in data gaps so as to aid in microplanning and strengthen vaccination campaigns (e.g. Mendes et al., 2021). As illustrated here, geospatial technologies and GIS is essential for making data actionable (e.g. responding to outbreaks in Lake Chad; Ajiri et al., 2021) and enabling capacity building through visualization (WHO_Africa, 2022).

Mobile Apps and Other Technologies for Health

During the 2020–2023 COVID-19 global pandemic, a proliferation of mobile apps and web-based dashboards were created (Visconti et al., 2022; Blanford et al., 2022; Guemes et al., 2021; Banskota et al., 2020; Pandit et al., 2022; Davalbhakta et al., 2020; Ming et al., 2020; Singh et al., 2020). In Europe alone, 40 apps were developed to manage and monitor COVID-19 (Blanford et al., 2022). The purpose of these digital tools ranged from active participatory surveillance (e.g. of symptoms globally ZOE – Bowyer et al., 2021; of symptoms in the US with the US Symptom app – Guemes et al., 2021); passive population-level tracking (e.g. outbreaks near me); individual risk assessment; or providing situational awareness by providing 24/7 updates (e.g. web-based dashboards keeping track of infected, recovered, vaccinated and those less fortunate who died); see (Pandit et al., 2022; Johns Hopkins University COVID-19 dashboard). As shown during the COVID-19 pandemic apps can be useful for providing information about health risks.

Imagery – Satellite/UAV Imagery to Medical Imaging

Remote sensing satellites provide continuous measurements of the earth and its environment. The availability of imagery has increased immensely in the past decades, making it much easier and more affordable to sense our environment not only through satellites and drones but also through medical imaging. In the next few paragraphs, I will provide a few examples of how imagery is being used for health-related studies.

Remotely sensed imagery of the earth's surface has been used in a variety of epidemiology and vector-borne disease studies (e.g. Hay et al., 2006; Kalluri et al., 2007; Rogers and Randolph, 2003; Hay et al., 2000). For example in the surveillance of vector-borne infectious diseases (Kalluri et al., 2007); aiding in vaccination campaigns by identifying settlements (Mendes et al., 2021); or determining Avian influenza disease reservoirs (Gulyaeva et al., 2020).

The Normalized Difference Vegetation Index (NDVI), an indicator of vegetation health, has been extensively used for monitoring the environment and used for determining the greenness of neighbourhoods (Reid et al., 2018) from which health-based assessments have been made. Monitoring air quality (Holloway et al., 2021) using satellites has been used for developing policies to improve air quality (Engel-Cox and Hoff, 2005). Thermal imaging and land surface temperatures have been useful for monitoring temperatures of the environment. With changing climates this type of information is of increasing importance for evaluating the potential effects heat has on the environment. This has been used for mapping local climate zones in Europe (Demuzere et al., 2019), monitoring temperatures across Europe (Hellings and Rienow, 2021), quantifying urban heat islands in European cities (Schwarz et al., 2011) and determining the effectiveness of mitigation strategies on heat island effects (Seeberg et al., 2022). Not only has this been useful for understanding heat-related effects in the built-up environment (e.g. urban heat island effects) but also in better understanding seasonal climate variations in urban areas (Qiu et al., 2019) and the variations in heat absorption capacities by different types of buildings and structures within an urban area (e.g. urban green spaces (Zeng et al., 2023)) and the role of different types of vegetation in absorbing and reflecting heat (Still et al., 2019).

These are but a few examples to highlight how remotely sensed information can be used in health-related studies. For more local studies, drones have also been used with a variety of sensors to capture information about the environment.

Global Health Challenges and Sustainable Development Goals (SDGs)

We are living in times of change and are currently facing many challenges that can affect the health of our planet and all living organisms, including us (KNAW, 2023; Folke et al., 2021; Almada et al., 2017; Iyer et al., 2021; Myers, 2017; Whitmee et al., 2015). We are currently facing a global water, food and energy crisis. By 2050, the number of people living in slums or informal settlements is expected to rise to 3 billion (Ross et al., 2021). These numbers are likely to increase with rising mobility and migration (IOM, 2015), leading to increasing health risk outcomes (Lilford et al., 2016).

Changes in the climate will further exacerbate health and socioeconomic disparities, resulting in an increase in susceptibility to diseases due to heat events and drought-induced water and food insecurities (Scovronick et al., 2015). Heat can affect cardiorespiratory diseases and result in increases in the number of deaths (Corburn and Sverdlik, 2018); temperature can affect the survival of bacterial pathogens (Corburn and Sverdlik, 2018); and climate change-induced weather can influence the adverse health impacts of aeroallergens and air pollution (Corburn and Sverdlik, 2018).

Air pollution is a leading contributor to mortality (Wallace-Wells, 2021; Egondi et al., 2016), with an estimated 4.3 million deaths occurring annually due to exposure to air pollution from solid fuels in households (Prüss-Ustün et al., 2016). Household air pollution is responsible for 33% of the total disease burden from respiratory infections (Prüss-Ustün et al., 2016), such as respiratory illness, asthma and acute respiratory infections (Egondi et al., 2013). With only 17% of the population having access to clean cooking and 2.5 million premature deaths annually from cooking smoke, there is a need for clean economical cooking solutions (EnvironBuzz, 2022). Knowing where these needs are will not only help in reducing air pollution but also reduce a reliance on wood and fossil fuels that continue to degrade the environment.

Interventions used to improve health outcomes are often hampered by a lack of reliable population- and place-based or geographic location exposure data (Corburn and Riley, 2016). Therefore, there is a need to improve the data we collect so that we can improve how we monitor health and disease (Corburn and Sverdlik, 2018; Thomson et al., 2020; Abascal et al., 2022) and how we access health services physically and digitally (e.g. mHealth; Lee et al., 2017).

Sustainable Development Goal (SDG) 3 (Good health and well-being) is *to ensure healthy lives and promote well-being for all at all ages.* The aim is to reduce global maternal mortality; end preventable deaths of newborns and children; end the epidemics of AIDS, tuberculosis, malaria and other communicable diseases; reduce mortality from non-communicable diseases; strengthen prevention and treatment of substance abuse; halve the number of deaths and injuries from road traffic accidents; ensure universal access to sexual and reproductive health-care services; achieve universal health coverage; and reduce the number of deaths and illnesses from hazardous chemicals and pollution.

SDG 3 is interwoven throughout the 2030 Agenda for Sustainable Development, linking to targets in other SDGs. These include access to clean indoor and outdoor energy (SDG 7); improving universal health access and health service to support economic growth (SDG 8), gender equity (SDG 5) and reducing poverty (SDG 1) and inequalities (SDG 10) (WHO, 2016); improved access to water and sanitation (SDG 6) and quality education (SDG 4) to help reduce disease transmission and support development of healthy sustainable cities and communities (SDG 11). Geographic Information and geospatial technologies play a central role in achieving the SDG's.

COVID-19 Pandemic 2020–2023

The impact of the COVID-19 pandemic has been disruptive and tragic for many. But the collective response to the pandemic, particularly from a GeoHealth point of view, has arguably been to shine a brighter light on the importance of geographic information and geospatial technologies and the role they play in disease monitoring, mapping, providing context, assessing relationships, impacts and interventions. The speed and enthusiasm with which the general public accepted and became adept at interpreting the numerous graphs, geographic outputs and summaries available in the media (e.g. Johns Hopkins dashboard – a daily view in our house) says something about how important it is to provide such information. It is also thanks to the geospatial industry for having such tools readily available to make it possible to put interactive web-based dashboards together in an instant.

The consequences of the pandemic were to drive people, by their own choice or government legislation, into isolation. If this had been a matter of a few days, the consequences of lockdowns might not have been highlighted. But the sheer length of travel bans and school closures and their repeated nature have forced us to re-evaluate many aspects of our health and well-being, as well as the health of our environment. Not only was the COVID-19 pandemic a health crisis, but it also has the potential to become one of the worst human and economic crises of our lifetime (WHO, 2023; see Chapter 9) from all of the post-pandemic effects, many of which are still unknown. As a result of the COVID-19 pandemic of 2020–2023, due to the disruptions in health services, many setbacks have occurred. Vaccination rates have fallen, and access to healthcare and medicines has resulted in an increase in disease prevalence (UN, 2023b; OECD, 2021).

The pandemic has shown us what changes are possible. Now more than ever, we need to build healthy spaces that are fit for the future. For this we need geospatial-technological relevant solutions.

Health in 2023 and Beyond

It is staggering that, in 2023, 400 million people had no basic healthcare, and more than 15 million people were waiting for HIV treatment (UN, 2023a). Each year, 829,000 people die from diarrhoea as a result of unsafe drinking water, sanitation and hand hygiene (UN, 2021). In 2022, 419 million people practiced open defecation; 2.2 billion people lacked access to safely managed drinking water, with 703 million without basic water service; and 2 billion lacking basic handwashing facilities with soap and water at home (UN, 2023b). Despite progress being made in the medical and health fields and in our societies, many households are being pushed into deprivation due to chronic and catastrophic diseases and the burdens associated with these.

Over the past four decades, a shift in the global disease burdens has occurred from infectious diseases, nutrition and communicable diseases to non-communicable diseases (NCDs) (see the Global Disease Burden Visualisation Hub (IHME, 2024)). The four main types of noncommunicable diseases are cardiovascular diseases (e.g. heart attacks and strokes), cancer, chronic respiratory diseases (e.g. chronic obstructed pulmonary disease and asthma) and diabetes (WHO, 2023). NCDs impose a large burden on human health worldwide. Sixty-three percent of all deaths worldwide stem from NCDs such as cardiovascular diseases, cancers, chronic respiratory diseases and diabetes. The cumulative economic losses from these four diseases are estimated to surpass US$ 7 trillion by 2025 (Fund, 2023).

The key global health priorities are summarized in Table 1.3.

TABLE 1.3

Summary of Key Global Health Priorities and Actions

Topic of Concern	Global Health Areas of Concern	Response and Actions
Population	• Demographic and socioeconomic statistics • Mortality and global health estimates	Crude birth and death rates
Infrastructure needs and policies	• Essential medicines (availability of generic medicines; consumer price ratio) • Universal health coverage (health systems strengthening, health financing, health equity monitor, noncommunicable diseases, global health workforce statistics and immunization coverage) • Governance and aid effectiveness • Health systems (health systems and financing, health service delivery, pharmaceuticals, drugs and supplies, health information system and health security) • Health workforce • Immunization (coverage estimates for BCG, diphtheria (DTP3), Hib3, HepB3, HepB, measles MCV1,2), neonates (tetanus – PAB), polio (Pol3), pneumococcal (PCV3), rotavirus (RotaC) and vaccination cards seen	• Priority health technologies (biomedical engineers and technicians, health infrastructure and medical devices) • Public health and environment (air pollution, radon, second-hand smoke, chemical safety, children environmental health, climate change, electromagnetic fields, occupational risk and UV, • Water, sanitation and hygiene • Resources for substance use disorders (governance, policy and financing; service organization and delivery; human resources; information systems; youth) • Tobacco control (policies for reducing demand; enhance prevention policies (taxes, advertising, health warnings, treatment solutions); air pollution – protection from smoke) • Road safety (road traffic mortality; national legislation; institutional framework; policy; post-crash response)

(Continued)

TABLE 1.3 (Continued)

Summary of Key Global Health Priorities and Actions

Topic of Concern	Global Health Areas of Concern	Response and Actions
Global surveillance	• Global dementia observatory • Global influenza virological surveillance • Global information system on alcohol and health • World health statistics	• International regulation monitoring framework • Health equity monitor • SDGs
Health and disease	Sexually transmitted infections	HIV/AIDS syphilis, STIs risk factors and WHO gonococcal AMR Surveillance Programme (WHO-GSP)
	Antimicrobial resistance	Global action plan
	Infectious diseases	• Cholera (cases, mortality, vaccine distribution and genetic) • Malaria (cases, death and tracking biological challenges to elimination) • Meningococcal meningitis (cases, mortality and vaccine distribution) • Tuberculosis (cases, mortality, drug resistance, treatment, coepidemics: TB and HIV)
	Vaccine-preventable diseases	Vaccine availability, costs, infrastructure requirements, delivery and waning immunity Congenital rubella syndrome, diphtheria Japanese encephalitis, yellow fever measles, mumps, pertussis (whooping cough), poliomyelitis, rubella, tetanus, neonatal tetanus, influenza and many STIs

(Continued)

TABLE 1.3 (Continued)

Summary of Key Global Health Priorities and Actions

Topic of Concern	Global Health Areas of Concern	Response and Actions
	Neglected tropical diseases	Diverse group of 20 conditions of parasitic, bacterial, viral, fungal and non-communicable origin Buruli ulcer, Chagas disease (American trypanosomiasis), dengue and chickungunya Dracunculiasis (guinea-worm disease) Echinococcosis, foodborne trematode infections, human African trypanosomiasis (sleeping sickness), leishmaniasis, leprosy, lymphatic filariasis, mycetoma, onchocerciasis, rabies, scabies, schistosomiasis, soil-transmitted helminthiases, snakebite envenoming, taeniasis and cysticercosis, trachoma and yaws
	Noncommunicable diseases	Heart disease, stroke, cancer, chronic respiratory diseases and diabetes Mortality, risk factors (include tobacco, harmful use of alcohol, unhealthy diet, insufficient physical activity, overweight/obesity (BMI), raised blood pressure, raised blood sugar and raised cholesterol) and national capacity) Air pollution (household, ambient and joint effect) Urban health and violence prevention
	Mental health	Suicide rates and mental health service availability
Women	Maternal and reproductive health Women and health Violence against women Child health	Maternal mortality, maternal and reproductive health and marriage/union age 15–18 Infant/child nutrition, preventing child mortality, causes of mortality, anaemia, breastfeeding and low birth weight Child malnutrition

Sources: WHO (2020); WHO (2021c); WHO (2021b); WHO (2021a).

Summary

> Everything we do during and after this crisis [COVID-19] must be with a strong focus on building more equal, inclusive and sustainable economies and societies that are more resilient in the face of pandemics, climate change, and the many other global challenges we face.
>
> *António Guterres, Secretary-General, United Nations.*
> *https://unstats.un.org/sdgs/report/2020/*

In summary, geographic information, spatial methods and geospatial technologies are useful for achieving more equal, inclusive, sustainable and resilient societies through:

- **Data collection and engineering**: collecting, managing and transforming data for operational use.
- **Visualization and communication**: mapping where and when disease and health risks are taking place. Static and interactive maps can be useful for informing the public and health officials.
- **Exploration**: understanding what is taking place where and when by interacting with and exploring the data; integrating different types of data, geographies and layers of information to provide context. Identifying the possible factors that might be contributing to the patterns of disease and developing hypothesis about the prevalence of the disease
- **Analysis and evaluation**: Performing different types of analysis to evaluate the spatial distribution of spatial patterns, relationships and gain insights into what maybe occurring.
- **Modelling and prediction**: model spatial relations over space and time between disease occurrence and different environmental and host factors and predict outcomes.

Now that you have some background about how geospatial technologies and how geographic information can be used in health studies, let's get started! In Chapter 2, we will talk about epidemiology and disease to introduce you to a framework that can be used to investigate the driving factors and processes of a disease and the patterns associated with their distribution, followed by details on methods and visualizations (Chapter 3), disasters (Chapter 4) and spatial data (Chapter 5) including the use of interactive web maps and dashboards to communicate what is going on. For the remainder of the chapters, we will move onto the fun stuff – mapping and modelling vector-borne diseases (Chapter 6), identifying hotspots and spatial disease clusters (Chapter 7), location-based analyses and accessibility to healthcare (Chapter 8), and lastly planetary health and exposure-related health risks (Chapter 9). At the end of each chapter there are a series of activities that you can work through to using a variety of methods. The methods have been purposefully selected from the methods captured in Figure 1.5 so that you are able to develop a broad range of skills that vary in complexity and use of different types of data. These are set around the elements summarized in Figure 1.3 (Ecology of Disease – Response – Communication); an integrated health approach.

Activity – Geospatial Information and Technologies for Health

Now that you have a broad overview of different ways that geographic information and geospatial technologies can be used for health-related studies. How useful do you think geospatial technologies and apps are for better understanding health risks and outcomes? How can geographic information and geospatial technologies be used to tackle global health challenges? How do you see yourself using geographic information and geospatial technologies?

References

Abascal, A., N. Rothwell, A. Shonowo, D. R. Thomson, P. Elias, H. Elsey & M. Kuffer (2022) "Domains of deprivation framework" for mapping slums, informal settlements, and other deprived areas in LMICs to improve urban planning and policy: A scoping review. *Computers, Environment and Urban Systems*, 93, 101770.

AGU (2021) Geohealth at AGU. https://news.agu.org/files/2020/10/GeoHealth-at-AGU.pdf (last accessed Sep 15 2023).

Ajiri, A., J. Okeibunor, S. Aiyeoribe, B. Ntezayabo, M. Mailhot, M. Nzioki, A. Traore, A. Khalid, M. Diallo, M. Ilboudo, B. M. Mikeyas, D. Samba, T. Mulunda, N. De Medeiros, B. Rabenarivo, F. Diomande & S. Okiror (2021) Response to poliovirus outbreaks in the lake chad sub-region: A GIS mapping approach. *Journal of Immunology Science*, 12(Special Issue), 1115.

Al-Huraibi, A., S. Amer & J. I. Blanford (2023) Cycling to get my vaccination: how accessible are COVID-19 vaccination centers in the Netherlands?. *AGILE: GIScience Series*, 4.

Almada, A. A., C. D. Golden, S. A. Osofsky & S. S. Myers (2017) A case for planetary health/geohealth. *Geohealth*, 1, 75–78.

Auchincloss, A. H., S. Y. Gebreab, C. Mair & A. V. Diez Roux (2012) A review of spatial methods in epidemiology, 2000–2010. *Annual Review of Public Health*, 33, 107–22.

Bailey, P. (1994) A Review of Statistical Spatial Analysis in Geographical Information Systems. In *Spatial Analysis and GIS*, eds. S. Fotheringham & D. Rogerson, 13–44. Bristol, PA: Taylor & Francis.

Banskota, S., M. Healy & E. M. Goldberg (2020) 15 smartphone apps for older adults to use while in isolation during the COVID-19 pandemic. *Western Journal of Emergency Medicine*, 21, 514–525

Barau, I., et al. (2014) Improving polio vaccination coverage in Nigeria through the use of geographic information system technology. *The Journal of Infectious Diseases* 210(suppl_1), S102–S110.

Beale, L., J. J. Abellan, S. Hodgson & L. Jarup (2008) Methodologic issues and approaches to spatial epidemiology. *Environmental Health Perspectives*, 116, 1105–1110.

Blanford, J., P. Kennelly, B. King, D. Miller & T. Bracken (2020) Merits of capstone projects in an online graduate program for working professionals. *Journal of Geography in Higher Education*, 44, 45–69.

Blanford, J. I., N. Beerlage-de Jong, S. Schouten, A. Friedrich & V. Araujo-Soares (2022) Navigating travel in Europe during the pandemic: From mobile apps, certificates, quarantine to traffic light system. *Journal of Travel Medicine*, 29(3), taac006.

Blanford, J. I. & A. Jolly (2021) Public Health Needs GiScience (Like Now!). AGILE: GIScience Series. Greece, Copernicus Publications. 2, 18.

Bowyer, R. C. E., T. Varsavsky, E. J. Thompson, C. H. Sudre, B. A. K. Murray, M. B. Freidin, D. Yarand, S. Ganesh, J. Capdevila, E. Bakker, M. J. Cardoso, R. Davies, J. Wolf, T. D. Spector, S. Ourselin, C. J. Steves & C. Menni (2021) Geo-social gradients in predicted COVID-19 prevalence in Great Britain: Results from 1,960,242 users of the COVID-19 Symptoms Study app. *Thorax*, 76, 723–725.

Brown, T. & G. Moon (2004) From Siam to New York: Jacques May and the 'foundation' of medical geography. *Journal of Historical Geography*, 30, 747–763.

Brownstein, J. S., B. Rader, C. M. Astley & H. Tian (2023) Advances in artificial intelligence for infectious-disease surveillance. *New England Journal of Medicine*, 388, 1597–1607.

CDC Defining Field Epidemiology. https://www.cdc.gov/eis/field-epi-manual/chapters/Defining-Field-Epi.html (last accessed Sep 16 2023).

—— (2014) Introduction to Public Health. In: *Public Health 101 Series*. https://www.cdc.gov/training/publichealth101/public-health.html (last accessed Sep 16 2023).

—— (2021) One Health Basics. https://www.cdc.gov/onehealth/basics/index.html (last accessed May 20 2021).

—— (2023) Disaster Epidemiology. https://www.cdc.gov/nceh/hsb/disaster/epidemiology.htm (last accessed Sep 16 2023).

Chetty, R., M. Stepner, S. Abraham, S. Lin, B. Scuderi, N. Turner, A. Bergeron & D. Cutler (2016) The association between income and life expectancy in the United States, 2001–2014. *JAMA*, 315, 1750–1766.

Chin, C. S., J. Sorenson, J. B. Harris, W. P. Robins, R. C. Charles, R. R. Jean-Charles, J. Bullard, D. R. Webster, A. Kasarskis, P. Peluso, E. E. Paxinos, Y. Yamaichi, S. B. Calderwood, J. J. Mekalanos, E. E. Schadt & M. K. Waldor (2011) The origin of the Haitian cholera outbreak strain. *The New England Journal of Medicine*, 364, 33–42.

Clements, A. C. A. & D. U. Pfeiffer (2009) Emerging viral zoonoses: Frameworks for spatial and spatiotemporal risk assessment and resource planning. *Veterinary Journal*, 182, 21–30.

Corburn, J. & L. Riley (2016) *Slum Health: From the Cell to the Street*. Oakland: California University of California Press.

Corburn, J. & A. Sverdlik (2018) Informal Settlements and Human Health. In *Integrating Human Health into Urban and Transport Planning*, eds. N. Mark & K. Haneen, 155–171. Cham: Springer.

Cromley, E. K. & S. L. McLafferty (2012) *GIS and Public Health*. New York: Guilford Press.

CSDGS_NRC_Ch6 (2010) Ch 6: How Does Where People Live Affect Their Health? In *Understanding the Changing Planet: Strategic Directions for the Geographical Sciences*, eds. Committee on Strategic Directions for the Geographical Sciences in the Next Decade & National Research Council, 67–74. Washington, DC: National Academy of Sciences.

Dangermond, J. & M. F. Goodchild (2020) Building geospatial infrastructure. *Geo-spatial Information Science*, 23, 1–9.

Davalbhakta, S., S. Advani, S. Kumar, V. Agarwal, S. Bhoyar, E. Fedirko, D. P. Misra, A. Goel, L. Gupta & V. Agarwal (2020) A systematic review of smartphone applications available for corona virus disease 2019 (COVID19) and the assessment of their quality using the mobile application rating scale (MARS). *The Journal of Medical Systems*, 44, 164.

de Smith, M., M. Goodchild & P. Longley (2022) Geospatial Analysis - A Comprehensive Guide to Principles, Techniques and Software Tools. https://www.spatialanalysisonline.com/ (last accessed Sep 16 2023).

Demuzere, M., B. Bechtel, A. Middel & G. Mills (2019) Mapping Europe into local climate zones. *PloS One*, 14, e0214474.

DiBiase, D., T. Corbin, T. Fox, J. Francica, K. Green, J. Jackson & K. Schuckman (2010) The new geospatial technology competency model: Bringing workforce needs into focus. *URISA Journal*, 22, 55–72.

Dong, E., H. Du & L. Gardner (2020) An interactive web-based dashboard to track COVID-19 in real time. *Lancet Infectious Diseases*, 20, 533–534.

Du, Y., Y. Ding, Z. Li & G. Cao (2015) The role of hazard vulnerability assessments in disaster preparedness and prevention in China. *Military Medical Research*, 2, 27.

Dummer, T. J. (2008) Health geography: Supporting public health policy and planning. *CMAJ*, 178, 1177–1180.

ECDC (2021) Weekly Maps in Support of the Council Recommendation on a Coordinated Approach to Travel Measures in the EU. https://www.ecdc.europa.eu/en/covid-19/situation-updates/weekly-maps-coordinated-restriction-free-movement (last accessed Jul 31 2021).

— (2023) Cholera Worldwide Overview: Monthly update as of 20 July 2023. https://www.ecdc.europa.eu/en/all-topics-z/cholera/surveillance-and-disease-data/cholera-monthly (last accessed Aug 3 2023).

Egondi, T., C. Kyobutungi, N. Ng, K. Muindi, S. Oti, S. van de Vijver, R. Ettarh & J. Rocklöv (2013) Community perceptions of air pollution and related health risks in Nairobi slums. *International Journal of Environmental Research and Public Health*, 10, 4851–4868.

Egondi, T., K. Muindi, C. Kyobutungi, M. Gatari & J. Rocklöv (2016) Measuring exposure levels of inhalable airborne particles (PM2.5) in two socially deprived areas of Nairobi, Kenya. *Environmental Research*, 148, 500–506.

Elliott, P., J. C. Wakefield, N. G. Best & D. J. Briggs (2000) *Spatial Epidemiology: Methods and Applications*. Oxford: Oxford University Press.

Elliott, P. & D. Wartenberg (2004) Spatial epidemiology: Current approaches and future challenges. *Environ Health Perspect*, 112, 998–1006.

el-Hamza, M. (2023) Mapping for Better Coverage: How GIS Microplanning is Revolutionising Immunisation Campaigns in Kano State. https://articles.nigeriahealthwatch.com/mapping-for-better-coverage-how-gis-microplanning-is-revolutionising-immunisation-campaigns-in-kano-state/ (last accessed May 10 2024).

Engel-Cox, J. A. & R. M. Hoff (2005) Science-policy data compact: Use of environmental monitoring data for air quality policy. *Environmental Science & Policy*, 8, 115–131.

English, D. R. (1992) Geographical Epidemiology and Ecological Studies. In *Geographical and Environmental Epidemiology: Methods for Small-Area Studies*, eds. P. Elliott, J. Cuzick, D. English & R. Stern, 3–13. Oxford: Oxford University Press.

EnvironBuzz (2022) Access to Energy in Sub-Saharan Africa. *EnvironBuzz Mag*. https://environbuzz.com/access-to-energy-in-sub-saharan-africa/ (last accessed 25 Sep 2022).

eo4geo (2024) A Body of Knowledge (BoK) for EO/GI. http://www.eo4geo.eu/bok/ (last accessed May 24 2024).

ESRI (2023) ArcGIS Pro Help. https://pro.arcgis.com/en/pro-app/latest/help/main/welcome-to-the-arcgis-pro-app-help.htm (last accessed Sep 15 2023).

EuropeanCommission (2021) A Common Approach to Travel Measures in the EU. https://ec.europa.eu/info/live-work-travel-eu/coronavirus-response/travel-during-coronavirus-pandemic/common-approach-travel-measures-eu_en#lifting-restrictions (last accessed Jul 31 2021).

Folke, C., S. Polasky, J. Rockström, V. Galaz, F. Westley, M. Lamont, M. Scheffer, H. Österblom, S. R. Carpenter, F. S. Chapin, K. C. Seto, E. U. Weber, B. I. Crona, G. C. Daily, P. Dasgupta, O. Gaffney, L. J. Gordon, H. Hoff, S. A. Levin, J. Lubchenco, W. Steffen & B. H. Walker (2021) Our future in the Anthropocene biosphere. *Ambio*, 50, 834–869.

Franch-Pardo, I., B. M. Napoletano, F. Rosete-Verges & L. Billa (2020a) Spatial analysis and GIS in the study of COVID-19. A review. *Science of the Total Environment*, 739, 140033.

Franch-Pardo, I., B. M. Napoletano, F. Rosete-Verges & L. Billa (2020b) Spatial analysis and GIS in the study of COVID-19. A review. *Science of the Total Environment*, 739, 1–10.

Freifeld, C. C., K. D. Mandl, B. Y. Reis & J. S. Brownstein (2008) HealthMap: Global infectious disease monitoring through automated classification and visualization of Internet media reports. *Journal of the American Medical Informatics Association*, 15, 150–157.

Fund, U. J. S. (2023) Goal 3: Good Health and Well-being. https://www.jointsdgfund.org/sustainable-development-goals/goal-3-good-health-and-well-being (last accessed Aug 5 2023).

Gammino, V. M., A. Nuhu, P. Chenoweth, F. Manneh, R. R. Young, D. E. Sugerman, S. Gerber, E. Abanida & A. Gasasira (2014) Using geographic information systems to track polio vaccination team performance: Pilot project report. *Journal of the Infectious Diseases*, 210(Suppl 1), S98–101.

GeoSpatialSense (2023) getting started: Spatial Data Science, GIScientist…https://geospatialsense.com/2023/04/10/getting-started-spatial-data-science-giscientist/ (last accessed May 21 2024).

--- (2024) the geospatial ecosystem: Software, Libraries and Data Platforms. https://geospatialsense.com/2024/02/03/the-geospatial-ecosystem-software-libraries-and-data-platforms/ (last accessed May 21 2024).

Gibbs, E. P. (2014) The evolution of one health: A decade of progress and challenges for the future. *Veterinary Record*, 174, 85–91.

GoodChild, M. F. (1992) Integrating GIS and spatial data analysis: Problems and possibilities. *International Journal of Geographical Information Systems*, 6, 407–423.

Guemes, A., S. Ray, K. Aboumerhi, M. R. Desjardins, A. Kvit, A. E. Corrigan, B. Fries, T. Shields, R. D. Stevens, F. C. Curriero & R. Etienne-Cummings (2021) A syndromic surveillance tool to detect anomalous clusters of COVID-19 symptoms in the United States. *Scientific Reports*, 11, 4660.

Gulyaeva, M., F. Huettmann, A. Shestopalov, M. Okamatsu, K. Matsuno, D.-H. Chu, Y. Sakoda, A. Glushchenko, E. Milton & E. Bortz (2020) Data mining and model-predicting a global disease reservoir for low-pathogenic Avian Influenza (AI) in the wider pacific rim using big data sets. *Scientific Reports*, 10, 16817.

Guo, H., M. F. Goodchild & A. Annoni (2020) *Manual of Digital Earth*. Heidelberg: Springer Nature.

Hadfield, J., C. Megill, S. M. Bell, J. Huddleston, B. Potter, C. Callender, P. Sagulenko, T. Bedford & R. A. Neher (2018) Nextstrain: Real-time tracking of pathogen evolution. *Bioinformatics*, 34, 4121–4123.

Haines, A. (2016) Addressing challenges to human health in the Anthropocene epoch-an overview of the findings of the Rockefeller/Lancet Commission on Planetary Health. *Public Health Reviews*, 37, 14.

—— (2017) Addressing challenges to human health in the Anthropocene epoch-an overview of the findings of the Rockefeller/Lancet Commission on Planetary Health. *The International Health Regulations*, 9, 269–271.

Haines, A., C. Hanson & J. Ranganathan (2018) Planetary health watch: Integrated monitoring in the Anthropocene epoch. *Lancet Planet Health*, 2, e141–e143.

Hay, S. I., A. Graham & D. J. Rogers (2006) Global mapping of infectious diseases: Methods, examples and emerging applications. *Advances in Parasitology*, 368(1614), 20120250.

Hay, S. I., D. J. Rogers & S. E. Randolph (2000) *Remote Sensing and Geographical Information Systems in Epidemiology*. The Netherlands: Elsevier.

He, D., S. T. Ali, G. Fan, D. Gao, H. Song, Y. Lou, ... & L. Stone (2022). Evaluation of effectiveness of global COVID-19 vaccination campaign. *Emerging Infectious Diseases*, 28(9), 1873.

Hellings, A. & A. Rienow (2021) Mapping land surface temperature developments in functional urban areas across Europe. *Remote Sensing*, 13, 2111.

Holloway, T., D. Miller, S. Anenberg, M. Diao, B. Duncan, A. M. Fiore, D. K. Henze, J. Hess, P. L. Kinney & Y. Liu (2021) Satellite monitoring for air quality and health. *Annual review of biomedical data science*, 4, 417–447.

Horney, J. (2017) *Disaster Epidemiology: Methods and Applications*. London: Academic Press.

Horton, R. & S. Lo (2015) Planetary health: A new science for exceptional action. *Lancet*, 386, 1921–1922.

Idris, U. K. (2020) AVADAR: How digital health fast-tracked Nigeria's drive to eradicate polio. https://articles.nigeriahealthwatch.com/avadar-how-digital-health-fast-tracked-nigerias-drive-to-eradicate-polio/ (last accessed 16 Sep 2023).

IHME (2024) Global Burden Disease Compare Visualization Hub. https://vizhub.healthdata.org/gbd-compare/. Accessed 27 April 2024

IJHG (2021) International Journal of Health Geographics. https://ij-healthgeographics.biomedcentral.com/about (last accessed April 26 2024).

IOM. 2015. *Migration Health: Annual Review 2014*, 104 p. Geneva, Switzerland: International Organization for Migration (IOM): Migration Health Division.

ISGH (2021) Geospatial Health: Health Application in Geospatial Science. https://geospatialhealth.net/index.php/gh (last accessed May 20 2021).

Iyer, H. S., N. V. DeVille, O. Stoddard, J. Cole, S. S. Myers, H. Li, E. G. Elliott, M. P. Jimenez, P. James & C. D. Golden (2021) Sustaining planetary health through systems thinking: Public health's critical role. *SSM Popul Health*, 15, 100844.

JHU (2021) COVID-19 Dashboard. https://coronavirus.jhu.edu/map.html (last accessed May 20 2021).

Jolly, A. M. & J. L. Wylie (2013) Sexual Networks and Sexually Transmitted Infections; "The Strength of Weak (Long Distance) Ties". In *The New Public Health and STD/HIV Prevention: Personal, Public and Health Systems Approaches*, eds. S. O. Aral, K. A. Fenton & J. A. Lipshutz. New York: Springer Science+Business Media.

Kalluri, S., P. Gilruth, D. Rogers & M. Szczur (2007) Surveillance of arthropod vector-borne infectious diseases using remote sensing techniques: A review. *PLoS Pathogens*, 3, e116.

Kamadjeu, R. (2009) Tracking the polio virus down the Congo River: A case study on the use of Google Earth in public health planning and mapping. *International Journal of Health Geographics*, 8, 4.

Keim, P. S., F. M. Aarestrup, G. Shakya, L. B. Price, R. S. Hendriksen, D. M. Engelthaler & T. Pearson (2011) Reply to "South Asia instead of Nepal may be the origin of the Haitian cholera outbreak strain". *mBio*, 2, e00245–11.

Kirby, R. S., E. Delmelle & J. M. Eberth (2017) Advances in spatial epidemiology and geographic information systems. *The Annals of Epidemiology*, 27, 1–9.

Klass, J. I., S. Blanford & M. B. Thomas (2007a) Development of a model for evaluating the effects of environmental temperature and thermal behaviour on biological control of locusts and grasshoppers using pathogens. *Agricultural and Forest Entomology*, 9, 189–199.

—— (2007b) Use of a geographic information system to explore spatial variation in pathogen virulence and the implications for biological control of locusts and grasshoppers. *Agricultural and Forest Entomology*, 9, 201–208.

KNAW (2023) Planetary Health. An emerging field to be developed., 137. https://www.knaw.nl/nl/publicaties/planetary-health-emerging-field-be-developed (last accessed June 18 2023).

Koplan, J. P., et al. (2009) Towards a common definition of global health. *The Lancet* 373(9679), 1993–1995.

Kwan, M.-P. (2016) Algorithmic geographies: Big data, algorithmic uncertainty, and the production of geographic knowledge. *Annals of the American Association of Geographers*, 106, 274–282.

Lawson, A. B. (2013) *Statistical Methods in Spatial Epidemiology*. New York: John Wiley & Sons.

Lee, S., Y. M. Cho & S. Y. Kim (2017) Mapping mHealth (mobile health) and mobile penetrations in sub-Saharan Africa for strategic regional collaboration in mHealth scale-up: An application of exploratory spatial data analysis. *Global Health*, 13, 63.

Li, S., S. Dragicevic, F. A. Castro, M. Sester, S. Winter, A. Coltekin, C. Pettit, B. Jiang, J. Haworth & A. Stein (2016) Geospatial big data handling theory and methods: A review and research challenges. *ISPRS journal of Photogrammetry and Remote Sensing*, 115, 119–133.

Lilford, R. J., O. Oyebode, D. Satterthwaite, et al. (2016) Improving the health and welfare of people who live in slums. *Lancet*, 389, 559–570.

Logan, J. J., A. M. Jolly & J. I. Blanford (2016) The sociospatial network: Risk and the role of place in the transmission of infectious diseases. *PLoS One*, 11, e0146915.

Mandl, K. D., M. Overhage, M. M. Wagner, W. B. Lober, P. Sebastiani, F. Mostashari, et al. (2004) Implementing syndromic surveillance: A practical guide informed by the early experience. *Journal of the American Medical Informatics Association*, 11, 141–150.

May, J. M. (1950) Medical geography: Its methods and objectives. *Geographical Review*, 40, 9–41.

--- (1959) The ecology of human disease. *The Ecology of Human Disease*, 84, 789–794.

--- (1978) History, definition, and problems of medical geography: A general review: Report to the Commission on Medical Geography of the International Geographical Union 1952. *Social Science & Medicine. Part D: Medical Geography*, 12, 211–219.

Mendes, A., T. Palmer, A. Berens, J. Espey, R. Price, A. Mallya, S. Brown, M. Martinez, N. Farag & B. Kaplan (2021) Mapathons versus automated feature extraction: A comparative analysis for strengthening immunization microplanning. *International Journal of Health Geographics*, 20, 27.

Miller, H. J. & M. F. Goodchild (2015) Data-driven geography. *GeoJournal*, 80, 449–461.

Miller, I. F., A. D. Becker, B. T. Grenfell, & C. J. E. Metcal (2020). Disease and healthcare burden of COVID-19 in the United States. *Nature Medicine*, 26(8), 1212–1217.

Ming, L. C., N. Untong, N. A. Aliudin, N. Osili, N. Kifli, C. S. Tan, K. W. Goh, P. W. Ng, Y. M. Al-Worafi & K. S. Lee (2020) Mobile health apps on COVID-19 launched in the early days of the pandemic: Content analysis and review. *JMIR mHealth and uHealth*, 8, e19796.

Morand, S. & C. Lajaunie (2017) *Biodiversity and Health: Linking Life, Ecosystems and Societies*. Amsterdam, The Netherlands: Elsevier.

Myers, S. S. (2017) Planetary health: Protecting human health on a rapidly changing planet. *Lancet*, 390, 2860–2868.

OECD (2021) Health at a Glance 2021: OECD Indicators. https://doi.org/10.1787/ae3016b9-en (last accessed Aug 5 2023).

OneHealth (2021) One Health Initiative will Unite Human and Veterinary Medicine. https://onehealthinitiative.com/ (last accessed May 20 2021).

Ostfeld, R. S., G. E. Glass & F. Keesing (2005) Spatial epidemiology: An emerging (or re-emerging) discipline. *Trends in Ecology & Evolution*, 20, 328–336.

Pandit, J. A., J. M. Radin, G. Quer & E. J. Topol (2022) Smartphone apps in the COVID-19 pandemic. *Nature Biotechnology*, 40, 1013–1022.

Pavlovsky, E. N. (1966) *Natural Nidality of Transmissible Diseases with Special Reference to the Landscape Epidemiology of Zooanthroponoses*. Urbana, IL: University of Illinois Press.

Prüss-Ustün, A., J. Wolf, C. Corvalán, R. Bos & M. Neira (2016) Preventing Disease Through Healthy Environments: A Global Assessment of the Burden of Disease from Environmental Risks. https://www.who.int/publications/i/item/9789241565196 (last accessed).

Purwanto, P., S. Utaya, B. Handoyo, S. Bachri, I. S. Astuti, K. S. B. Utomo & Y. E. Aldianto (2021) Spatiotemporal analysis of COVID-19 spread with emerging hotspot analysis and space-time cube models in East Java, Indonesia. *ISPRS International Journal of Geo-Information*, 10, 133.

Qiu, C., L. Mou, M. Schmitt & X. X. Zhu (2019) Local climate zone-based urban land cover classification from multi-seasonal Sentinel-2 images with a recurrent residual network. *ISPRS Journal of Photogrammetry and Remote Sensing*, 154, 151–162.

Reid, C. E., L. D. Kubzansky, J. Li, J. L. Shmool & J. E. Clougherty (2018) It's not easy assessing greenness: A comparison of NDVI datasets and neighborhood types and their associations with self-rated health in New York City. *Health & Place*, 54, 92–101.

RIVM (2021) Weekcijfers COVID-19/Weekly COVID-19 Figures https://www.rivm.nl/en/coronavirus-covid-19/weekly-covid-19-figures (last accessed May 21 2021).

Robertson, C. (2017) Towards a geocomputational landscape epidemiology: Surveillance, modelling, and interventions. *GeoJournal*, 82, 397–414.

Rogers, D. J. & S. E. Randolph (2003) Studying the global distribution of infectious diseases using GIS and RS. *Nature Reviews Microbiology*, 1, 231–237.

Ross, A. G., M. Alam, M. Rahman, F. Qadri, S. S. Mahmood, K. Zaman, T. N. Chau, A. Chattopadhyay & S. P. Gon Chaudhuri (2021) Rise of informal slums and the next global pandemic. *The Journal of Infectious Diseases*, 224, S910–S914.

Rubin, D. H. F., F. G. Zingl, D. R. Leitner, R. Ternier, V. Compere, S. Marseille, D. Slater, J. B. Harris, F. Chowdhury, F. Qadri, J. Boncy, L. C. Ivers & M. K. Waldor (2022) Reemergence of Cholera in Haiti. *The New England Journal of Medicine*, 387, 2387–2389.

Rushton, G. (2003) Public health, GIS, and spatial analytic tools. *Annual Review of Public Health*, 24, 43–56.

Safaei, M., E. A. Sundararajan, M. Driss, W. Boulila & A. Shapi'i (2021) A systematic literature review on obesity: Understanding the causes & consequences of obesity and reviewing various machine learning approaches used to predict obesity. *Computers in Biology and Medicine*, 136, 104754.

Salathé, M. (2018) Digital epidemiology: What is it, and where is it going. *Life Sciences, Society and Policy*, 14, 1.

Schwarz, N., S. Lautenbach & R. Seppelt (2011) Exploring indicators for quantifying surface urban heat islands of European cities with MODIS land surface temperatures. *Remote Sensing of Environment*, 115, 3175–3186.

Scovronick, N., S. J. Lloyd & R. S. Kovats (2015) Climate and health in informal urban settlements. *Environment and Urbanization*, 27, 657–678.

Seeberg, G., A. Hostlowsky, J. Huber, J. Kamm, L. Lincke & C. Schwingshackl (2022) Evaluating the potential of landsat satellite data to monitor the effectiveness of measures to mitigate urban heat Islands: A case study for stuttgart (Germany). *Urban Science*, 6, 82.

Singh, H. J. L., D. Couch & K. Yap (2020) Mobile health apps that help with COVID-19 management: Scoping review. *JMIR Nursing*, 3, e20596.

Snow, J. (1855) *On the Mode of Communication of Cholera*. London: John Churchill.

Sorre, M. (1948) Fondements de la géographie humaine. *Cahiers Internationaux de Sociologie*, 5, 21–37.

Still, C., R. Powell, D. Aubrecht, Y. Kim, B. Helliker, D. Roberts, A. D. Richardson & M. Goulden (2019) Thermal imaging in plant and ecosystem ecology: Applications and challenges. *Ecosphere*, 10, e02768.

Thomson, D. R., M. Kuffer, G. Boo, B. Hati, T. Grippa, H. Elsey & C. Kabaria (2020) Need for an integrated deprived area "Slum" mapping system (IDEAMAPS) in low- and middle-income countries (LMICs). *Social Sciences*, 9, 80.

Touray, K., P. Mkanda, S. G. Tegegn, P. Nsubuga, T. B. Erbeto, R. Banda, A. Etsano, F. Shuaib & R. G. Vaz (2016) Tracking vaccination teams during polio campaigns in Northern Nigeria by use of geographic information system technology: 2013–2015. *Journal of Infectious Diseases*, 213(Suppl 3), S67–72.

Tunstall, H. V. Z., M. Shaw & D. Dorling (2004) Places and health. *Journal of Epidemiology and Community Health*, 58, 6–10.

UCGIS (2023) *Geographic Information Science & Technology Body of Knowledge (BoK)*. https://gistbok.ucgis.org/. Accessed 27 April 2024.

UN (2021) The United Nations World Water Development Report 2021: Valuing Water, 206. https://unesdoc.unesco.org/ark:/48223/pf0000375724 (last accessed Aug 5 2023).

—— (2023a) The SDGs in Action. https://www.undp.org/sustainable-development-goals#good-health (last accessed Aug 5 2023).

—————— (2023b) The Sustainable Development Goals Report 2023: Special edition. Towards a Rescue Plan for People and Planet, 80 p. https://unstats.un.org/sdgs/report/2023/The-Sustainable-Development-Goals-Report-2023.pdf (last accessed Aug 5 2023).

Visconti, A., B. Murray, N. Rossi, J. Wolf, S. Ourselin, T. D. Spector, E. E. Freeman, V. Bataille & M. Falchi (2022) Cutaneous manifestations of SARS-CoV-2 infection during the Delta and Omicron waves in 348 691 UK users of the UK ZOE COVID Study app. *British Journal of Dermatology*, 187, 900–908.

Wallace-Wells, D. (2021) Ten Million a Year, 43 p. https://www.lrb.co.uk/the-paper/v43/n23/david-wallace-wells/ten-million-a-year (last accessed 25 Sep 2022).

Whitmee, S., A. Haines, C. Beyrer, F. Boltz, A. G. Capon, B. F. de Souza Dias, A. Ezeh, H. Frumkin, P. Gong, P. Head, R. Horton, G. M. Mace, R. Marten, S. S. Myers, S. Nishtar, S. A. Osofsky, S. K. Pattanayak, M. J. Pongsiri, C. Romanelli, A. Soucat, J. Vega & D. Yach (2015) Safeguarding human health in the Anthropocene epoch: Report of the rockefeller foundation-lancet commission on planetary health. *Lancet*, 386, 1973–2028.

WHO (2023) Noncommunicable Disease Fact Sheet. https://www.who.int/mediacentre/factsheets/fs355/en/ (last accessed April 27 2024).

—— (2016) Burning Opportunity: Clean Household Energy for Health, Sustainable Development, and Wellbeing of Women and Children. https://apps.who.int/iris/handle/10665/204717 (last accessed Feb 18 2022).

—— (2017) One Health. https://www.who.int/news-room/q-a-detail/one-health (last accessed April 27 2024).

—— (2020) 10 Global Health Issues to Track in 2021. https://www.who.int/news-room/spotlight/10-global-health-issues-to-track-in-2021 (last accessed April 5 2021).

—— (2021a) Control of Neglected Tropical Diseases. https://www.who.int/teams/control-of-neglected-tropical-diseases (last accessed April 5 2021).

—— (2021b) The Global Health Observatory. https://www.who.int/data/gho (last accessed April 5 2021).

—— (2021c) Monitoring Health for Sustainable Development Goals. https://www.who.int/data/gho/data/themes/sustainable-development-goals (last accessed April 5 2021).

WHO_Africa (2022) Data for Action and Visualisation for Capacity Building. https://polioeradication.org/wp-content/uploads/2022/08/Data-for-action-and-visualization-capacity-building.pdf (last accessed Sep 16 2023).

You, Y. A., M. Ali, S. Kanungo, B. Sah, B. Manna, M. Puri, G. B. Nair, S. K. Bhattacharya, M. Convertino, J. L. Deen, A. L. Lopez, T. F. Wierzba, J. Clemens & D. Sur (2013) Risk map of cholera infection for vaccine deployment: The eastern Kolkata case. *PLoS One*, 8, e71173.

Yuki, K., M. Fujiogi, & S. Koutsogiannaki (2020). COVID-19 pathophysiology: A review. *Clinical Immunology*, 215, 108427.

Zavattaro, S. M. (2023). Death managers, public health, and COVID-19: An exploratory study. *Public Administration Review* 83(5), 1339–1350.

Zeng, Y., J. Guo & X. X. Zhu (2023) Spatial-Temporal Analysis of Urban Green Space and the Impact on Land Surface Temperature of Beijing, China, 1–4. Heraklion, Greece: IEEE.

Appendix A

Appendices for Introduction

TABLE A.1

Summary of Terms Used for the Application of Geography and Spatial Analysis Methods for studying Health and Disease

Spatial epidemiology	• The study of spatial variation in disease risk or incidence and is devoted to understanding the causes and consequences of spatial heterogeneity in infectious diseases, particularly in zoonoses (Elliott et al., 2000) • The analysis of the spatial/geographical distribution of the incidence of disease (Lawson, 2013) • Incorporates the spatial perspective into the design and analysis of the distribution, determinants and outcomes of all aspects of health and well-being from prevention to treatment (Kirby et al., 2017) • Used to assess health risks associated with environmental hazards (Beale et al., 2008) Risk patterns tend to have both a temporal and a spatial component Combine methods from epidemiology, statistics and geographic information science Includes the use of smoothing in risk maps to create an interpretable risk surface The extension of spatial models to incorporate the time dimension, and the combination of individual- and area-level information
Geographical epidemiology	The description of spatial patterns of disease incidence and mortality (English, 1992)
Landscape epidemiology	**Landscape epidemiology** consists of three basic Observations (Pavlovsky, 1966): • Diseases tend to be limited geographically. • This spatial variation arises from underlying variation in the physical and/or biological conditions that support the pathogen and its vectors and reservoirs. • If those abiotic and biotic conditions can be delimited on maps, then both contemporaneous risk and future change in risk should be predictable. Uses statistical associations between environmental variables and diseases to study and predict their spatial distributions (Clements and Pfeiffer, 2009) and geocomputational methods for examining outcomes and mapping interdependencies (Robertson, 2017)

(Continued)

TABLE A.1 (*Continued*)

Summary of Terms Used for the Application of Geography and Spatial Analysis
Methods for studying Health and Disease

Health geography	• Health geography is a subdiscipline of human geography that deals with the interaction between people and the environment.
	• Health geography views health from a holistic perspective encompassing society and space, and it conceptualizes the role of place, location and geography in health, well-being and disease.
	• Although health geography is closely aligned with epidemiology, its distinct primary emphasis is on spatial relations and patterns. Whereas epidemiology is predicated on the biomedical model and focuses on the biology of disease, health geography seeks to explore the social, cultural and political contexts for health within a framework of spatial organization (Dummer, 2008).
Medical geography	• Environmental influences on health was described by (Sorre, 1948) and the concept of pathogenic complex emerged where disease depends on physical, biological and social factors such as climate, natural biological environment and anthropo-geographical environment (Morand and Lajaunie, 2017).
	• Jacques May (May 1950, May 1978, May 1959), founder of modern medical geography, focused on the role the environment has on human diseases and the importance of geography in mapping pathological trends. May provided a theoretical framework for studying the environment and geographical factors ("geogenetics") of pathogen emergence (Brown and Moon, 2004).
Health Geographics	• Interdisciplinary study of geospatial information systems and science applications in health and healthcare (IJHG, 2021). Health geographics improves our understanding of the important relationships between people, location (and its characteristics: for example, environmental or socio-economic), time and health and assists with discovering and eliminating disease, in public health tasks like disease prevention and health promotion, and better healthcare service planning and delivery (IJHG, 2021).
Geohealth	• Is used for advancing interdisciplinary research that highlights issues at the intersection of the Earth and environmental sciences and health sciences (AGU, 2021)
	• It is a multidisciplinary field that investigates and highlights *how* environmental exposures and factors impact public health;
	• Connects the environment and health sciences to better protect populations and build a more resilient and healthy planet (AGU, 2021)
	• Advances our understanding of the complex interaction between the environment (including earth, water, soils and air) and the health, well-being, and continued progress of human populations (AGU, 2021).

(Continued)

TABLE A.1 (*Continued*)

Summary of Terms Used for the Application of Geography and Spatial Analysis Methods for studying Health and Disease

Geospatial health	• Focuses on all aspects of the application of geographic information systems, remote sensing, global positioning system and other geospatial tools in human and veterinary health (ISGH, 2021)
Precision health	• Is a population-based strategy that uses advanced tools and technology to discover how genetic, lifestyle and environmental factors influence a population's health. Moving beyond the traditional practice of medicine, Precision Health is a research, education and service initiative that uses big data, computational science, genetics, biology and social factors to better understand and prevent disease, *promote wellness* and *develop better treatment options* that allow patients to *improve* their health and wellness (UMichigan)
Digital epidemiology	• Digital epidemiologists conduct traditional epidemiological studies and health-related research using new data sources and digital methods from data collection to analysis (Salathé, 2018)
Field epidemiology	• Describes investigations initiated in response to urgent public health problems and requires going to the field to assess the context and possible factors contributing to the problem (CDC).
Disaster epidemiology	Disaster epidemiology is defined as the epidemiologic investigation of disaster forecasting and warning, emergency responses according to the different phases of disasters, and the short- and long-term adverse health effects of disasters on the population (Horney, 2017). It provides situational awareness; that is, information that helps us understand what the needs are, plan the response and gather the appropriate resources (Du et al., 2015). The main objectives of disaster epidemiology are to (CDC, 2023): • Prevent or reduce the number of deaths, illnesses and injuries caused by disasters • Provide timely and accurate health information for decision-makers • Improve prevention and mitigation strategies for future disasters by collecting information for future response preparation
OneHealth	**One Health** is a collaborative, multisectoral, and transdisciplinary approach—working at the local, regional, national, and global levels—with the goal of achieving optimal health outcomes recognizing the interconnection between people, animals, plants, and their shared environment. It is a collaborative, multisectoral, and transdisciplinary approach—working at the local, regional, national, and global levels—with the goal of achieving optimal health outcomes by recognizing the interconnection between people, animals, plants, and their shared environment (CDC, 2021; Gibbs, 2014)

(Continued)

TABLE A.1 (*Continued*)

Summary of Terms Used for the Application of Geography and Spatial Analysis Methods for studying Health and Disease

OneHealth	• It requires the cooperation of human, animal, and environmental health partners (CDC, 2021) worldwide (OneHealth, 2021) and is
	• Useful for
	   • Designing and implementing programmes, policies, legislation and research in which multiple sectors communicate and work together to achieve better public health outcomes (WHO, 2017).
	   • Further understanding the drivers of human, animal and environmental health by combining approaches and knowledge from medicine, biology and fields beyond.
	One Health' is an integrated, unifying approach to balance and optimize the health of people, animals and the environment.

Planetary health	In 2015 The Rockefeller Foundation-Lancet Commission on Planetary Health recognized the link between human health and the health of our planet (Horton and Lo, 2015; Almada et al., 2017; Myers, 2017; Horton and Lo, 2015; Haines, 2016; Haines, 2017; Haines et al., 2018). The premise of planetary health is that human well-being over the long-term depends on the well-being of the earth, including both its living and non-living systems. In other words the health of all living organisms is dependent on the functioning of healthy natural systems.
Public health	Public health is "the science and art of preventing disease, prolonging life, and promoting health through the organized efforts and informed choices of society, organizations, public and private communities, and individuals."—CEA Winslow (CDC, 2014) Public health emphasizes the health of populations. Planetary health recognizes the health of the planet *as a system*, and that even its non-living components are wrapped up in that state of well-being or disease.
Global health	Global health is about worldwide health improvement, reduction of disparities and protection against global threats "Global health is an area for study, research, and practice that places a priority on improving health and achieving equity in health for all people worldwide. Global health emphasizes transnational health issues, determinants, and solutions; involves many disciplines within and beyond the health sciences and promotes interdisciplinary collaboration; and is a synthesis of population-based prevention with individual-level clinical care." (Koplan et al., 2009)

TABLE A.2

Cholera Reports from around the World, July 2023

Region	Country	Report Dates (2023)	New Cases	Num Deaths	Total Cases Since 1 Jan 2023	Total Deaths	Additional Comments
Asia	Afghanistan	10 Jun–9 Jul	23,298	16	91,052	43	
Asia	India	23 Apr–19 May	71		616		
Asia	Pakistan	15 May–10 Jul	2,029		9,343		
Asia	Philippines	29 Apr–3 Jun	679		1,911	10	
Asia	Syria	20 May–15 Jun	34,161	614	114,064	621	
Asia	Taiwan	9 Jul	1				First cholera case reported since 2022
Asia	Yemen	7 May–11 Jun	864	1	3,878	4	
Asia	Bangladesh	9-Jul-23					No updates
Asia	Iraq	9-Jul-23					No updates
Asia	Thailand	9-Jul-23					No updates
Africa	Burundi	28 May–9 Jul	124	2	574	9	
Africa	Cameroon	7 May–2 Jul	3,067	95	3,787	138	
Africa	Congo	14-Jul	15				First cholera report since 2018
Africa	DRC	7 May–2 Jul	8,469	74	27,263	178	
Africa	Ethiopia	13 May–2 Jul	5,974	65	11,425	142	
Africa	Kenya	7 May–2 Jul	1,397	28	8,735	137	
Africa	Malawi	20 Jun–17 Jul	64	5	41,493	1190	
Africa	Mozambique	29 May–16 Jul	2,017	3	32,983	137	

(Continued)

TABLE A.2 (*Continued*)

Cholera Reports from around the World, July 2023

Region	Country	Report Dates (2023)	New Cases	Num Deaths	Total Cases Since 1 Jan 2023	Total Deaths	Additional Comments
Africa	Nigeria	30 Apr–28 May	222	4	1,851	52	
Africa	Somalia	4 Jun–2 Jul	1,295	2	10,686	30	
Africa	South Africa	2 Jun–3 Jul	722	32	1,265	47	
Africa	Zambia	25 May–22 Jun	69	1	757	14	
Africa	Zimbabwe	28 May–9 Jul	1,781	34	3,430	78	
Africa	Benin	2023					No updates
Africa	Eswatini	2023					No updates
Africa	South Sudan	2023					No updates
Africa	United Republic of Tanzania	2023					No updates
Americas	Dominican Republic	20 Mar–15 Jun	8		99		
Americas	Haiti	11 Jun–10 Jul	6,701	42	33,058	405	
Americas	Mexico	4-Jul	1		1		First new case since 2016

Source: Data extracted from (ECDC, 2023).

2

Epidemiology and Disease Ecology

Overview

In the past 40–70 years, we have seen changes in the intensity, duration and types of diseases that are affecting the health and well-being of populations around the world. COVID-19 is not the first pandemic and certainly will not be the last pandemic as we continue to have outbreaks (e.g. global outbreak of monkeypox in 2022; Karim et al., 2022; Del Rio and Malani, 2022) (Figure 2.1).

In the end, we are all interested in staying healthy and minimizing our risk of getting sick. We want to avoid getting ill and as such, we want to know not only *how* to avoid getting ill but also *where* a disease may be located and *when* it is present, if there is a temporal component to its prevalence, and *what* precautions we can take to stop getting sick.

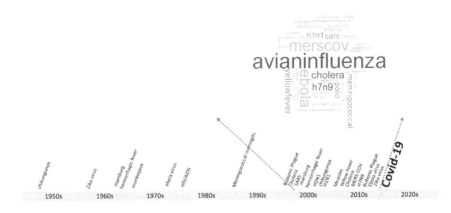

FIGURE 2.1
Overview of diseases and outbreaks recorded between 1950 and 2020 (adapted from Blanford and Jolly, 2021). In the last 20 years, we have seen the emergence and re-emergence of many diseases, some of which have led to outbreaks of varying intensities and durations (Blanford and Jolly, 2021); many occurring locally for decades before spreading to new locations (e.g. Zika, Theze et al., 2018; chikungunya, Villero-Wolf et al., 2019).

DOI: 10.1201/9781003435082-2

Diseases: An Expression of Change

Patterns influencing health and disease in the environment are complex and may lead to the emergence of new diseases in new locations or the establishment and re-emergence of diseases in places where they may have been absent (Table 2.1).

Diseases can be an expression of change. The change can be a change in the environment, behaviour or pathogen. In general, the main drivers of disease fall into one of four types of change: changes in *local environment, host behaviour, inadequate policies and governance* and/or changes in the *pathogen* (Table 2.2) (Woolhouse and Gowtage-Sequeria, 2005; Myers, 2017; Baker et al., 2022), which are captured by the epidemiologic triad (Figure 2.2) and governed by several factors as summarised in Table 2.4 for several diseases.

Patterns influencing health and disease in the environment are complex and require an understanding of the ecology of the disease (e.g. Figure 2.2). That is the agent (health risk or pathogen), a host (and host reservoirs) and how these interact in and with dynamically changing environments

TABLE 2.1

Categories of Infectious Diseases

Category	Description	Examples
Established infectious diseases	Endemic diseases that have been around for a sufficient amount of time to allow for a relatively stable and predictable levels of morbidity and mortality.	Examples are common diarrheal pathogens, drug-susceptible malaria, tuberculosis, helminthic and other parasitic diseases.
Newly emerging infectious diseases	Diseases that have been detected in the human host for the first time.	Nipah virus, severe acute respiratory syndrome virus (SARS), human metapneumovirus (hMPV) and new influenza subtypes (swine H1N1, avian H5N1 or avian H7N9) are examples in this category. Often RNA viruses causing respiratory diseases are in this category.
Re-emerging infectious diseases	Diseases that historically have infected humans but continue to • Reappear either in new locations or • In resistant forms or • Reappear after apparent control or elimination or • Under unusual circumstances	• Appear in new regions (e.g. West Nile virus (WNV) in the Americas) • e.g. drug-resistant bacterial infections (penicillin-resistant pneumococcal pneumonia, carbapenem-resistant hospital acquired infections); drug-resistant malaria and oseltamivir-resistant influenza. • Polio in parts of Africa, Dengue in Florida. • Release of anthrax in the USA in 2001.

Source: Fauci and Morens (2012); Horby et al. (2014).

TABLE 2.2

Main Drivers Associated with the Emergence and Reemergence of Human Pathogens

Type	Driver
Local environment	• Changes in land use or agricultural practices • Changes in human demographics and society • Changes in climate
Behavior (lifestyle) (social and economic)	• Mobility (international travel and/or international trade) • Events (Morens et al., 2004) • Anti-vaccination (Dube et al., 2015)
Governance and policies	• Hospital and medical procedures • Failure of public health programs • Contaminated food sources or water supplies • Poor population health (e.g. HIV and malnutrition)
Evolution	• Pathogen evolution and adaptation (e.g. antimicrobial drug resistance; Kennedy and Read, 2018), change in virulence (e.g. COVID-19; van Oosterhout et al., 2021); waning immunity (e.g. whooping cough; Borba et al., 2015)

Source: Adapted from Woolhouse and Gowtage-Sequeria (2005); Myers (2017); Baker et al. (2022).

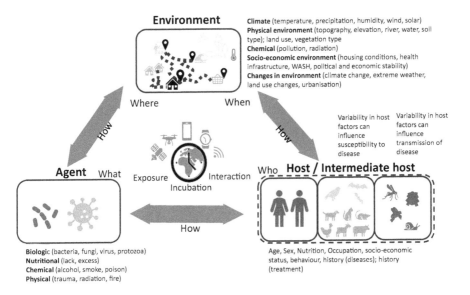

FIGURE 2.2

The Epidemiological Triad. Factors that can influence disease include an agent, a host and the environment in space and time. In the centre of the triangle is time. Time can represent an incubation period (the time between when the host is infected and when disease symptoms occur or when a vector can start to transmit a pathogen); or the duration of the illness or the amount of time a person can be sick before death or recovery occurs; or the period from an infection to the threshold of an epidemic for a population.

(space-time interactions). Thus, a better understanding of the processes associated with the emergence of new diseases, the re-emergence of known health risks in new locations or the management of established health risks requires a transdisciplinary approach, an idea promoted by the One Health concept (Trueba, 2014) and Planetary Health (Horton and Lo, 2015; Almada et al., 2017; Myers, 2017; Horton and Lo, 2015; Haines, 2016; Haines, 2017; Haines et al., 2018). More on Planetary Health will be provided in chapter 9, but first let's delve into some background on epidemiology.

Epidemiology

Epidemiology is the **study** of the **distribution** and **determinants** of **health-related states or events** in **specified populations** and the **application** of this study to the control of health problems (Last, 2001). It is concerned not only with the **frequency** (that is, the number of health events) but also with the relationship of that number to the size of the population and **pattern** (that is, the occurrence of health-related events by time, place and person) of health events in a population (Table 2.3).

Factors Influencing Disease

As already highlighted, health and disease risks and outcomes are affected by a variety of factors that include place-based influences such as **physical circumstances** (e.g. altitude, temperature regimes and pollutants), **social context** (e.g. social networks, behaviours, access to health services and perception of risk behaviours) and **economic conditions** (e.g. quality

TABLE 2.3

Definition of Epidemiology

Term	Explanation
Study	Includes: surveillance, observation, hypothesis testing, analytic research and experiments.
Distribution	Refers to analysis of: times, persons, places and classes of people affected.
Determinants	Include factors that influence health: biological, chemical, physical, social, cultural, economic, genetic and behavioural.
Health-related states and events	Refer to: diseases, causes of death, behaviours such as use of tobacco, positive health states, reactions to preventive regimes and provision and use of health services.
Specified populations	Include those with identifiable characteristics, such as occupational groups.
Application to prevention and control	The aims of public health-to promote, protect and restore health.

Source: Table adapted **from Bonita** et al. (2006).

of nutrition and ability to purchase services) (CSDGS_NRC_Ch6, 2010). Many of these are constantly shifting as people themselves move between each of these factors. Therefore, one of the most challenging components of understanding health impacts is understanding how where people live (geographical/environmental component) affects their health (CSDGS_NRC_Ch6, 2010).

Chain of infection (CDC, 2012): Disease occurs when an agent leaves its **reservoir/host** through a **portal of exit** and is conveyed by a **mode of transmission** to enter through an appropriate **portal of entry** to infect a **susceptible host**. Transmission may be **direct** (direct contact host-to-host, droplet spread from one host to another) or **indirect** (the transfer of an infectious agent from a reservoir to a susceptible host by suspended air particles, inanimate objects (vehicles or fomites), or animate intermediaries (vectors)).

Epidemiologic triad (Bonita et al., 2006; CDC, 2012): the traditional model of infectious disease causation has three components: an external agent, a susceptible host and an environment that brings the host and agent together so that disease occurs. The agent, host and environmental factors interrelate in a variety of complex ways that can result in the development of a disease. We will revisit this again throughout the book. In the meantime, let us examine Figure 2.2, as it provides a summary of different agents, environmental factors and host characteristics that might affect disease outcomes. Examples of different factors that can influence the occurrence of a range of diseases are provided in Table 2.4.

An Integrated Health Approach: Managing and Planning Health and Disease Risks

The ultimate goal of healthcare officials is to minimize health risks by promoting, protecting and restoring health through the management and planning of public health resources. This consists of several components that include:

- **Ecology of disease (monitoring, surveillance, research, scientific experiments, and surveys)**: to better understand the interaction of the disease agent and host in the environment through the collection of data and observations to enable the identification of characteristics that might be important in the occurrence of a disease. This may include the source(s) of infection, the causative agent(s), disease reservoirs, mode of transmission, susceptible host(s) and environmental factors (see Epidemiologic Triad) (Figure 2.2).

- **Response (diagnosis and treatment)**: provide the necessary healthcare infrastructure to enable the diagnosis and treatment of disease and the recovery from disease when possible with timely and relevant treatment.

TABLE 2.4

Factors Important in the Occurrence of a Range of Infectious Diseases and Major Noncommunicable Diseases

Disease	Key Factors
Vector-borne diseases Mosquito: dengue, malaria, WNV, chikungunya, Zika, yellow-fever ticks (Lyme)	Transmission via insect vectors; biting rate; influence of climate and environment on vector and pathogen development; animal reservoirs; interaction between strains within-host and between-host; exposure; drug-resistance
Childhood infectious diseases e.g. measles, whooping cough	Immunizing infections; spatial and temporal heterogeneity; demography; socio-economic; age structure; household structure.
Sexually transmitted infections e.g. HIV, gonorrhoea	High/low risk groups; nonrandom contact structure; partnerships; within-host strain diversity and evolution; space-time interaction; drug-resistance
Novel emerging infections e.g. OSCARS, Nipah virus, MERS and other coronaviruses	Behaviour change; global interconnectedness and international cooperation in control; responses in absence of biomedical measures; animal reservoirs.
Influenza including avian influenza	Distribution of prior immunity; within-population and species strain differences, virus evolution and interaction; role of wildlife and farm animals.
Bacterial infections Pneumococcal disease, MRSA and tuberculosis	Antibiotic/drug-resistance; adaptive dynamics.
Veterinary outbreaks e.g. BSE and FMD. Mad cow disease and foot and mouth	Fixed spatial locations with changing contact networks.
Macroparasites	Clumped infections, multistrain and multispecies infections, cross immunity, concurrent infections.
Neglected tropical diseases (NTDs) Podoconiosis, leprosy, schistosomiasis	Influence of the environment (e.g. direct contact with soil); socio-economic factors, education, access to treatment and healthcare.
Noncommunicable diseases (NCDs)	
Lifestyle, exposure to environments (physical or human-induced) Chronic diseases: cardiovascular diseases (e.g. heart attacks and stroke), cancers, chronic respiratory diseases (e.g. chronic obstructed pulmonary disease and asthma) Diabetes/Obesity overdoses – opioid Environmental diseases – heat (heat-related deaths; skin cancer), safety (crime, crashes), malnutrition Poverty-related diseases (e.g. Noma)	Social structure (Social position): Social class, age, sex, ethnicity Lifestyle influences (individual behaviours) Smoking nutrition, physical activity Psychosocial factors, physiological influences (the body) Blood pressure, cholesterol, obesity, blood glucose Environmental influences (places) Geographic location, housing conditions, occupational risks, access to services, availability of food, environment (climate); electricity and cooking type

(Continued)

TABLE 2.4 (*Continued*)

Factors Important in the Occurrence of a Range of Infectious Diseases and Major Noncommunicable Diseases

Disease	Key Factors
Mental health	Variety of reasons ranging from changes in personal circumstances, environmental influences, etc. Access to services, individual, social and community
Women's health	Variety of reasons ranging from environment, behavior, cultural, etc. Access to services, individual, social and community

Sources: **Table Adapted from** Heesterbeek et al. (2015) Additional sources: Bonita et al. (2006); WHO (2015);. Podoconiosis: Deribe et al., 2013, 2018; Harter, 2022; Leprosy: Gabbatt, 2023).

- **Response (control and prevention)**: interventions used for the control of and prevention of disease so as to minimize further risks and enhance the well-being of the public.
- **Communication**: provide the public with relevant information about a disease or health risk that is easily accessible, up to date and easy to understand (Figure 2.3).

In summary, when thinking about health and disease, we need to think of all of these elements so that we can understand what is taking place and start to think about how to respond. Who is affected, where, by what and when? Why are they being affected and how? What is needed to prevent further infections? (Table 2.5)

Investigating, Monitoring and Tracking an Outbreak

Epidemic and pandemic-prone diseases threaten public health security and can be responsible for high levels of morbidity and mortality, which can have a devastating impact on economies. Therefore, when an outbreak occurs, the primary goal is to control the outbreak and prevent additional cases. To do so, it may be relevant to think about *who, what, where, when, how and why* so that control and intervention efforts can be directed against one or more segments in the chain of transmission (agent, source, mode of transmission, portal of entry or host) that are susceptible to intervention. A lot of sleuthing is required to identify the different segments in the chain of transmission: the agent, source of infection, mode of transmission, portal of entry, host and incubation period, as well as who iss affected (e.g. age, race and sex). This is an iterative process that requires the collection of data, visualising and analysing the data as captured by Figure 2.4.

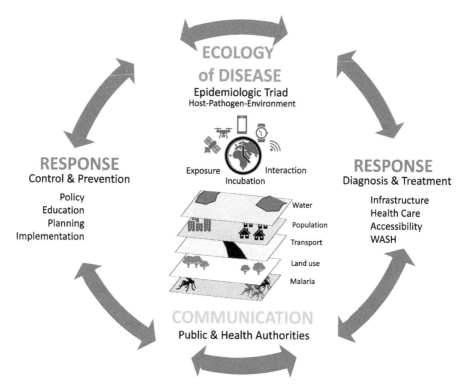

FIGURE 2.3

Integrated Health Approach. Ecology of Disease through monitoring, surveillance and analysis and integrating the epidemiologic triad (host-pathogen-environment); Response through directed control and prevention strategies to minimize future risks; and Communication through different channels to disseminate up-to-date information about risks and preventative methods.

Eight Steps for Investigating and Managing an Outbreak

Several steps are required for investigating and managing an outbreak. These can be broken down into about eight steps that include (CDC, 2012; Smith et al., 2015):

1. Establish the existence of an outbreak
2. Confirm diagnosis
3. Define and identify outbreak cases
4. Describe cases and develop hypotheses
5. Evaluate hypotheses and draw conclusions
6. Compare with established facts
7. Execute prevention measures
8. Communicate findings

TABLE 2.5

Breakdown of Descriptive and Analytic Elements Important for Determining Associations between Health and Disease Risks in Time, Place and Persons

	Descriptive
Where	Where did the exposure occur?
	Place – exposure environment and activity space (home, work, or leisure activity)
Who	Who is at risk?
	How many people are affected?
	What are the attributes of the person(s) affected?
What	What are the health and disease problems of the community?
	What are the attributes of the illness?
When	When are they at risk?
	At a specific time of year/season?
	Over what period of time? (latency of onset of disease)
	Analytic
Why	What are the causal agents?
	What factors affect outcomes?
How	How are they at risk? (mode of transmission)
	Intervention and priority
Intervention	**How to Intervene** and minimize the occurrence and spread? (identify factors or conditions that can be manipulated to modify or prevent disease)
Priority	**How to Prioritize** and respond quickly and effectively to minimize the occurrence of an outbreak?

Source: CDC (2012); Bonita et al. (2006).

During each of the eight steps, a variety of visualizations along with descriptive and spatial analysis to understand what interactions are taking place, how the disease maybe spreading/diffusing and what actions or responses may be necessary to protect the population (Figure 2.5).

Communication of information during an ongoing pandemic is vital for keeping everyone informed and may occur through a variety of channels (e.g. local public health authorities, social media, outbreak reports, etc.), as shown during the COVID-19 pandemic 2020–2023. Not only is it useful for communicating risk but also for communicating of interventions and disease prevention measures that may be necessary for the general public to adhere to (e.g. travel restrictions, vaccination requirements, isolation requirements, and curfews (Blanford et al., 2022)). Communication may take many forms and evolve over time, as has taken place during the COVID-19 pandemic, with weekly news updates and regular broadcasts by leaders at a country level and the development of digital tools to provide information in various forms through statistical figures, interactive maps, and a variety of dashboards that can be updated at set time intervals, as shown in Figure 2.6. Each of these

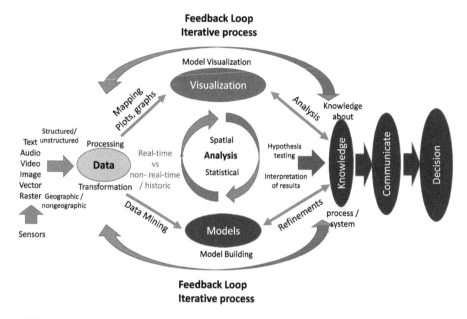

FIGURE 2.4

Geographic Information and Health Analysis: an iterative process (adapted from (Keim et al. 2010); (Blanford et al. 2019).

examples provides different types of information. For example, Figure 2.6a captures COVID-19 patients at a municipality level for a 2-week time interval. Data were updated once a week. Figure 2.6b captures the risk levels throughout Europe were compiled using a variety of information described in Figure 1.13. Data were aggregated to administrative level (Provincial, State) and also updated once a week. For the Johns Hopkins COVID-19 Dashboard, data were updated daily (Figure 2.6c).

Interventions

An important way to combat disease risks is to understand the transmission dynamics of the ecology of the disease and break the chain of transmission. This can be done in a number of different ways and maybe:

- **Medical**: treatment of the infected and vaccinations
- **Environment**: adjusting the environment and improving infrastructure

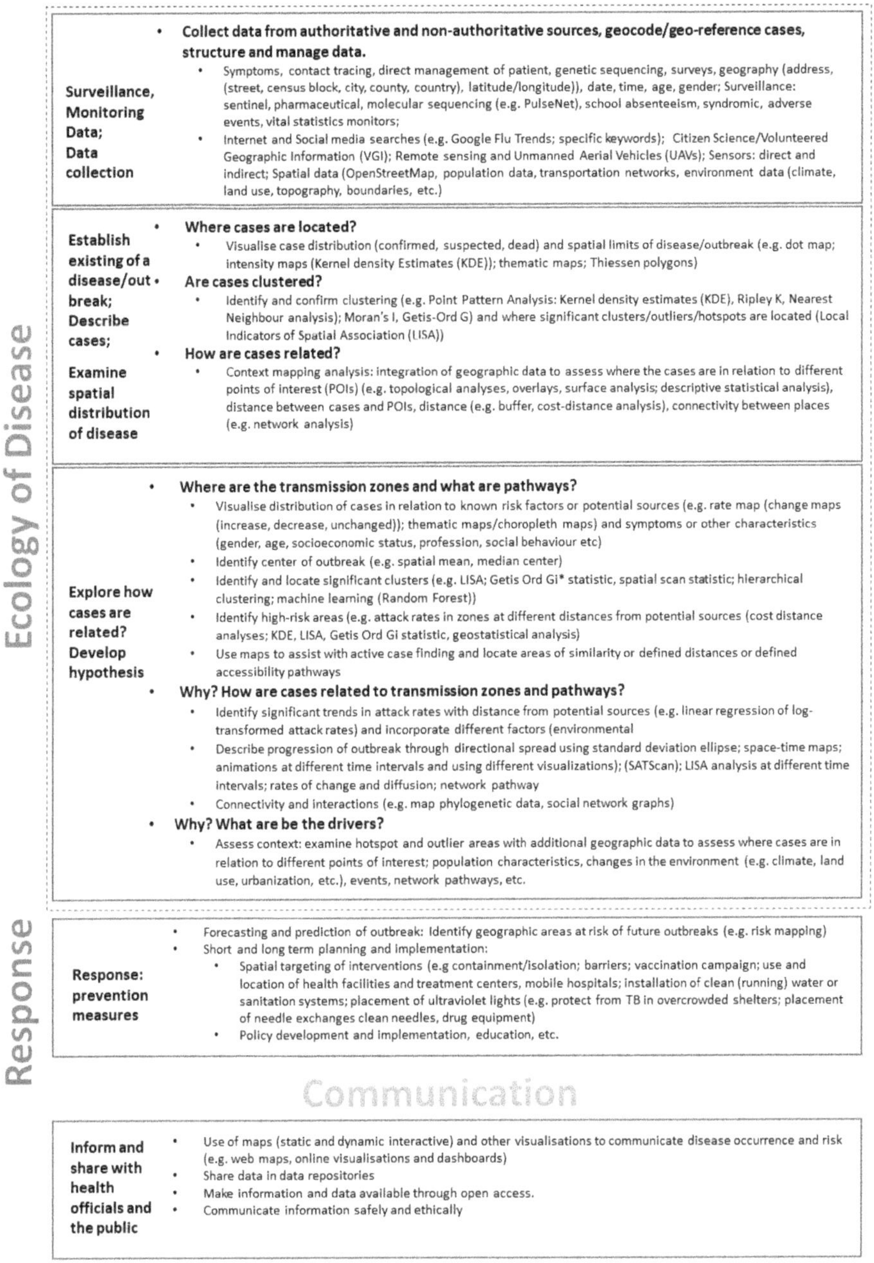

Ecology of Disease

Surveillance, Monitoring Data; Data collection

- Collect data from authoritative and non-authoritative sources, geocode/geo-reference cases, structure and manage data.
  - Symptoms, contact tracing, direct management of patient, genetic sequencing, surveys, geography (address, (street, census block, city, county, country), latitude/longitude)), date, time, age, gender; Surveillance: sentinel, pharmaceutical, molecular sequencing (e.g. PulseNet), school absenteeism, syndromic, adverse events, vital statistics monitors;
  - Internet and Social media searches (e.g. Google Flu Trends; specific keywords); Citizen Science/Volunteered Geographic Information (VGI); Remote sensing and Unmanned Aerial Vehicles (UAVs); Sensors: direct and indirect; Spatial data (OpenStreetMap, population data, transportation networks, environment data (climate, land use, topography, boundaries, etc.)

Establish existing of a disease/outbreak; Describe cases; Examine spatial distribution of disease

- Where cases are located?
  - Visualise case distribution (confirmed, suspected, dead) and spatial limits of disease/outbreak (e.g. dot map; intensity maps (Kernel density Estimates (KDE)); thematic maps; Thiessen polygons)
- Are cases clustered?
  - Identify and confirm clustering (e.g. Point Pattern Analysis: Kernel density estimates (KDE), Ripley K, Nearest Neighbour analysis); Moran's I, Getis-Ord G) and where significant clusters/outliers/hotspots are located (Local Indicators of Spatial Association (LISA))
- How are cases related?
  - Context mapping analysis: integration of geographic data to assess where the cases are in relation to different points of interest (POIs) (e.g. topological analyses, overlays, surface analysis; descriptive statistical analysis), distance between cases and POIs, distance (e.g. buffer, cost-distance analysis), connectivity between places (e.g. network analysis)

Explore how cases are related? Develop hypothesis

- Where are the transmission zones and what are pathways?
  - Visualise distribution of cases in relation to known risk factors or potential sources (e.g. rate map (change maps (increase, decrease, unchanged)); thematic maps/choropleth maps) and symptoms or other characteristics (gender, age, socioeconomic status, profession, social behaviour etc)
  - Identify center of outbreak (e.g. spatial mean, median center)
  - Identify and locate significant clusters (e.g. LISA; Getis Ord Gi* statistic, spatial scan statistic; hierarchical clustering; machine learning (Random Forest))
  - Identify high-risk areas (e.g. attack rates in zones at different distances from potential sources (cost distance analyses; KDE, LISA, Getis Ord Gi statistic, geostatistical analysis)
  - Use maps to assist with active case finding and locate areas of similarity or defined distances or defined accessibility pathways
- Why? How are cases related to transmission zones and pathways?
  - Identify significant trends in attack rates with distance from potential sources (e.g. linear regression of log-transformed attack rates) and incorporate different factors (environmental
  - Describe progression of outbreak through directional spread using standard deviation ellipse; space-time maps; animations at different time intervals and using different visualizations; (SATScan); LISA analysis at different time intervals; rates of change and diffusion; network pathway
  - Connectivity and interactions (e.g. map phylogenetic data, social network graphs)
- Why? What are be the drivers?
  - Assess context: examine hotspot and outlier areas with additional geographic data to assess where cases are in relation to different points of interest; population characteristics, changes in the environment (e.g. climate, land use, urbanization, etc.), events, network pathways, etc.

Response

Response: prevention measures

- Forecasting and prediction of outbreak: Identify geographic areas at risk of future outbreaks (e.g. risk mapping)
- Short and long term planning and implementation:
  - Spatial targeting of interventions (e.g containment/isolation; barriers; vaccination campaign; use and location of health facilities and treatment centers, mobile hospitals; installation of clean (running) water or sanitation systems; placement of ultraviolet lights (e.g. protect from TB in overcrowded shelters; placement of needle exchanges clean needles, drug equipment)
- Policy development and implementation, education, etc.

Communication

Inform and share with health officials and the public

- Use of maps (static and dynamic interactive) and other visualisations to communicate disease occurrence and risk (e.g. web maps, online visualisations and dashboards)
- Share data in data repositories
- Make information and data available through open access.
- Communicate information safely and ethically

FIGURE 2.5

An integrated public health response approach for investigating a disease outbreak and other health risks. A breakdown of the steps and associated spatial analysis methods that are useful at each stage of investigating a disease outbreak. (Adapted from Blanford and Jolly, 2021; CDC, 2012; Smith et al., 2015.)

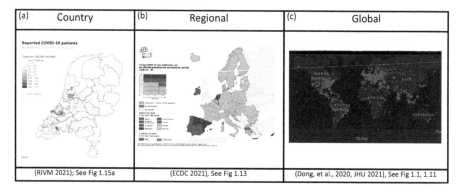

(a) Country	(b) Regional	(c) Global
(RIVM 2021); See Fig 1.15a	(ECDC 2021), See Fig 1.13	(Dong, et al., 2020, JHU 2021), See Fig 1.1, 1.11

FIGURE 2.6
Examples of different maps used to communicate risk of COVID-19 during the 2020–2023 pandemic at (a) country; (b) regional; and (c) global level.

- **Mobility/Contact**: reducing direct contact, quarantine and isolation
- **Behaviour**: education, incentives and government policies or restrictions

During the COVID-19 pandemic of 2020–2023, a variety of interventions were used to reduce the interaction of infected populations with uninfected populations. This was accomplished using all four of the interventions to enhance prevention:

- **Environment**: disinfect surfaces and air, ventilation of closed spaces and distancing measures
- **Medical**: vaccination and testing (testing centres and self-tests)
- **Mobility/Contact**: reduction in mobility of populations by closing borders, implementation of lockdowns and curfews, quarantine/ isolation measures and distancing measures to reduce contact and interaction
- **Behaviour**: educating the population on how to reduce transmission by wearing masks, washing hands and distancing; policies and restrictions implemented; and advertisements to inform the public

Often multiple approaches are needed. Bradley (1974) reclassified diseases based on their transmission characteristics (Table 2.6). By taking this approach, a multi-phased and multi-layered preventative strategy can be developed and implemented, each tackling a different aspect of the transmission route. For diseases with multiple transmission routes, it may be necessary to use one or many preventative strategies to be effective.

As seen by a variety of intervention programs (e.g. polio (Chapter 1), eliminating malaria in Suriname (see vector-borne disease chapter) and leprosy

TABLE 2.6

Reclassification of Water-Related Diseases Based on Their Transmission Route

Transmission Route	Description	Disease Group Example	Preventative Strategy
Water-borne	Pathogen is in the water that is ingested	Faeco-oral e.g. diarrhoea, dysentery, typhoid fever, leptospirosis	Improve water quality; alert of risk in water body and avoid contact with water (e.g. swimming)
Water-washed	Person-to-person transmission due to lack of water for hygiene	Skin and eye infections e.g. Scabies, trachoma; COVID-19	Improve water quantity and accessibility; Improve hygiene
Water-based	Transmission via an aquatic intermediate host	Water-based e.g. snails (schistosomiasis); parasite-infected water fleas (guinea worm)	Decrease need for water contact; control snail populations; improve quality
Water-related insect vector	Transmission by insects that breed in water or bite near water	Water-related insect vectors e.g. mosquitoes (dengue, malaria); flies (trypanosomiasis)	Improve surface water management; destroy breeding sites of insects; decrease need to visit breeding sites.

Source: Compiled from Bradley (1974); Feachem (1977); Cairncross and Feachem (1993); Morgan et al. (2002).

in different parts of the world), to be effective, they require constant surveillance and multi-phased and levelled strategies targeting different transmission routes. Other diseases can also benefit from taking a variety of approaches (e.g. sexually transmitted diseases and cholera) that are based on various host-environment-pathogen-related characteristics that may include those related to behavioural or socio-economic characteristics and vulnerabilities (e.g. Kienberger and Hagenlocher, 2014).

Summary

Tackling global health challenges through local actions is vital. To be able to do so requires us to be able to examine health and disease processes across dynamically changing environments at a local and global level. Often, the scale at which we can analyse some of these processes is driven not only by what we know about how a disease system works but also by what data is available, as illustrated by Figure 2.7. Having data available at the level that John Snow collected during the cholera outbreak of 1854 would be ideal (Granular data at a local spatial level), however this is likely the exception.

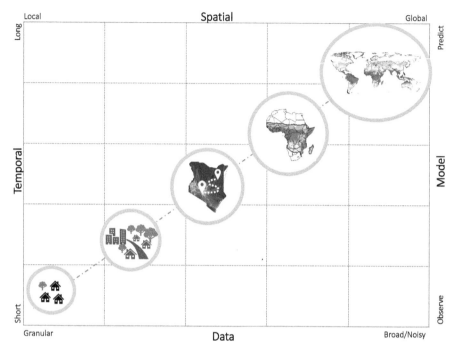

FIGURE 2.7

Spatial-temporal analysis framework for health and disease mapping and modelling. (i) **Temporal scale**: short-term (e.g. hourly or daily) to long-term (e.g. several weeks to multiple years or longer to enable inter-annual comparisons); (ii) **Spatial scale**: local (e.g. individual, household or village level) to global ; (iii) **Data** can be granular (e.g. GPS point location) to broad (e.g. aggregated to an areal unit such as a municipality or country) or noisy (e.g. social media posts about a topic using a variety of geographic scales); or (iv) **Model**: map observations and model or predict what is taking place across space and time.

Activity – Mapping Cholera and Examining the Extent of an Outbreak

Background

Cholera is still a large problem today in many parts of the world. Geospatial data and spatial analysis are being used to assess cholera prevalence and create maps to develop targeted vaccination campaigns so as to minimize disease risks (e.g. vaccine deployment for cholera in Kolkata; You et al., 2013).

During the mid-19th century, many European cities, including London, suffered from a series of cholera outbreaks. Dr John Snow, regarded as one of the founding fathers of modern epidemiology, theorized that cholera was spread through contaminated water. At the time, London's water supply system consisted of shallow public wells where people could pump their own water to carry home or have it delivered by water companies that drew water from the Thames using a series of water pumps.

During the 1854 cholera outbreak, Snow mapped the 13 public wells and all of the known cholera deaths around Soho and noted the spatial clustering of cases around one particular water pump on the southwest corner of the intersection of Broad (now Broadwick) Street and Cambridge (now Lexington) Street (Snow, 1855). To prevent further deaths, Snow had the pump handle removed from the Broad Street pump after examining water samples, which confirmed the presence of an unknown bacterium in the Broad Street samples.

Snow published a map of the epidemic to support his theory. The map showed the locations of the 13 public water pumps and the 578 cholera deaths mapped by home address, marked as black bars stacked perpendicular to the streets. To learn more about cholera in London, read **On the Mode of Communication of Cholera** by John Snow or the short overview Bynum (2013). In this activity you too can analyze the cholera outbreak of 1854 using a variety of spatial analysis and statistical methods.

Overview of Activity

- Visualize the cholera outbreak of 1854
- Examine the relationship of cholera deaths by pump
- *Where* was the outbreak concentrated?
    - Where was the mean centre of the outbreak?
    - Where was the highest density of cholera deaths?
- Are the cholera deaths really clustered?
- Conclusion. *What* did you conclude from your analysis?

Data

Data Set	Description	Date	Source
Pump	Location of water pumps Pumps.shp	1854	Blanford
Cholera deaths	Total no. of deaths per address. Each point represents the location of the cholera outbreak of 1854 Cholera_deaths.shp	1854	Blanford

Data of pump locations and cholera death locations were digitized from the John Snow Cholera image. **Software**: ArcGISPro will be used to perform the analysis.

	Points 	Kernel Density Estimate 	Nearest Neighbour Analysis 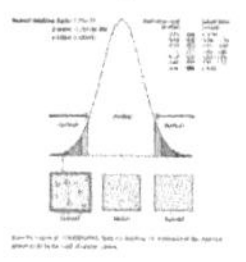
PPA Method			
	Measuring geographic distributions	Exploratory Data Analysis	Distance-based analysis
Description	Descriptive spatial statistics -Mean Center; -Central/Median Center; -Standard distance; -Standard Deviational Ellipse	Used to "visually enhance" a point pattern by showing where features are concentrated. Captures disease intensity or concentrations through a continuous surface.	Measures spatial dependence based on distances of points from one another. Calculates a nearest neighbour index based on the average distance from each feature to its nearest neighbouring feature.
Limitations		Sensitive to change in bandwidth.	Very dependent on the the area of the study region. Based on the mean distance to the nearest neighbour.

FIGURE 2.8
Summary of different point pattern analysis methods available for analysing point data.

Point Pattern Analysis

Point pattern analysis is a suite of analyses available for analysing points. The analyses range from qualitative to quantitative methods. As humans, we are predisposed to seeing patterns when there may not necessarily be any existing pattern; therefore, it is important to use statistical methods to determine if a spatial pattern really exists or not. The qualitative methods allow us to visualize and describe the spatial pattern, while the quantitative methods allow us to evaluate whether a pattern really exists or not by comparing an *observed* pattern with an *expected* pattern (O'Sullivan and Unwin, 2010). Thus, point pattern analyses are useful as they can help us determine whether a pattern is clustered or not, providing us with statistical confirmation of what we may have assumed when we first mapped the disease incidence. A summary of these methods is provided in Figure 2.8. More details can be found in Chapter 7.

Visualizing the Cholera Outbreak of 1854

Map the cholera outbreak that occurred in London in 1854 and add the pumps to the map (Figure 2.9a). You may also want to change the symbology of the points to graduated symbols (right-click layer, symbology, select graduated symbols, field=count).

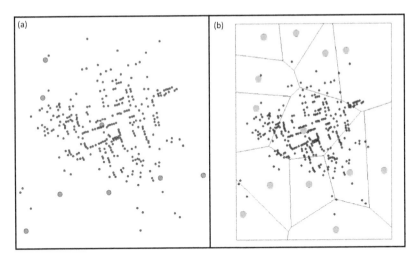

FIGURE 2.9
Map of (a) distribution of cholera deaths (black dots) and water pumps (pink dots) (b) Thiessen polygon based on water pumps (green dots).

Next, you may want to see how many deaths occurred within a certain area. To create data summaries, use the summary statistics tool (**Toolboxes – Analysis Tools – Statistics – Summary Statistics – you may need to add the Toolbox (Insert – Toolbox)**). Add Cholera_deaths_prj to the project and summarize the number of deaths by streetname.

What did you find? Were the deaths concentrated in a particular area? Were there any outliers? What pumps were closest to the outbreak area?

Examine the Relationship of Deaths by Pump

In Chapter 1, the spatial analysis functions of GIS were introduced. To recap, these fall into five classes: measurement, topological analysis, network analysis, surface analysis and statistical analysis. To analyse the relationship of cholera deaths to pump locations, we will use some of the measurement capabilities to examine spatial distances.

Create Thiessen Polygons around each water pump (Navigate to **Analysis Tools – Proximity – Create Thiessen Polygon**) to delineate areas closest to the associated water pump than to any other water pump. When you create the Thiessen Polygon, set the processing extent in the environments to the study boundary. The outputs will look something like Figure 2.9b.

Calculate the number of deaths associated with each pump and summarize your findings. First, we will perform a spatial join to add the pumpname to each cholera death location (Navigate to **Analysis Tools – Overlay – Spatial Join**).

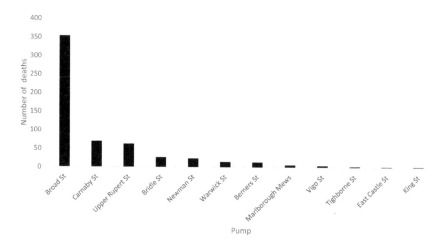

FIGURE 2.10
Figure capturing the total number of deaths by pump.

Target Features: **Cholera_deaths_prj**
Join Features: **ThiessenPolygon**
Match Option: **Completely Within**
Output Table: **CholeraDeathPump**

Keep the default settings for the rest of the options.

Secondly, we will use the summary statistics tool to calculate the total number of deaths by pump. Once you have your table, again copy and paste the results to excel and create a graph by pump as shown in Figure 2.10 (Navigate to **Analysis Tools – Statistics – Summary Statistics**).

Input Table: **CholeraDeathPump**
Case Field: **pumpname**
Statistics Field: **Count** and Statistics Type: **Sum**

How useful was this in assessing the concentration of the outbreak and identifying the potential source of the deaths?

Where Was the Outbreak Concentrated?

To assess where the centre of the outbreak was and where the highest density of the outbreak was you will use a variety of methods.

Where Was the Mean Centre of the Outbreak?

To find the centre of concentration (or geographic centre) of the outbreak, calculate the mean centre of all of the cholera deaths using the mean centre tool (Navigate to **Spatial Statistics Tools – Mean Centre**) and input **Cholera_deaths_prj**.

Where Was the Highest Density of Cholera Deaths?

Kernel density estimate (KDE) is an "exploratory spatial data analysis" (ESDA) that can be used to assess the intensity of an event based on the mean number of events occurring within a specified distance. In other words, a density surface shows where features are concentrated, providing a visual representation of the location(s) of potential hotspots. In this case it is useful for showing where disease concentrations are located through a continuous surface of 'peaks' and 'valleys'. To create a continuous surface in kernel estimation, a spatial window or *kernel* is moved across the study area, and the density of events is computed within the kernel or spatial window. The window is a circle with a constant radius or *bandwidth.* Events within the window are weighted according to their distance from the centre of the window, the point at which the density is being estimated. The kernel function describes mathematically how those weights vary over a distance. This method is useful for mapping the uneven and/or irregular shaped spatial distribution of health events.

Bandwidth Size Considerations

A key issue with the kernel estimation method is the selection of the bandwidth. Generally speaking, large bandwidths smooth the data more, removing local variation, while, in contrast, small bandwidths do very little smoothing, producing an irregular 'spiky' map. When selecting a bandwidth, a compromise is reached between these two extremes, depending on the objectives of the analysis. See the chapter on spatial clustering for additional information.

When a density surface is created, individual cell values are calculated by dividing the number of features that fall within the search area (e.g., cholera deaths) by the size of the search area. The resulting value is then assigned to the cell. If two or more kernels overlap at a cell centre, the value for that cell is the sum of the overlapping kernel values divided by the area of the search radius.

In summary, a KDE surface is created by fitting a smoothly curved surface over each point, and the number of points that fall within a kernel is summed. The total is then divided by the area of the kernel to provide the grid cell value. As shown in the figure above, the surface value will be highest at the location of the point and diminish with increasing distance from the point (at the centre), reaching zero at the outer boundary edge of the kernel.

A "neighbourhood" or kernel is defined around each grid cell, consisting of all grid cells with centres within the specified kernel (search) radius.

Create a kernel density surface to show where the key area(s) of the cholera outbreak were concentrated. Before you settle on a specific output, explore how the bandwidth affects the density surface.

Start with the default (Navigate to **Spatial Analyst Tools – Density – Kernel Density**)

Overlay the mean centre of the outbreak with the KDE surface output (e.g. Figure 2.11).

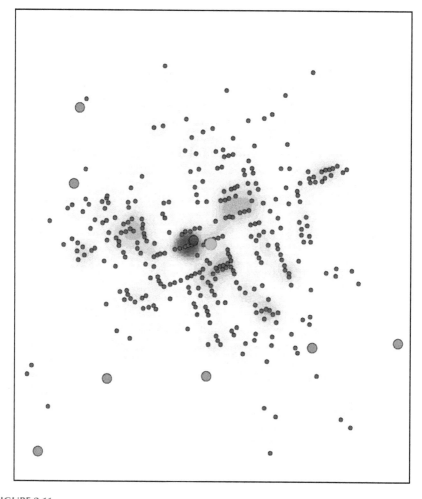

FIGURE 2.11

Map capturing the concentration of cholera deaths, the mean center of deaths (green dot) in relation to the water pumps (pink dots).

- How does the location of the mean centre of the outbreak compare with the KDE surface?
- Where was the mean centre of the outbreak in comparison to the highest concentration of deaths?
- What was the closest water pump?
- What did you find when you explored the different search radius settings? What are the limitations of using the KDE method?

Are the Cholera Deaths Really Clustered?

Based on the KDE outputs, it certainly looks as if the cholera outbreak was clustered. However, to be sure, we will use a distance-based statistical analysis method such as the nearest neighbour index that measures spatial dependence by examining the distance of points from one another.

The Nearest Neighbour Index (Navigate to **Spatial Statistics Tools – Analysing patterns – Average Nearest Neighbour**) method calculates a nearest neighbour index based on the average distance from each observation to its nearest neighbouring observation and compares this with the expected mean distance if the distribution was random. If NNI=0, the distribution is clustered; if NNI=1, the distribution is random; and if NNI>2.14, then the distribution is considered to be uniform.

Ensure that you tick the **Generate Report**. To view the report, go to RESULTS, click on Average Nearest Neighbour Analysis, and click on Report File. An HTML file will open in your browser with the following information

Although this method provides a quick statistical assessment of whether the observations are clustered or not, there are some limitations. Calculations are dependent on the size of the study area. The NNI is based only on the overall mean distance to the nearest neighbour and misses local variations that may exist. Furthermore, this analysis only uses the geographic location of the observation when making an assessment; therefore, quantities that may be associated with each observation are not taken into consideration.

- What conclusions did you come to about the cholera outbreak?
- How useful were the point pattern analysis methods for evaluating the cholera outbreak?
- Highlight any limitations with the data and methods used. Were there any challenges you encountered? What, if any improvements would you make?

References

Allen, L. (2017) Are we facing a noncommunicable disease pandemic? *Journal of Epidemiology and Global Health*, 7, 5–9.

Almada, A. A., C. D. Golden, S. A. Osofsky & S. S. Myers (2017) A case for Planetary Health/GeoHealth. *Geohealth*, 1, 75–78.

Baker, R. E., A. S. Mahmud, I. F. Miller, M. Rajeev, F. Rasambainarivo, B. L. Rice, S. Takahashi, A. J. Tatem, C. E. Wagner, L. F. Wang, A. Wesolowski & C. J. E. Metcalf (2022) Infectious disease in an era of global change. *Nature Reviews Microbiology*, 20, 193–205.

Blanford, J. I. & A. Jolly (2021) Public Health Needs GiScience (Like Now!). In *AGILE Conference*. Greece: AGILE.

Blanford, J. I., N. Beerlage-de Jong, S. Schouten, A. Friedrich & V. Araujo-Soares (2022) Navigating travel in Europe during the pandemic: From mobile apps, certificates, quarantine to traffic light system. *Journal of Travel Medicine*, 29(3), taac006.

Bonita, R., R. Beaglehole & T. Kjellstrom (2006) *Basic Epidemiology*. Geneva, Switzerland: World Health Organization (WHO). https://apps.who.int/iris/bitstream/10665/43541/1/9241547073_eng.pdf (last accessed).

Borba, R. C. N., V. M. Vidal & L. O. Moreira (2015) The re-emergency and persistence of vaccine preventable diseases. *Annals of the Brazilian Academy of Sciences*, 87, 1311–1322.

Bradley, D. J. (1974) Human Rights in Health. In *Ciba Foundation Symposium 23*, 81–98. Amsterdam: Associated Scientific Publishers.

Bynum, W. (2013) In retrospect: On the mode of communication of cholera. *Nature*, 495, 169–170.

Cairncross, S. & R. Feachem (1993) *Environmental Health Engineering in the Tropics*. Chichester, UK: John Wiley & Sons.

CDC (2012) *Principles of Epidemiology in Public Health Practice*. U.S. Department of Health and Human Services, Centers for Disease Control and Prevention. https://www.cdc.gov/ophss/csels/dsepd/SS1978/SS1978.pdf (last accessed April 10 2024).

CSDGS _NRC_Ch6 (2010) Ch 6: How Does Where People Live Affect Their Health? In *Understanding the Changing Planet: Strategic Directions for the Geographical Sciences*, eds. Committee on Strategic Directions for the Geographical Sciences in the Next Decade & National Research Council, 67–74. Washington, DC: National Academy of Sciences.

Del Rio, C. & P. N. Malani (2022) Update on the monkeypox outbreak. *JAMA*, 328, 921–922.

Deribe, K., S. J. Brooker, R. L. Pullan, A. Hailu, F. Enquselassie, R. Reithinger, M. Newport & G. Davey (2013) Spatial distribution of podoconiosis in relation to environmental factors in Ethiopia: a historical review. *PLoS One*, 8, e68330.

Deribe, K., J. Cano, M. L. Trueba, M. J. Newport & G. Davey (2018) Global epidemiology of podoconiosis: A systematic review. *PLoS Neglected Tropical Diseases*, 12, e0006324.

Dong, E., H. Du & L. Gardner (2020) An interactive web-based dashboard to track COVID-19 in real time. *Lancet Infectious Diseases*, 20, 533–534.

Dube, E., M. Vivion & N. E. MacDonald (2015) Vaccine hesitancy, vaccine refusal and the anti-vaccine movement: influence, impact and implications. *Expert Review of Vaccines*, 14, 99–117.

ECDC (2021) Weekly Maps in Support of the Council Recommendation on a Coordinated Approach to Travel Measures in the EU. https://www.ecdc.europa.eu/en/covid-19/situation-updates/weekly-maps-coordinated-restriction-free-movement (last accessed Jul 31 2021).

Fauci, A. S. & D. M. Morens (2012) The perpetual challenge of infectious diseases. *The New England Journal of Medicine*, 366(5), 454–461.

Feachem, R. G. (1977) Infectious disease related to water supply and excreta disposal facilities. *Ambio*, 6, 55–58.

Gabbatt, A. (2023) CDC Warning as Leprosy Cases Increase in Florida. https://www.theguardian.com/us-news/2023/aug/01/florida-leprosy-cases-increase-cdc (last accessed Sep 24 2023).

Haines, A. (2016) Addressing challenges to human health in the Anthropocene epoch-an overview of the findings of the Rockefeller/Lancet Commission on Planetary Health. *Public Health Reviews*, 37, 14.

—— (2017) Addressing challenges to human health in the Anthropocene epoch-an overview of the findings of the Rockefeller/Lancet Commission on Planetary Health. *International Health*, 9, 269–271.

Haines, A., C. Hanson & J. Ranganathan (2018) Planetary Health Watch: integrated monitoring in the Anthropocene epoch. *Lancet Planet Health*, 2, e141–e143.

Harter, F. (2022) A Common Condition: How Wearing Shoes Could Eliminate One of the World's Most Neglected Tropical Diseases. https://www.theguardian.com/global-development/2022/nov/17/how-wearing-shoes-tropical-diseases-podoconiosis-acc (last accessed 3 Sep 2023).

Heesterbeek, H., R. M. Anderson, V. Andreasen, S. Bansal, D. De Angelis, C. Dye, K. T. Eames, W. J. Edmunds, S. D. Frost, S. Funk, T. D. Hollingsworth, T. House, V. Isham, P. Klepac, J. Lessler, J. O. Lloyd-Smith, C. J. Metcalf, D. Mollison, L. Pellis, J. R. Pulliam, M. G. Roberts, C. Viboud & I. D. D. C. Isaac Newton Institute (2015) Modeling infectious disease dynamics in the complex landscape of global health. *Science*, 347, aaa4339.

Horby, P. W., N. T. Hoa, D. U. Pfeiffer & H. F. L. Wertheim (2014) *Drivers of Emerging Zoonotic InfectiousDiseases*. Toyko, Japan: Springer.

Horton, R. & S. Lo (2015) Planetary health: A new science for exceptional action. *Lancet*, 386, 1921–1922.

IHME (2024) Global Burden Disease Compare Visualization Hub. https://vizhub.healthdata.org/gbd-compare/. Accessed 27 April 2024

JHU (2021) COVID-19 Dashboard. https://coronavirus.jhu.edu/map.html (last accessed May 20 2021).

Karim, A. M., J. E. Kwon, M. A. Karim, H. Iftikhar, M. Yasir, I. Ullah & S. C. Kang (2022) Comprehensive update on the monkeypox outbreak. *Frontiers in Microbiology*, 13, 1037583.

Kennedy, D. A. & A. F. Read (2018) Why the evolution of vaccine resistance is less of aconcern than the evolution of drug resistance. *PNAS*, 115, 12878–12886.

Kienberger, S. & M. Hagenlocher (2014) Spatial-explicit modeling of social vulnerability to malaria in East Africa. *International Journal of Health Geographics*, 13, 29.

Morens, D. M., G. K. Folkers & A. S. Fauci (2004) The challenge of emerging and re-emerging infectious diseases. *Nature*, 430, 242–249.

Morgan, J., S. L. Bornstein, A. M. Karpati, M. Bruce, C. A. Bolin, C. C. Austin, C. W. Woods, J. Lingappa, C. Langkop, B. Davis, D. R. Graham, M. Proctor, D. A. Ashford, M. Bajani, S. L. Bragg, K. Shutt, B. A. Perkins, J. W. Tappero & G. Leptospirosis Working (2002) Outbreak of Leptospirosis among Triathlon participants and community residents in Springfield, Illinois, 1998. *Clinical Infectious Diseases*, 34, 1593–1599.

Myers, S. S. (2017) Planetary health: Protecting human health on a rapidly changing planet. *Lancet*, 390, 2860–2868.

O'Sullivan, D. & D. Unwin (2010) *Geographic Information Analysis*. Chichester, UK: John Wiley & Sons.

RIVM (2021) Weekcijfers COVID-19/Weekly COVID-19 Figures https://www.rivm.nl/en/coronavirus-covid-19/weekly-covid-19-figures (last accessed May 21 2021).

Smith, C. M., S. C. Le Comber, H. Fry, M. Bull, S. Leach & A. C. Hayward (2015) Spatial methods for infectious disease outbreak investigations: Systematic literature review. *European Surveillance*, 20, 21.

Snow, J. (1855) *On the Mode of Communication of Cholera*. London: John Churchill.

Theze, J., T. Li, L. du Plessis, J. Bouquet, M. U. G. Kraemer, S. Somasekar, G. Yu, M. de Cesare, A. Balmaseda, G. Kuan, E. Harris, C. H. Wu, M. A. Ansari, R. Bowden, N. R. Faria, S. Yagi, S. Messenger, T. Brooks, M. Stone, E. M. Bloch, M. Busch, J. E. Munoz-Medina, C. R. Gonzalez-Bonilla, S. Wolinsky, S. Lopez, C. F. Arias, D. Bonsall, C. Y. Chiu & O. G. Pybus (2018) Genomic epidemiology reconstructs the introduction and spread of Zika Virus in Central America and Mexico. *Cell Host Microbe*, 23, 855–864e7.

Trueba, G. (2014) *The Origin of Human Pathogens*. Toyko, Japan: Springer.

van Oosterhout, C., N. Hall, H. Ly & K. M. Tyler (2021) COVID-19 evolution during the pandemic - Implications of new SARS-CoV-2 variants on disease control and public health policies. *Virulence*, 12, 507–508.

Villero-Wolf, Y., S. Mattar, A. Puerta-Gonzalez, G. Arrieta, C. Muskus, R. Hoyos, H. Pinzon & D. Pelaez-Carvajal (2019) Genomic epidemiology of Chikungunya virus in Colombia reveals genetic variability of strains and multiple geographic introductions in outbreak, 2014. *Scientific Reports*, 9, 9970. WHO (2023) Noncommunicable Disease Fact Sheet. https://www.who.int/mediacentre/factsheets/fs355/en/ (last accessed April 27 2024).

Woolhouse, M. E. J. & S. Gowtage-Sequeria (2005) Host range and emerging and reemerging pathogens. *Emerging Infectious Diseases*, 11, 1842–1847.

You, Y. A., M. Ali, S. Kanungo, B. Sah, B. Manna, M. Puri, G. B. Nair, S. K. Bhattacharya, M. Convertino, J. L. Deen, A. L. Lopez, T. F. Wierzba, J. Clemens & D. Sur (2013) Risk map of cholera infection for vaccine deployment: The eastern Kolkata case. *PLoS One*, 8, e71173.

3

Statistics, Analysis and Visualizations

Overview

Where we live plays a fundamental role in shaping our health, be it through direct contact with our environment (e.g., altitude, climate, greenness, hazards and pollutants), our behaviour (e.g., social networks, risk behaviours and lifestyles), or economic status (e.g., access – to health care; healthy environments – e.g., quality of food and physical activity spaces). Knowing how different factors affect our health and the health of our environment is important for determining how to respond, be it through education/ policies, or identifying the necessary services needed. To do so, we need to understand how these factors affect health outcomes. In this chapter, you will be introduced to a number of different analytical methods (Figure 3.1) that will help you explore, analyse, synthesise and communicate. Before we continue with the fun stuff, a few essentials are needed. In this chapter I provide an overview of different measures, essential statistics and an overview of spatial analysis methods and different visualisations to consider for communication.

Measures: A Variety of Measures Are Used to Assess Health Risk

A number of measures of risk are used. You are not expected to remember all of these. These have been provided as a source of reference. Much of what is provided in this section was compiled from the following sources CDC (2012) and Shantikumar, Barratt and Kirwan (2018).

DOI: 10.1201/9781003435082-3

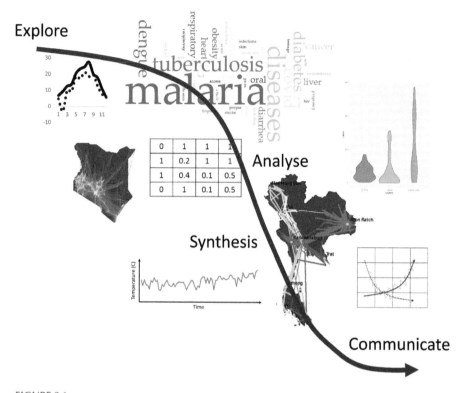

FIGURE 3.1
Visualization and analytics play an important role in scientific research and understanding. Visualizations such as wordclouds, graphs and maps facilitate visual thinking and insights, helping to communicate research findings to a wider audience.

Measures of risk (Table 3.1) are standardized measures that are used as follows:

- **Counts** to capture the number of cases or health events and are used to describe them in terms of time, place and person;
- **Divides** the number of cases by an appropriate denominator to calculate rates; and
- **Compares** these rates over time or for different groups of people.

If you are curious to see some of these in action, look at the information provided by the WHO (find mortality and global health estimates, world health statistics or life expectancy statistics) or visit the Global Burden Disease Compare Visualization Hub. https://vizhub.healthdata.org/gbd-compare/.

TABLE 3.1

Summary of Different Measures of Risk

Ratios, Proportions and Rates	Morbidity (disease), Mortality (deaths) and Natality (births)	These frequency measures are based on the formula: **Ratio, proportion, rate $= x/y \times 10^{n}$** where, x and y are the two quantities that are being compared; 10^{n} is a constant used to transform the results into a uniform quantity. The size of 10^{n} may equal 1, 10, 100, 1,000 and so on depending on the value of n.	CDC (2012)
Incidence rates	Incidence rate; attack rate; secondary attack rate	Incidence rates are used for measuring and comparing the frequency of disease in populations. Three types of incidence rates are used to measure risk and include incidence rate, attack rate and secondary attack rate.	CDC (2012)
Prevalence rates	Point prevalence; period prevalence	Prevalence = all new and pre-existing cases during a time period / population during the same time period * 10^{n}	CDC (2012)
Measures of mortality	Crude death rate; causes specific death rate; proportionate mortality; death-to-case ratio; neonatal mortality rate; postneonatal morality rate; infant mortality rate; maternal mortality rate	Mortality Rate = deaths occurring during a given time period / size of the population * 10^{n}. The most commonly used values for 10^{n} are 1,000 and 100,000.	CDC (2012)
Mortality rate	Crude mortality rate; cause-specific mortality rate; other specific mortality rate	Mortality rate from all causes of death for a population during a specified time period; Mortality rate from a specified cause for a population during a specified time period.	CDC (2012)
Case fatality rate		A measure of the severity of an infectious disease. It is the proportion of known cases who died of the disease in question.	CDC (2012)

(Continued)

TABLE 3.1 *(Continued)*

Summary of Different Measures of Risk

Disability-adjusted life year (DALY)	A measure of disease burden. The DALY for disease or injury causes is calculated as the sum of the years of life lost (YLL) due to premature mortality in the population and the years lost due to disability (YLD) for incident cases of the disease or injury. $$DALY = YLL + YLD$$ One DALY = one lost year of healthy life	WHO (2013)
YLL	One YLL represents the loss of one year of life. YLLs are calculated from the number of deaths multiplied by a global standard life expectancy at the age at which death occurs. $$YLL(c, s, a, t) = N(c, s, a, t) \times L(s, a)$$ where: $N(c, s, a, t)$ is the number of deaths due to the cause c for the given age a and sex s in year t $L(s, a)$ is a standard loss function specifying years of life lost for a death at age a for sex s For example, if there were ten deaths at the age of 1, the YLL contribution of this cause of death and age group would be: Number of deaths at the age of 1 year $\times$ The number of years lost had each individual lived to the age of $75 = 10 \times 74$ years $= 740$ years	https://www.who.int/data/gho/indicator-metadata-registry/imr-details/159 Number of deaths are from the WHO Global Health Estimates, and the standard loss function is based on the frontier national life expectancy projected for the year 2050 by the World Population Prospects 2012 (UN Population Division, 2013), with a life expectancy at birth of 92 years.
Years lost due to disability	Time lived in states of less than optimal health, loosely referred to as "disability" $$YLD(c, s, a, t) = I(c, s, a, t) \times DW(c, s, a) \times L(c, s, a, t)$$ where: $I(c, s, a, t) =$ number of incident cases for cause c, age a and sex s $DW(c, s, a) =$ disability weight for cause c, age a and sex s $L(c, s, a, t) =$ average duration of the case until remission or death (years)	https://cdn.who.int/media/docs/default-source/gho-documents/global-health-estimates/ghe2019_daly-methods.pdf?sfvrsn=31b25009_7
Burden of disease	In a population is the gap between the current years of healthy life and years of healthy life that would be lived if everyone in the population survived the full life spam free of disease or disability.	Cromley and McLafferty (2012)
Measures of natality	Crude birth rates; crude fertility rates; crude rate of natural increase; low-birth weight ratio	Birth rate = births occurring during a given time period/size of the population * 10^n. The most commonly used values for 10^n are 1,000.	CDC (2012)

Ratios, Proportions and Rates

Three kinds of frequency measures are used in infectious disease epidemiology to describe **morbidity** (disease), **mortality** (death) and **natality** (birth) (CDC, 2012) (Table 3.2).

These frequency measures are based on the formula: ratio, proportion, rate $= x/y \times 10^n$

where, x and y are the two quantities that are being compared; 10^n is a constant used to transform the results into a uniform quantity. The size of 10^n may equal 1, 10, 100, 1,000 and so on, depending on the value of n.

- **Ratio's** are used: to compare two independent groups
- **Proportions** are used to compare one group with a larger one to which it belongs; usually expressed as a percentage
- **Rates** are used to measure an event in a population over time
  - Rates are always specific to a particular population. They reflect groupings of people based on time (year, month, week, day or hour), place (country, state, county, township, school, institution and area) and person (age, sex and membership in some group or class).
  - Comparisons across different population groups can be made since rates take into account the size of the population
  - Rates may be harder to calculate, because accurate denominator data may not be available for small, localized population groups.

TABLE 3.2

Summary of Different Measures

Condition	Ratio	Proportion	Rate
Morbidity (Disease)	Risk ratio (Relative risk) Rate ratio Odds ratio Period prevalence	Attack rate (Incidence proportion) Secondary attack rate Point prevalence Attributable proportion	Person-time incidence rate
Mortality (Death)	Death-to-case ratio	Proportionate mortality	Crude mortality rate Case-fatality rate Cause-specific mortality rate Age-specific mortality rate Maternal mortality rate Infant mortality rate
Natality (Birth)		Crude birth rate Crude fertility rate	

Morbidity Frequency Measures

Several standard measures are used to measure and describe the frequency of disease and include **incidence** and **prevalence rates** (CDC, 2012)

Incidence Rates

Incidence rates are used for measuring and comparing the frequency of disease occurrence in a population. Three types of incidence rates are used to measure risk and include the incidence rate, attack rate and secondary attack rate (Table 3.3).

Disease incidence rates imply a change over time so **the period of time must be specified**. For surveillance purposes, this is usually one calendar year, but any time period may be used as long as it is stated.

Prevalence Rates

Prevalence is the proportion of people in a population who have a particular disease at a specified point in time or over a specified period of time (Table 3.4). The numerator includes not only new cases, but also old cases (people who remained ill during the specified point or period in time). A case is counted in prevalence until death or recovery occurs. Prevalence is most useful for measuring the burden of chronic diseases such as tuberculosis, malaria and HIV in a population (CDC 2012).

The formula for calculating prevalence is: **Prevalence**=all new and pre-existing cases during a time period/population during the same time period $* 10^n$

Point vs. period prevalence: The amount of disease present in a population obviously changes over time. Sometimes, we want to know how much

TABLE 3.3

Incidence, Attack and Secondary Attack Rates

Measure	Numerator (x)	Denominator (y)	Expressed per Number at Risk (10^n)
Incidence rate	No. new cases of a specified disease reported during a given time interval	Average population during time interval	Varies:10^n where $n=2, 3, 4, 5, 6$
Attack rate	No. new cases of a specified disease reported during an epidemic period	Population at start of the epidemic period	Usually a percentage: 10^n where $n=2$
Secondary attack rate	No. new cases of a specified disease among contacts of known cases	Size of contact population at risk	Usually a percentage: 10^n where $n=2$

of a particular disease is present in a population at a single point in time (Table 3.4).

- **Point prevalence**: Point prevalence is useful in comparing different points in time to help determine whether an outbreak is occurring.
- **Example**: What is the prevalence of TB in Community A today? A review of patients reported to the tuberculosis registry revealed that as of July 1, 2005, there were 35 cases that had not yet completed therapy. The most recent population estimate was 57,763. The prevalence of TB on July 1, 2005 was: **35/57,763 * 10,000=6.1 per 10,000 people**, where the numerator would include all known TB patients who lived in Community A that day. The denominator would be the population of Community A that day.
- **Period prevalence**: To assess how much of a particular disease is present in a population over a longer time period. Period prevalence is calculated in exactly the same way as point prevalence, except the numerator is the number of people who had the disease at any time during a specified time period. Period prevalence can be calculated for a week, month, year, decade or any other specified length of time.

Ensure that the length of the time period is the same. Prevalence may be compared among different diseases or different populations, as long as the same length of time is used.

Mortality Frequency Measures

Mortality rates measure the frequency of the occurrence of death in a defined population during a specified interval (Table 3.5).

Mortality Rate=deaths occurring during a given time period/size of the population * 10^n. The most commonly used values for 10^n are 1,000 and 100,000 (Table 3.5).

TABLE 3.4

Point and Period Prevalence Rates

Measure	Numerator (x)	Denominator (y)	Expressed per Number at Risk (10^n)
Point prevalence	Number of current cases (new and pre-existing) at a specified point in time	Population at the same specified point in time	10^n
Period prevalence	Number of current cases (new and pre-existing) over a specified period of time	Average or mid-interval population	10^n

TABLE 3.5

Frequently Used Measures of Mortality

Measure	Numerator	Denominator	10^n
Crude death rate	Total number of deaths during a given time interval	Mid-interval population	1,000 or 100,000
Cause-specific death rate	Number of deaths assigned to a specific cause during a given time interval	Mid-interval population	100,000
Proportionate mortality	Number of deaths assigned to a specific cause during a given time interval	Total number of deaths from all causes during the same time interval	100 or 1,000
Death-to-case ratio	Number of deaths assigned to a specific cause during a given time interval	Number of new cases of same disease reported during the same time interval	100
Neonatal mortality rate	Number of deaths among children < 28 days of age during a given time interval	Number of live births during the same time interval	1,000
Postneonatal mortality rate	Number of deaths among children 28–364 days of age during a given time interval	Number of live births during the same time interval	1,000
Infant mortality rate	Number of deaths among children < 1 year of age during a given time interval	Number of live births during the same time interval	1,000
Maternal mortality rate	Number of deaths assigned to pregnancy-related causes during a given time interval	Number of live births during the same time interval	100,000
Crude mortality rate	The crude mortality rate is the mortality rate from all causes of death for a population during a specified time period		CDC (2012)
Cause-specific mortality rate	This is the mortality rate from a specified cause for a population during a specified time period		CDC (2012)
Other specific mortality rates	Mortality rates calculated for population subgroups defined by age, sex, race or other demographic factors.	For example, the mortality rate attributed to HIV among 25- to 44-year-olds in the US in 1987 was: HIV mortality rate = 9,280 deaths/77,600,000 aged 25–44 years * 100,000 = 12/100,000 This is an example of a cause- and age-specific mortality rate.	CDC (2012)
Case fatality rate	The case fatality rate is often used as a measure of the severity of an infectious disease. It is the proportion of known cases who died of the disease in question.	For example, in an outbreak of 20 cases of invasive meningococcal disease on a large university campus, three patients died. The case fatality rate was 3/20 or 15%.	CDC (2012)

Source: CDC (2012).

Natal Frequency Measures

Birth rates measure the frequency of the occurrence of births in a defined population during a specified interval (Table 3.6).

Birth Rate=births occurring during a given time period/size of the population $* 10^n$. The most commonly used values for 10^n are 1,000 (Table 3.6).

Odds Ratio

Odds Ratio (OR) is a measure of the association between exposure and an outcome. The OR represents the odds that an outcome will occur given a particular exposure, compared to the odds of the outcome occurring in the absence of that exposure (Figure 3.2).

An OR is a comparison of the odds between people who were exposed and people who were not exposed (Goodwin and Ryu, 2023). It is an indicator of the effect of exposure on the likelihood of becoming ill (CDC). Essentially, the OR is the "measure of association". It quantifies the relationship between an exposure and a disease. The OR tells us how much higher the odds of exposure are compared with those not exposed. The results can be interpreted as follows (Szumilas, 2010; Goodwin and Ryu, 2023; CDC):

- **OR=1**: An OR of 1.0 (or close to 1.0) indicates that the exposure is NOT associated with the disease since the odds of exposure are the same as, or similar to, the odds of exposure among controls.
- **OR>1**: Greater than 1.0 indicates that the odds of exposure among case-patients are greater than the odds of exposure among controls. The exposure might be a risk factor for the disease.
- **OR<1**: Less than 1.0 indicates that the odds of exposure among case-patients are lower than the odds of exposure among controls. The exposure might be a protective factor against the disease.

TABLE 3.6

Frequently Used Measures of Natality

Measure	Numerator	Denominator	10^n
Crude birth rate	Number of live births during a specified time interval	Mid-interval population	1,000
Crude fertility rate	Number of live births during a specified time interval	Number of women ages 15–44 years at mid-interval	1,000
Crude rate of natural increase	Number of live births minus number of deaths during a specified time interval	Mid-interval population	1,000
Low-birth weight ratio	Number of live births <2,500 grams during a specified time interval	Number of live births during the same time interval	100

Source: CDC (2012).

	Disease Status		
Risk Factor	**Disease**	**No Disease**	**Total**
Exposed	a	b	a+b
Not exposed	c	d	c+d
Total	a+c	b+d	N=a+b+c+d

Risk = (a+c) /N

Exposure-specific Risk (exposed to risk factors) = a /(a+b)

Exposure-specific Risk (not exposed to risk factors) = c /(c+d)

$$\textbf{Relative Risk } = \frac{a/(a+b)}{c/(c+d)} = \frac{a(c+d)}{b(a+b)}$$

$$\textbf{Odds ratio} \text{ (comparison of those exposed to not exposed risk factor)} = \frac{a/b}{c/d} = \frac{ad}{bc}$$

FIGURE 3.2

The association between risk factors and disease incidence. Calculating the Odds Ratio, risk, exposure-specific risk and relative risk in a population. Adapted from Cromley and McLafferty (2012).

The magnitude of the OR provides information about the "strength of the association." The further away an OR is from 1.0, the more likely it is that the relationship between the exposure and the disease is causal.

Population Data

Census data can provide the denominators for many of these types of analyses. However, there are limitations to using these data. For many countries, censuses are obtained every 10 years. Not all countries conduct regular censuses; therefore, obtaining up-to-date population estimates is not always available for all countries and can become out-of-date in areas where populations change rapidly. For potential data sources, see chapter on data.

Types of Variables

A lot of the information we need is stored as attribute data and will fall into one of four types (Table 3.7).

TABLE 3.7

Four Different Types of Variables

Types of Variables Scale	Example	Values	
Nominal ordinal	\ "categorical" or / "qualitative	disease status ovarian cancer	yes/no
Interval ratio	\ "continuous" or / "quantitative"	date of birth tuberculin skin test	Any date from recorded time

Source: CDC (2012).

TABLE 3.8

Variable Types and How Used

Scale	Ratio or Proportion	Measure of Central Location	Measure of Spread
Nominal	Yes	No	No
Ordinal	Yes	No	No
Interval	Yes, but might need to group first	Yes	Yes
Ratio	Yes, but might need to group first	Yes	Yes

Source: CDC (2012).

- A *nominal-scale variable* is one whose values are categories without any numerical ranking, such as county of residence. A nominal variable with two mutually exclusive categories is sometimes called a dichotomous variable.
- An *ordinal-scale variable* has values that can be ranked but are not necessarily evenly spaced.
- An *interval-scale variable* is measured on a scale of equally spaced units, but without a true zero point, such as the date of birth.
- A *ratio-scale variable* is an interval variable with a true zero point, such as height in centimetres or duration of illness.

Nominal- and ordinal-scale variables are considered **qualitative** or **categorical** variables, whereas interval- and ratio-scale variables are considered **quantitative** or **continuous** variables. Sometimes the same variable can be measured using both a nominal scale and a ratio scale.

Many variables used by epidemiologists are categorical (e.g., exposed yes/no, test positive/negative, case/control, etc.). Categorical variables are summarized with frequency measures such as ratios, proportions and rates. Continuous variables are often further summarized with measures of central location and measures of spread (Table 3.8).

In the data chapter, you will be introduced to different definitions of variables for data that may be stored in tables associated with geographic features.

Getting to Know Your Data

This is often overlooked, as we just want to start analysing our data and see what interesting patterns may exist, if any. But to do so, we really need to take the time to get to know what is in the data. Are there missing values? What is the distribution of the data? Often, when we really start to get to know the data and look more closely at what is in the data, many times the data does not have what we need in the format we need it in or there are a lot of gaps.

Descriptive Statistics: Summarizing Data

In the chapter on data, I talk about data, including data sources, data types and some of the challenges associated with analysing spatial data. Once you have obtained the data you need, it is necessary to become familiar with it. One way to do so is to use descriptive statistics (Table 3.9–3.10) as well as apply different visualizations (Figure 3.3).

Visualizations and Descriptive Measures

Visualizations play an important role in exploring data as well as telling a story about the data. There are many types of graphs and plots that you can use, as illustrated by the R gallery (Figure 3.3).

Although R/RStudio has plenty to offer, also check Microsoft Excel, Tableau and Orange (https://orangedatamining.com/). Microsoft offers an increased set of visualization capabilities (e.g., sparklines, conditional formatting to create heatmaps, treemaps and sunburst charts).

Keep in mind the purpose of the visualization and select a visualization that may be suitable for that purpose (comparison, distribution, relationship and composition).

Visualizing the Data Distribution

Frequency distributions and histograms reveal three features (Figure 3.4):

1. where the distribution has its peak (*central location*),
2. how widely dispersed it is on both sides of the peak (*spread*), and
3. whether it is more or less symmetrically distributed on the two sides of the peak.

   - **Central location:** The clustering at a particular value is known as the *central location* or *central tendency* of a frequency distribution. Measures of central location can be in the middle or off to one side or the other.

   - **Spread (also known as variation or dispersion):** Spread refers to the distribution away from a central value. Two measures of spread commonly used are *range* and *standard deviation*.

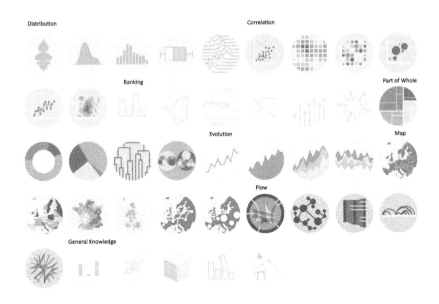

FIGURE 3.3
Overview of a variety of visualizations (distributions, relationships, ranking, evolution/trends, flows, maps) to explore, compare and evaluate the data. *Source*: R https://www.r-graph-gallery.com/index.html.

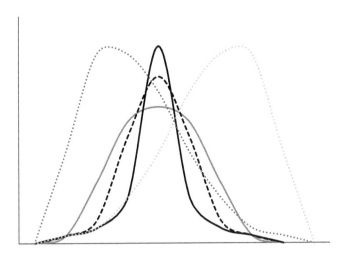

FIGURE 3.4
Illustration of a symmetric and skewed distribution.

- **Shape**: Shape can be symmetrical or skewed. A distribution that has a central location to the left and a tail off to the right is said to be *positively skewed* **or** *skewed to the right*. A distribution that has a central location to the right and a tail to the left is said to be *negatively skewed* **or** *skewed to the left*.

Measures of central location include the *mode, median, arithmetic mean, midrange* **and** *geometric mean*. Selecting the best measure to use for a given distribution depends largely on (i) the **shape or skewness** of the distribution and (ii) the intended **use** of the measure.

Means, Median and Mode

See Table 3.9 for more information about these measurements and how they are calculated.

Three Centres: Spatial Distributions of the Data

Equivalent descriptive measures for examining geographic distributions are provided in Table 3.10. These measures can be used to identify the centre, shape and orientation of features or how dispersed they are in space.

To demonstrate the usefulness of the spatial descriptive measures, I have performed a variety of analyses using the 2014–2016 Ebola outbreak data for each week of the outbreak (Figure 3.5) using the mean center and the standard deviational ellipse measures (Table 3.10).

Maps, Graphs and Colour

While analysing your data, it may be necessary to use different visualizations to represent your findings and also to view the data. Maps are useful, but sometimes the data may be better presented in other formats or by combining a variety of different visualizations. Of course, colour is an important consideration. Use the colorbrewer palettes and the tool http://colorbrewer2.org/ to explore colour combinations (Figure 3.6).

Choropleth Maps and Classification Schemes

When mapping data, there are a number of classification schemes that can be used to reduce the complexity of data and enhance the information within the data. When mapping the information consider the following:

- For unevenly distributed data, use *natural breaks*.
- For evenly distributed data (and the emphasis is to highlight differences between entities), use *equal interval or standard deviation*.
- For evenly distributed data (and interested in highlighting relative differences), use *quantiles*.
- For set defined classifications use the *manual* option.

Examine the different classifications available next time you create a choropleth map (Figure 3.7).

TABLE 3.9

Measures of Central Tendency

Measures of Central Tendency		
Measures of Central Tendency	Mean	• Sample average • Arithmetic mean is the value that is closest to all the other values in a distribution • Also called the *centre of gravity* of a frequency distribution • Best descriptive measure for data that are normally distributed • Because the arithmetic mean uses all of the observations in the distribution, it is affected by any extreme value (outliers)
	Median	• The middle value of a set of data that has been put into rank order • Good descriptive measure, particularly for data that are skewed • Not generally affected by one or two extreme values (outliers)
	Mode	• The value that occurs most often in a set of data • Not affected by one or two extreme values (outliers)
	Geometric mean	• Is the mean or average of a set of data measured on a logarithmic scale • Used when the logarithms of the observations are distributed normally (symmetrically) rather than the observations themselves • Dampens the effect of extreme values and is always smaller than the corresponding arithmetic mean • Is less sensitive than the arithmetic mean to one or a few extreme values • Measure of choice for variables measured on an exponential or logarithmic scale and when levels can range over several orders of magnitude
Measures of Variability	Variance	The differences between members for all possible pairs of measurements in the sample
	Standard deviation	The measure of spread The difference between each observation in the sample and the sample mean Conveys how widely or tightly the observations are distributed from the centre Calculated only when the data are more-or-less "normally distributed" For normally distributed data, approximately two-thirds (68.3%) of the data fall within one standard deviation of either side of the mean; 95.5% of the data fall within two standard deviations of the mean; and 99.7% of the data fall within three standard deviations. Exactly 95.0% of the data fall within 1.96 standard deviations of the mean
	Range	Difference between the largest (maximum) value and smallest (minimum) value
	Midrange	The midrange is the half-way point or the midpoint of a set of observations
	Interquartile range	Is a measure of spread commonly used with the median Useful for characterizing the central location and spread of any frequency distribution, but particularly those that are skewed It represents the central portion of the distribution, from the 25th percentile to the 75th percentile

Bar Graphs, Epi Curves and Sparklines

A variety of graphs can be used to capture case numbers. Some familiar ones include histograms (Figure 3.8a) or cumulative graphs (Figure 3.8b) or line graphs (Figures 3.9 and 3.10) or sparklines (Figure 3.12). These can easily be created in Excel or R.

TABLE 3.10

Measures the Geographic Distributions

	Spatial Descriptive Statistic	Description	Calculation
Measures of Central Tendency	Central	Identifies the most centrally located feature for a set of points, polygon(s) or line(s)	Point with the shortest total distance to all other points is the most central feature $$D = \sum_{i=1}^{n} \sum_{j=1}^{n-1} \sqrt{(x_j - x_i)^2 + (y_j - y_i)^2}$$ $D_{central} = $ minimum (D)
	Mean	Identifies the geographic centre (or the centre of concentration) for a set of features ****Mean sensitive to outliers****	Simply the mean of the X coordinates and the mean of the Y coordinates for a set of points $$\bar{X} = \frac{\sum_{i=1}^{n} x_i}{n}, \bar{Y} = \frac{\sum_{i=1}^{n} y_i}{n}$$
	Weighted mean		Produced by weighting each X and Y coordinate by another variable (W_i) $$\bar{X} = \frac{\sum_{i=1}^{n} wx_i}{\sum_{i=1}^{n} w_i}, \bar{Y} = \frac{\sum_{i=1}^{n} wy_i}{\sum_{i=1}^{n} w_i}$$
Measures of Variability	Standard distance	Measures the degree to which features are concentrated or dispersed around the geometric mean centre The greater the standard distance, the more the distances vary from the average, thus features are more widely dispersed around the centre Standard distance deviation is a good single measure of the dispersion of the points around the mean centre, but it doesn't capture the shape of the distribution.	Represents the standard deviation of the distance of each point from the mean centre: $$SD = \sqrt{\frac{\sum_{i=1}^{n} (x_i - \bar{X})^2}{n} + \frac{\sum_{i=1}^{n} (y_i - \bar{Y})^2}{n}}$$ Where x_i and y_i are the coordinates for a feature and $\bar{X}$ and $\bar{Y}$ are the mean centre of all the coordinates **Weighted SD** $$SDw = \sqrt{\frac{\sum_{i=1}^{n} w_i(x_i - \bar{X})^2}{n} + \frac{\sum_{i=1}^{n} w_i(y_i - \bar{Y})^2}{n}}$$ where x_i and y_i are the coordinates for a feature and $\bar{X}$ and $\bar{Y}$ are the mean centre of all the coordinates. w_i is the weight value
	Standard deviational ellipse	Captures the shape of the distribution.	Creates standard deviational ellipses to summarize the spatial characteristics of geographic features: Central tendency, dispersion and directional trends

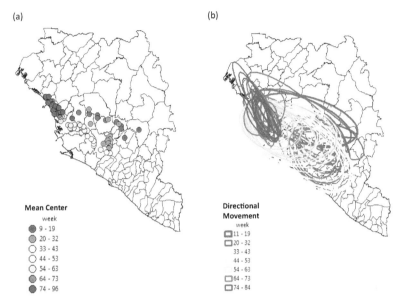

(a)   (b)

Mean Center
week
● 9 - 19
● 20 - 32
○ 33 - 43
○ 44 - 53
○ 54 - 63
● 64 - 73
● 74 - 96

Directional Movement
week
□ 11 - 19
□ 20 - 32
  33 - 43
  44 - 53
  54 - 63
□ 64 - 73
□ 74 - 84

FIGURE 3.5
Distribution of Ebola cases during the outbreak of 2014–2016. (a) Mean centre of the outbreak by week and (b) directional movement of the outbreak by week. *Data Source*: WHO. Images created by Blanford (2023) and also published in Blanford and Jolly (2021). Analysis performed using ArcGIS Pro.

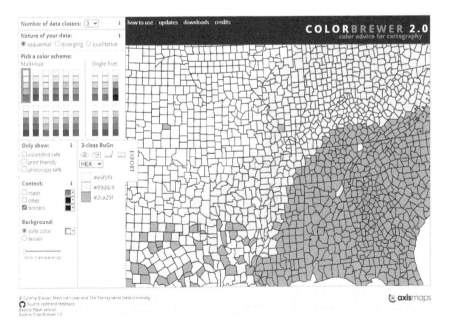

FIGURE 3.6
ColorBrewer 2.0: the colour advice interactive web application for cartography (source: https://colorbrewer2.org)

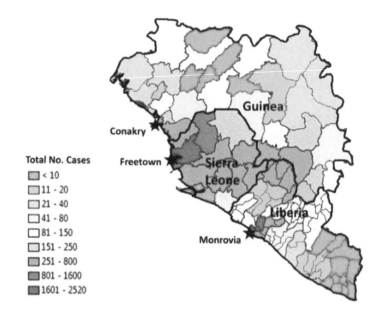

FIGURE 3.7
The spatial distribution confirmed Ebola cases reported during the Ebola outbreak 2014–2016 for Guinea, Sierra Leone and Liberia. *Data Source*: WHO. *Image Source*: Blanford and Jolly (2021). Map created in ArcGIS Pro.

Frequency and Cumulative Graphs

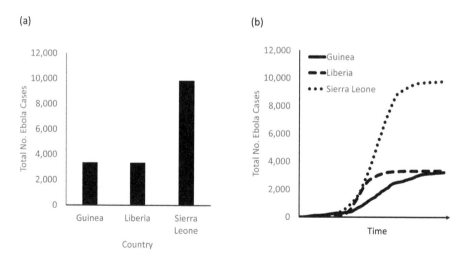

FIGURE 3.8
(a) Total number of Ebola cases reported by country; and (b) the total number of cumulative Ebola cases reported by country in Guinea, Liberia and Sierra Leone *Data Source*: WHO. Image created Blanford (2023).

Epidemic Curve (Epi Curve)

The **epidemic curve** (**epi curve** for short) provides a simple visual display of the outbreak's magnitude and time trend. The epi curve shows the number of illnesses over time during an outbreak and can be useful for characterizing or assessing an outbreak (Figures 3.9 and 3.10).

Benefits of an Epi Curve

There are many benefits to creating an epidemic curve (CDC, 2012). The shape of the epidemic curve provides clues about the pattern of spread (e.g., point vs intermittent sources (Figure 3.11)).

- The curve shows where you are in the course of the epidemic, e.g., increasing (on the upswing), decreasing (on the down slope) or after the epidemic has ended.
- This information is useful for predicting the number of cases that are likely to occur in the near future.
- Identifying outliers.
- To deduce the time of exposure.
- Evaluation of intervention methods and response time.

Epi curves are useful for understanding the dynamics of an outbreak. To do so, consider the shape of the curve. This is governed by the epidemic pattern (e.g., point source versus propagated), the period of time over which susceptible persons are exposed and the incubation periods of the disease (minimum, mean and maximum). Some examples include (Figure 3.11):

- **Common point source**: Persons are exposed to the same source over a relative brief period. Any sudden rise in the number of cases suggests sudden exposure to a common source. All cases occur within one incubation period.
- **Common persistent source**: Similar to common point source except that the duration of exposure is prolonged resulting in a continuous and persistent epidemic event.
- **Common intermittent source**: Exposure to the causative agent is sporadic over time. This usually produces an irregularly jagged epidemic curve, reflecting the intermittence and duration of exposure.
- **Propagated source**: The spread occurs from person to person with the number of cases increasing with each generation.

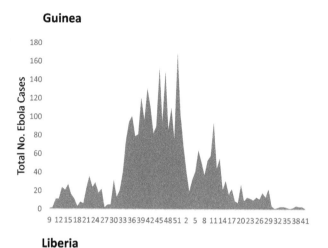

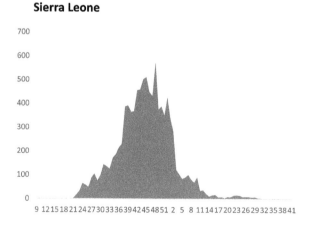

FIGURE 3.9

Epi curve capturing the total weekly number of confirmed Ebola cases for Guinea, Liberia and Sierra Leone, 2014–2016. Data obtained from World Health Organization. *Data Source*: WHO. Image created Blanford (2023).

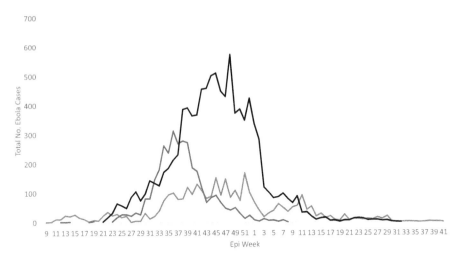

FIGURE 3.10
Total weekly number of confirmed Ebola cases for Liberia, Sierra Leone and Guinea, 2014–2016.
Data obtained from World Health Organization. *Data Source*: WHO. Image created Blanford
(2023).

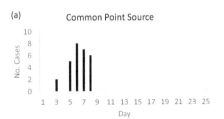

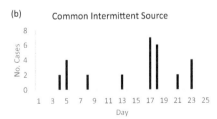

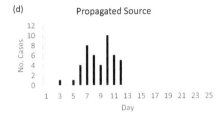

FIGURE 3.11
Examples of four typical epidemic curves for different types of disease spread that include
(a) common point source, (b) common intermittent source, (c) common persistent source
and (d) propagated source. *Source*: CDC (2012). Adapted from: European Programme for
Intervention Epidemiology Training [Internet]. Solna, Sweden: Smittskyddsinstitutet [updated
2004 Sep 27; cited 2006 Sep 22].

Epi weeks: An epidemiological week (an epi week) is a standardized method of counting weeks to allow for the comparison of data year after year. The first epi week of the year ends, by definition, on the first Saturday of January, as long as it falls at least four days into the month. Each epi week begins on a Sunday and ends on a Saturday. Note – not all countries calculate epi weeks in the same way. Excel also has a function that calculates the week number.

The Epi Curve When Dealing with Lots of Information and for Multiple Locations

In some cases, you may want to view the shape of the epi curve for many different data locations to see if they hold the same distribution. To do this, 20–100 times becomes tedious. Instead of creating a line or bar graph for each location or having to program something in R, you can use sparklines in Excel or create a heatmap (in Excel) or R.

Sparklines or Ridgelines

Sparklines are useful for exploring trends in a series of values as shown in Figure 3.12. The sparklines captures the different Ebola trends within each district across different countries.

Heatmaps

Heatmaps may be useful for looking at the intensity of events occurring in both space and time. Although there are a number of different ways

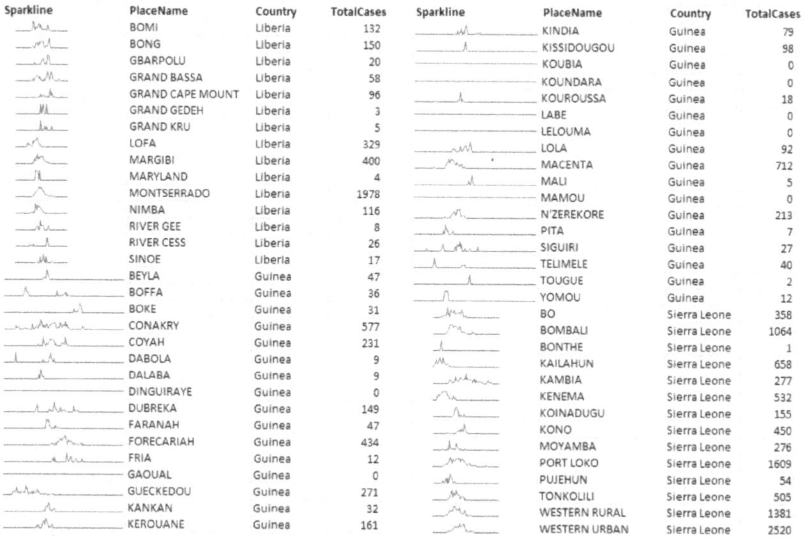

Sparkline	PlaceName	Country	TotalCases	Sparkline	PlaceName	Country	TotalCases
	BOMI	Liberia	132		KINDIA	Guinea	79
	BONG	Liberia	150		KISSIDOUGOU	Guinea	98
	GBARPOLU	Liberia	20		KOUBIA	Guinea	0
	GRAND BASSA	Liberia	58		KOUNDARA	Guinea	0
	GRAND CAPE MOUNT	Liberia	96		KOUROUSSA	Guinea	18
	GRAND GEDEH	Liberia	3		LABE	Guinea	0
	GRAND KRU	Liberia	5		LELOUMA	Guinea	0
	LOFA	Liberia	329		LOLA	Guinea	92
	MARGIBI	Liberia	400		MACENTA	Guinea	712
	MARYLAND	Liberia	4		MALI	Guinea	5
	MONTSERRADO	Liberia	1978		MAMOU	Guinea	0
	NIMBA	Liberia	116		N'ZEREKORE	Guinea	213
	RIVER GEE	Liberia	8		PITA	Guinea	7
	RIVER CESS	Liberia	26		SIGUIRI	Guinea	27
	SINOE	Liberia	17		TELIMELE	Guinea	40
	BEYLA	Guinea	47		TOUGUE	Guinea	2
	BOFFA	Guinea	36		YOMOU	Guinea	12
	BOKE	Guinea	31		BO	Sierra Leone	358
	CONAKRY	Guinea	577		BOMBALI	Sierra Leone	1064
	COYAH	Guinea	231		BONTHE	Sierra Leone	1
	DABOLA	Guinea	9		KAILAHUN	Sierra Leone	658
	DALABA	Guinea	9		KAMBIA	Sierra Leone	277
	DINGUIRAYE	Guinea	0		KENEMA	Sierra Leone	532
	DUBREKA	Guinea	149		KOINADUGU	Sierra Leone	155
	FARANAH	Guinea	47		KONO	Sierra Leone	450
	FORECARIAH	Guinea	434		MOYAMBA	Sierra Leone	276
	FRIA	Guinea	12		PORT LOKO	Sierra Leone	1609
	GAOUAL	Guinea	0		PUJEHUN	Sierra Leone	54
	GUECKEDOU	Guinea	271		TONKOLILI	Sierra Leone	505
	KANKAN	Guinea	32		WESTERN RURAL	Sierra Leone	1381
	KEROUANE	Guinea	161		WESTERN URBAN	Sierra Leone	2520

FIGURE 3.12

Sparklines used to view multiple epi curves for confirmed Ebola cases for each district in Liberia, Guinea and Sierra Leone (2014–2016). *Data Source*: WHO. Image by Blanford (2023).

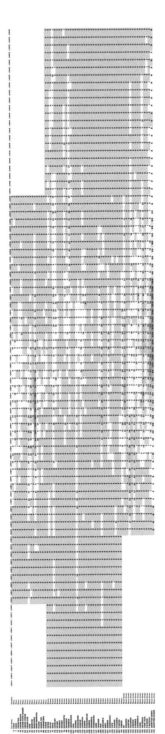

FIGURE 3.13
A heatmap of confirmed Ebola cases by district for each country, Guinea, Sierra Leone and Liberia, using conditional formatting in Excel. Image created by Blanford (2023).

to create heatmaps in R, the example below (Figure 3.13) was created in Excel using the conditional formatting option. In this case, each column represents a week, and each row represents a unique geographic location, such as a district. Each cell is coloured to show low intensity (green), high intensity (red) and values laying somewhere in between (yellow). By viewing the data in this way, it is easy to see the intensity and duration of each event across numerous locations. This is not as easy to view on a single map.

Boxplots

A boxplot or a box and whiskers plot is also useful for examining the different distributions. The Ebola data were aggregated by country by week to create the boxplot (Figure 3.14).

Violins

The Ebola data were aggregated by week for each country to create the violins (Figure 3.15).

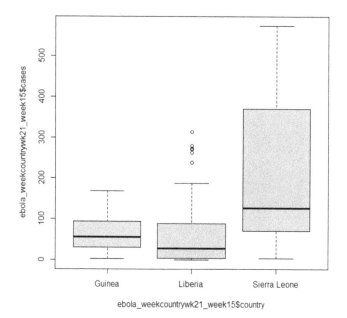

FIGURE 3.14
A boxplot for confirmed Ebola cases for each country (Guinea, Sierra Leone, and Liberia) to show how weekly outbreaks varied between the three countries. Boxplot was created in R using boxplot. Image created by Blanford (2023).

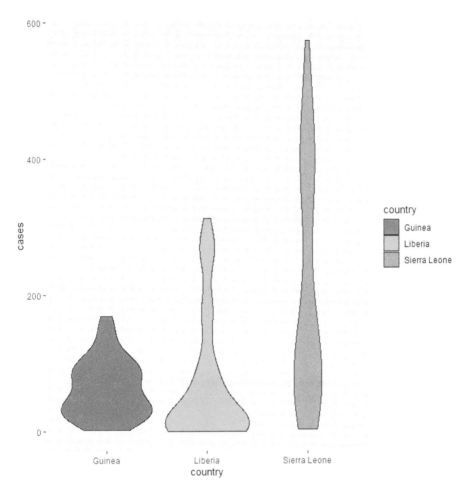

FIGURE 3.15
Violin plots were used to show how weekly outbreaks varied between each of the three countries (Guinea, Liberia, Sierra Leone) during the Ebola outbreak week 21, 2014 to week 15, 2015. Violin plots were created in R using ggplot. Image created by Blanford (2023).

Wordclouds

Visualizations can also be performed on text, as illustrated with the wordcloud in Figure 3.16. A wordcloud is a visual representation of text, in which the words appear bigger the more often they are mentioned. Wordclouds can be useful for gaining insights about trends and patterns in text data, particularly unstructured text data. To create the wordcloud in Figure 3.16 the total number of cases reported for each district was used as the frequency

WESTERN AREA URBAN
MONTSERRADO

MOYAMBA
COYAH FARANAH
GRAND GEDEH   YOMOU
GBARPOLU SINOE CONAKRY
KOINADUGU
BEYLA   KISSIDOUGOU TONKOLILI
WESTERN AREA RURAL
KANKAN KOUROUSSA   SIGUIRI LOLA   MALI   BO
FRIA BOKE NZEREKORE
PORT LOKO
DUBREKA RIVER GEE   PITA BOFFA DALABA GUECKEDOU
KINDIA NIMBA KAILAHUN   MARYLAND
PUJEHUN
BOMBALI   KONO
MACENTA
GRAND CAPE MOUNT
BONG FORECARIAH GRAND BASSA
MARGIBI LOFA BOMI
KENEMA   KAMBIA
KEROUANE
TELIMELE

FIGURE 3.16

A word cloud created to highlight where the outbreak was concentrated. The words are Admin Level 2 (District) and the frequency is the total number of cases reported during the Ebola outbreak 2014–2016. Word cloud created in R. Data from WHO. Image created by Blanford (2023).

of mentions. More complex types of text-based analyses are available, such as topic modelling or sentiment-based analyses, but are beyond the scope of this course.

Combinations: Using Multiple Visualizations to Tell a Story

Each of the visualization examples (Figure 3.5, 3.7–3.10, 3.12–3.16) uses the same data and provides different insight about what took place during the Ebola outbreak of 2014–2016. A variety of visualizations are useful for understanding the spatial and temporal distribution of a disease or an outbreak (Figure 3.17).

DO: Review Figure 3.17 and the other examples provided about the Ebola outbreak. How useful are the visualizations when presented together in a figure rather than individually? What insights were you able to gain about the Ebola outbreak? Are there visualizations missing from Figure 3.17? Which visualizations do you find useful and which do you not find useful?

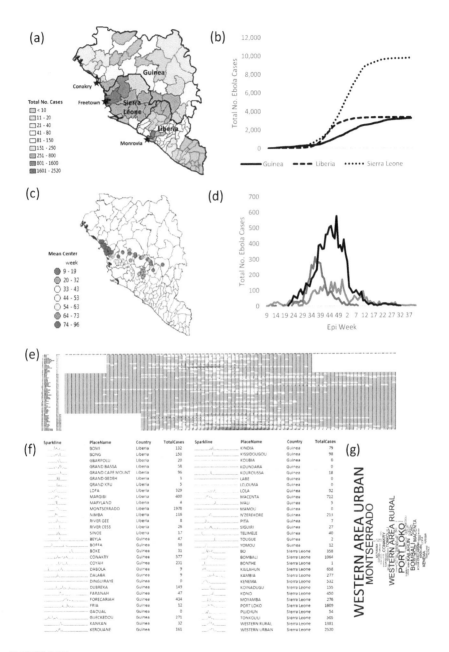

FIGURE 3.17

The Ebola outbreak 2014–2016. (a) The spatial distribution of Ebola cases mapped using a choropleth capturing the total number of cases by admin level 2; (b) Cumulative cases by country over time; (c) The mean centre of the outbreak for each week; (d) The epi curve capturing the total number of confirmed cases reported by week for each country; (e) A heatmap showing the intensity of cases reported across space and time. Each cell represents the total number of cases reported; (f) Epi curves for each district; (g) A wordcloud capturing the district where the outbreak was concentrated. *Data Source*: WHO. Image created by Blanford (2023).

Confidence Limits (Intervals)

A common way of determining a measurement's precision is by calculating the confidence interval. A narrow confidence interval indicates high precision; a wide confidence interval indicates low precision. Confidence intervals are one of the most useful tools in epidemiology. In general, a confidence interval uses these concepts to create reasonable bounds for the population mean, based on information from a sample (see Bonita et al., 2006; CDC, 2012).

The confidence interval for a mean is based on the mean itself and some multiple of the standard error of the mean. Fortunately, regardless of how the data are distributed, means (particularly from large samples) tend to be normally distributed (Table 3.11).

Understanding the Why?

Understanding why a disease is occurring in a particular location or in a population is important for determining how to respond and what interventions may be needed, such as changing the environment, policy changes, new diagnostic equipment or medicines, etc. Correlations enable us to examine the relationship between variables and how strong those relationships are, and regression analysis enables us to describe the relationship. Before we get into the details, here are some examples of how these have been used in health studies to understand why?

TABLE 3.11

Variability of the Data

Standard error of the mean	How similar the sample means are to each other Standard error of the mean refers to variability we might expect in the arithmetic means of repeated samples taken from the same population.	$SE = \dfrac{s}{\sqrt{n}}$
95% confidence interval for a mean		**Step 1.** Calculate the mean and its standard error. **Step 2.** Multiply the standard error by 1.96. **Step 3.** Lower limit of the 95% confidence interval=mean minus 1.96 × standard error. Upper limit of the 95% confidence interval=mean plus 1.96 × standard error.

Source: CDC (2012).

- **Healthcare expenditure**: Estimate annual health care expenditures among adults (Wee et al., 2005) http://www.ncbi.nlm.nih.gov/pmc/articles/PMC1449869/; (An, 2015)
- **Socio-economic factors**: Examining socio-economic status affecting access to health care (Comber, Brunsdon and Radburn, 2011) https://ij-healthgeographics.biomedcentral.com/articles/10.1186/1476-072
- **Accessibility to parks**: And physical activity sites in New York City (Maroko et al., 2009)
- **Pollution and mortality**: Examines the association between air pollution and mortality using a Cox's Regression analysis (Dockery et al., 1993) http://www.ncbi.nlm.nih.gov/pubmed/8179653/
- **Correlations of disease and different factors**: Malaria and environmental factors (Hakre et al., 2004); cancer mortality and socio-economic inequality (Vinnakota and Lam, 2006)
- **Health disparities**: The association between income and life expectancy (Chetty et al., 2016)

Correlation Analysis

Correlation quantifies the degree to which two variables vary together. If two variables are independent, then the value of one has no relationship to the value of the other. If they are correlated, then the value of one is related to the value of the other, either high when the other is high (positive correlation) or high when the other is low (negative correlation).

A commonly used measure is Pearson's r. Pearson' r has the following characteristics:

1. **Non-unit tied**: allows for comparisons between variables measured using different units
2. **Strength of relationship**: Assess the strength of the relationship
3. **Direction of relationship**: Provides an indication of the direction of that relationship
4. **Statistical measure**: Provides a statistical measure of that relationship

Pearson's correlation coefficient measures the linear association between two variables and ranges between $-1 \leq r \leq 1$.

When **r is near −1**, then there is strong linear negative association, that is, when a low value for x tends to imply a high one for y.

When **r = 0**, there is no linear association.

When **r is near +1**, then there is a strong positive linear association, that is, when a low value of x tends to imply a low value for y (Figure 3.18)

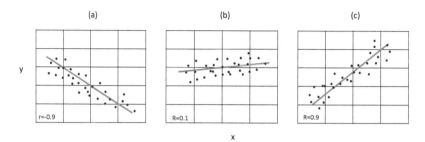

FIGURE 3.18
Examples of scatterplots capturing the correlations between two variables ((a) negative, (b) little to no association and (c) positive association). Image created by Blanford (2023).

Remember that correlation measures the *strength of the relationship* (or "association") between *two* variables. Just because you get a **correlation** between two variables does **NOT** necessarily imply that one **causes** the other.

Regression Analysis

Regressions are used to evaluate relationships between two or more feature attributes. Identifying and measuring relationships lets you better understand what's going on in a place, predict where something is likely to occur or begin to examine the causes of why things occur where they do (ESRI). For example, you might use regression analysis to explain childhood obesity using a set of related variables such as income, education and accessibility to healthy food. Typically, regression analysis helps you answer these why questions so that you can do something about them. If, for example, you discover that childhood obesity is lower in schools that serve fresh fruits and vegetables at lunch, you can use that information to guide policy and make decisions about school lunch programs (ESRI).

Regression analysis is a type of statistical evaluation that employs a model that describes the relationships between the dependent variables and the independent variables in a simplified mathematical form. It provides three things (Schneider et al., 2010):

- **Description**: Relationships among the dependent variables and the independent variables can be statistically described by means of regression analysis.
- **Estimation**: The values of the dependent variables can be estimated from the observed values of the independent variables.
- **Prognostication**: Risk factors that influence the outcome can be identified, and individual prognoses can be determined.

Some regression analysis include (Table 3.12):

TABLE 3.12

List of Regression Models

Regression Models	Application	Dependent Variables	Independent Variables
Linear regression	Description of a linear relationship	Continuous (weight, blood pressure)	Continuous and/or categorical
OLS – Ordinary least square	Global regression equation – to generate predictions or to model a dependent variable in terms of its relationships to a set of explanatory variables		
Logistic regression	Prediction of the probability of belonging to groups (outcome: yes/no)	Dependent variable is derived from the presence or absence of a phenomena, typically represented by 0 (absent) or 1 (present) or success of treatment (yes/no)	Continuous and/or categorical
Proportional hazard regression (Cox regression)	Modelling of survival data	Survival time (time from diagnosis to event)	Continuous and/or categorical
Poisson regression	Modelling of counting processes	Counting data: whole numbers representing events in temporal sequence (e.g., the number of times a woman gave birth over a certain period of time)	Continuous and/or categorical
GWR – Geographically Weighted Regression	A local form of linear regression used to model spatially varying relationships.		

Source: Table adapted from Schneider et al. (2010); Scott (2009).

In summary, regression models are used extensively in epidemiological research, are a SIMPLIFICATION of reality and provide us with:

- A simplified view of the relationship between two or more variables
- A way of fitting a model to our data
- Means for evaluating the importance of the variables and the fit (correctness) of the model
- A way of trying to "explain" the variation in y across observations using another variable (x) or variables
- Concerned with "predicting" one variable (Y – the dependent variable) from another variable (X – the independent variable)

Spatial regression analysis: However, when working with spatial data, we must be careful since some of the assumptions required for traditional nonspatial statistical methods are violated (ESRI, 2014; Scott, 2009).

Spatial Interpolation Methods

When you analyse malaria, you will explore an interpolation method. We do not have time to go into these methods in much detail but it is worth mentioning them since they are quite useful for filling in data gaps and are useful for creating continuous surfaces of the landscape from a set of points such as elevation or environmental information such as temperature.

Interpolation methods are based on the principle of spatial autocorrelation or spatial dependence, which measures the degree of relationship/dependence between near and distant objects (Childs, 2004). Spatial autocorrelation determines if values are interrelated. If values are interrelated, it determines if there is a spatial pattern and is used to measure (Childs, 2004):

- Similarity of objects within an area
- Determine the degree to which a spatial phenomenon is correlated to itself in space
- The level of interdependence between the variables
- Nature and strength of the interdependence

A number of different interpolation methods are available (e.g., natural neighbours, trend method, inverse distance weighting, spline and a variety of geostatistical/kriging methods (Table 3.13). For a quick overview, refer to Childs (2004) or the ESRI Help documentation (ESRI, 2021).

The geostatistical methods have been useful for creating the world malaria maps by the Malaria Atlas Project (MAP) (Hay et al., 2009).

Text, Natural Language Processing and Sentiment Analysis

A lot of information is captured as text in documents or a range of microblogs, such as social media posts. Being able to extract information from these documents and map them can be useful for gaining insights about different phenomena. These can include, for example, insights into perceptions and behaviour through sentiment-based analyses (e.g., women's march – Felmlee et al., 2020; vaccination attitudes – Liu and Liu, 2021; Ye et al., 2023) or topic-related mapping

TABLE 3.13

A Summary of Different Interpolation Methods

Deterministic	Deterministic techniques use parameters that control either the extent of similarity of the values or the degree of smoothing in the surface. There are a number of different interpolators that include: • Global polynomial • Local polynomial • Inverse distance weighted (IDW) • Radial basis functions • Spline • Natural neighbour • Topo to raster based on the ANUDEM (Hutchinson et al.)
Geostatistical Methods/ Kriging	Geostatistics methods are used to produce predictions and related measures of uncertainty of the predictions at unsampled locations. There are several methods available that include: • Kriging/co-kriging • Empirical Bayesian kriging • Areal interpolation

Source: ESRI (2021); Childs (2004).

(e.g., public behaviours and emergency responses to a disaster such as a tornado; Blanford et al., 2014) or text-mining for sense-making to improve our understanding about disease hotspots (e.g., measles in Niger; Tomaszewski et al., 2011) or the impact of storms (e.g., hurricane; Hu, 2018).

Clustering Methods

There are a variety of clustering methods that can be used for analysing spatial and non-spatial data. In Chapter 7 more information is provided on spatial clustering methods, but in short, clustering methods are useful for making sense of data by identifying patterns and trends or for extracting information. Clustering methods are also extensively used in the classification of imagery.

Image Analysis and Feature Classification and Extraction

More and more sensors are being used to capture information about our environment and ourselves. Artificial Intelligence, Machine Learning and Deep Learning are all part of the many tools and methods available to analyse the data captured through different sensors. Here is a quick overview about these different methods.

- **Machine learning** is a branch of artificial intelligence in which structured data is processed with an algorithm. Traditional structured data requires a person to label the data. For example, a picture of an animal such as flamingos or elephants so that specific features for each animal type can be understood within the algorithm and used to identify these animals in the image or other pictures.
- **Deep learning** is a subset of machine learning that uses several layers of algorithms in the form of neural networks. Input data are analysed through different layers of the network, with each layer defining specific features and patterns in the data. These are useful for identifying features such as buildings, fields or roads. To achieve this, the deep learning model can be trained with images of different buildings and roads, process the images through layers within the neural network and then find the identifiers required to classify a building or road.

Machine learning and deep learning methods are used in many GIS systems and methods to perform image classification, clustering data and modelling spatial relationships. These can include:

- **Image classification** involves assigning a label or class to a digital image. Object classification or image recognition can be used to categorize features in an image.
- **Object detection** is the process of locating features in an image such as animals. This process involves drawing a bounding box around the features of interest.
- **Semantic segmentation** occurs when each pixel in an image is classified as belonging to a class, such as a particular type of land use. This is also referred to as pixel classification, image segmentation or image classification.
- **Instance segmentation** is a more precise object detection method in which the boundary of an object is drawn (e.g., the roof of a house). This type of deep learning application is also known as object segmentation.
- **Panoptic segmentation** combines semantic segmentation and instance segmentation to characterize unique objects.
- **Image translation or image-to-image translation** is used to translate an image from one representation to another and is useful for noise reduction or super-resolution.

And also **change detection** methods for detecting changes in features between images obtained from different dates (e.g., Cheng et al., 2023; Asokan and Anitha, 2019; Hussain et al., 2013; Blaschke, 2010; Lu et al., 2004; Blaschke, 2005; Tewkesbury et al., 2015).

Spatial Data and Some Considerations

> Everything is related to everything else but near things are more related
> than those further away.
>
> *Tobler's First Law of Geography (Tobler 1970)*

Processes that happen in space are *not independent* observations and can be influenced by **first-order effects** (due to the local environment) and/or **second-order effects** (due to local interactions) (Cromley and McLafferty, 2012; O'Sullivan and Unwin, 2010)

Spatial Autocorrelation

Spatial autocorrelation is the presence of systematic spatial variation in a mapped variable. Adjacent observations with similar values can be **positive** (features that are similar in their location, as well as in their attributes); **negative** (features that are dissimilar in their location); or have **zero** autocorrelation (the location of a feature has no influence).

Modifiable Areal Unit Problem (MAUP)

Tha Modifiable Areal Unit Problem was first addressed by Openshaw (1984) who highlighted that "the areal units (zonal objects) used in many geographical studies are arbitrary, modifiable and subject to the whims and fancies of whoever is doing, or did, the aggregating." – Thus, MAUP is related to the aggregation fallacy present in geographic data. It is a statistical bias that occurs when data is aggregated. This results in two types of biases. One that has a scale effect and one that has a zonal effect. Standard statistical techniques may be sensitive to the chosen unit scale, aggregation and zonation.

Scale effect. The scale effect occurs when the analytical result differs at different levels of aggregation (e.g. size of the units used). For example, the results obtained at one geographic scale (e.g. district) differs to another geographic scale (e.g. neighbourhood or postcode level analysis) resulting in patterns that may be clustered at one scale but dispersed at another.

Zone effect occurs when the shape of the aggregation unit changes. For example, the zone effect is observed when the scale of analysis is fixed, but the shape of the aggregation unit has changed.

Boundary and Edge Effects

Edges of the map, beyond which there is no data, can affect the results of an analysis, as can the boundary selected in which to conduct the analysis (Figure 3.19).

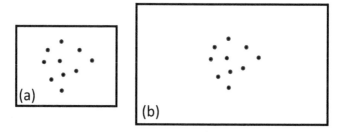

FIGURE 3.19

Boundary and edge effects. Both pattern are identical and yield identical measures of central tendency or dispersion, However, (a) represents a pattern that is dispersed and (b) represents a pattern that is clustered. Image credit Blanford (2021).

Summary

In this chapter you were introduced to statistical and visualization methods that are useful for evaluating health risks and disease outcomes. This included a variety of measures used to assess health risks, basic statistics and a variety of visualizations. However, there are some considerations to take into account when working with spatial data that include spatial autocorrelation, the modifiable areal unit problem and boundary and edge effect. Keep in mind there are many methods needed when performing an analysis that start with acquiring the data (what data is needed), getting to know the data and lastly identifying the methods needed to conduct the analysis (what methods to use to conduct the analysis) (Figure 3.20).

Activity – Working in R

This activity is to provide you with an opportunity to work in R/R Studio if you are not yet familiar with R/R Studio.

1. Familiarizing yourself with RStudio and getting comfortable with statistics!
   a. Obtaining and installing R packages into RStudio
2. Perform the analysis and tell a story about what you found
   a. Loading the data into R
   b. Getting to know your data:

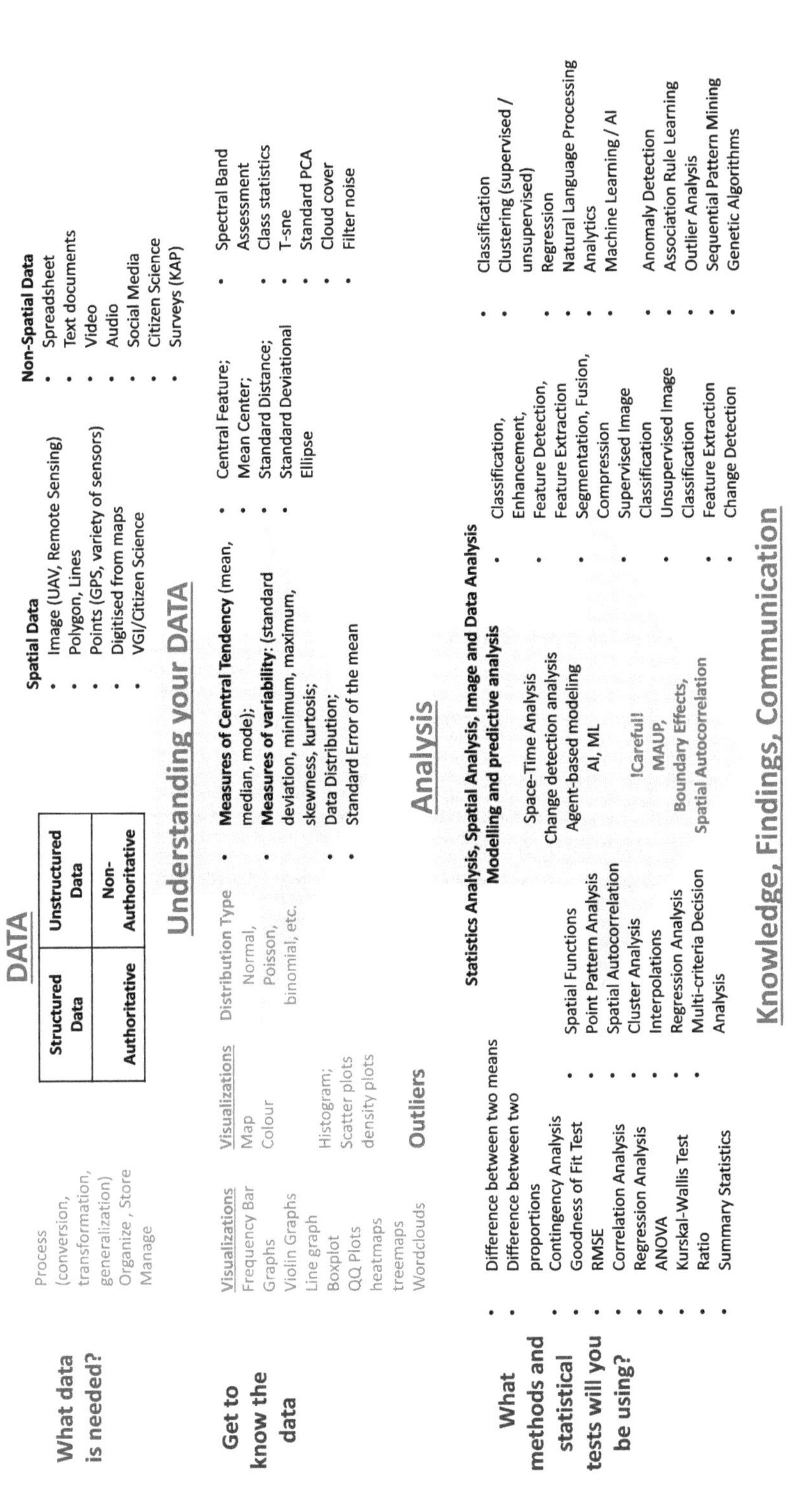

DATA

Structured Data	Unstructured Data
Authoritative	Non-Authoritative

Spatial Data
- Image (UAV, Remote Sensing)
- Polygon, Lines
- Points (GPS, variety of sensors)
- Digitised from maps
- VGI/Citizen Science

Non-Spatial Data
- Spreadsheet
- Text documents
- Video
- Audio
- Social Media
- Citizen Science
- Surveys (KAP)

Understanding your DATA

What data is needed?

Process (conversion, transformation, generalization)
Organize , Store
Manage

Get to know the data

Visualizations
Map
Colour

Histogram;
Scatter plots
density plots

Distribution Type
Normal,
Poisson,
binomial, etc.

Visualizations
Frequency Bar Graphs
Violin Graphs
Line graph
Boxplot
QQ Plots
heatmaps
treemaps
Wordclouds

Outliers

- Measures of Central Tendency (mean, median, model);
- Measures of variability: (standard deviation, minimum, maximum, skewness, kurtosis;
- Data Distribution;
- Standard Error of the mean

- Central Feature;
- Mean Center;
- Standard Distance;
- Standard Deviational Ellipse

Analysis

What methods and statistical tests will you be using?

- Difference between two means
- Difference between two proportions
- Contingency Analysis
- Goodness of Fit Test
- RMSE
- Correlation Analysis
- Regression Analysis
- ANOVA
- Kurskal-Wallis Test
- Ratio
- Summary Statistics

Statistics Analysis, Spatial Analysis, Image and Data Analysis

Modelling and predictive analysis

- Spatial Functions
- Point Pattern Analysis
- Spatial Autocorrelation
- Cluster Analysis
- Interpolations
- Regression Analysis
- Multi-criteria Decision Analysis

- Space-Time Analysis
- Change detection analysis
- Agent-based modeling

AI, ML

!Careful!
MAUP,
Boundary Effects,
Spatial Autocorrelation

- Classification, Enhancement,
- Feature Detection,
- Feature Extraction
- Segmentation, Fusion, Compression
- Supervised Image Classification
- Unsupervised Image Classification
- Feature Extraction
- Change Detection

- Classification
- Clustering (supervised / unsupervised)
- Regression
- Natural Language Processing
- Analytics
- Machine Learning / AI
- Anomaly Detection
- Association Rule Learning
- Outlier Analysis
- Sequential Pattern Mining
- Genetic Algorithms

- Spectral Band Assessment
- Class statistics
- T-sne
- Standard PCA
- Cloud cover
- Filter noise

Knowledge, Findings, Communication

FIGURE 3.20
Data, process and methods used to conduct an analysis. Image created by Blanford (2023).

    i. Descriptive statistics
1. What type of distribution does your data have?
2. What is a typical value in this dataset? (the average)
3. How widely do values in the dataset vary?
4. Are there any unusually high or low values in this dataset? (Outliers)
   ii. Visualizing the data
1. Create some graphs:
   a. Is there a place or year when disease incidence occurs more than others?
   b. Has disease incidence increased or decreased over time?
   c. Etc.
  iii. Discuss your findings

Getting Started in R and RStudio

RStudio/R is a standard statistical analysis package that is free, extensible and contains many new methods and analysis that are contributed by researchers across the globe.

You'll install R and RStudio at this link: https://www.rstudio.com/products/rstudio/download/. Versions are available for all the major platforms. Once installed, start the program.

The command window appears, as shown in Figure 3.21

Obtaining and Installing R Packages

R contains many packages that researchers around the world have created. To learn more about what is available, look through the following links:

https://cran.r-project.org/web/packages/available_packages_by_name.html

https://www.rstudio.com/products/rpackages/

https://www.datacamp.com/community/tutorials/r-packages-guide

Installing packages by command line

```
> install.packages("ggplot2")
> library(ggplot2)
```

FIGURE 3.21
The RStudio console window.

OR load packages in RStudio

To do so, click the **Packages** tab and click **Install**. A window will popup. Select the repository to install from, and then type in the package(s) you want. To install the package(s), click Install. Once the package has been installed, you will see it listed in the packages library. To turn a package on or off, click on the box next to the package listed in the User Library on the packages tab. Packages are updated, so check for updates regularly.

Install the following packages:
Notebook and interactive apps

- rmarkdown (book: https://bookdown.org/yihui/rmarkdown/)
- shiny: creating interactive apps https://www.rstudio.com/products/shinyapps/

Learning tools in R

- swirl: Learning R in R – http://swirlstats.com/
- rcmdr: GUI for different statistical methods.

Various packages used for mapping, spatial analysis, statistics and visualizations.

- leaflet: mapping elements – http://rstudio.github.io/leaflet/morefeatures.html

- maptools
- spatstat
- sp
- rgdal
- dplyr
- ggplot2
- doBy
- ggmap
- gridExtra
- RColorBrewer

R Markdown: A Quick Overview

Since typing commands into the console can get tedious, a notebooking approach is useful for tracking what you did and embedding the results into a single document. R Markdown is a package where you can execute R code, display the results in a single file and add comments and notes.

To get started. **Install the RMarkdown library**. Create a new file. To do so, go to **File – New File – R Markdown**. For the output format and document, **select html**.

Opening the RMD File and How to Code the File

Load the rmd file, **File – Open File – select the file.Rmd** you saved on your computer.

The file will open in a window in the top-left quadrant of RStudio. This is essentially a text file where you will add your R code, run your analysis and write up your results.

The template contains several pieces of information. The header is where you should add a title, date and your name as the author.

R code should be written in the grey areas between the ``` and ``` comments. {r, echo=TRUE) will include the *r* code into the output. If you do not want to include the *r* code, then set echo=FALSE.

For additional information on formatting, see the RStudio cheat sheet https://rmarkdown.rstudio.com/lesson-15.html.

To execute the R code and create a formatted document, use **Knit**. **Knit** integrates the pieces in the file to create the output document, in this case the html_document. The **Run** button will execute the R code and integrate the results into the notebook. Or you can run chunks of code.

A Brief Introduction to R: The Essentials

If you are new to R, at the end of the document there are some essential commands to get you going. This is not a definitive resource since it would end up as a book in its own right. Instead, it provides a variety of commands that will allow you to load, explore and visualize your data.

Analysing Data in R and Documenting What You Have Done

Adding the R Packages That Will Be Needed for the Analysis

Many of the analyses you will be using require commands contained in different packages.

To install and load packages here, add the following code. Note that #install.packages will be ignored in this instance because of the # in front of the command. If you do need to install a package, then remove the # when you run your code.

```
#install and load packages here
library(ggplot2)
library(spatstat)
library(leaflet)
library(maptools)
library(rgdal)
library(dplyr)
library(doBy)
library(rgdal)
library(ggmap)
library(gridExtra)
library(sp)
```

Set the Working Directory

Before starting, set the working directory:

```
#check and set the working directory
# If you do not set the working directory ensure that you
use the full directory pathname to where you saved the
file.
getwd()

#set directory to where the data is saved.
filedir <-"C:/project/data/disease/"
#to view the files in the directory
list.files(filedir)
```

Loading and Viewing the Data File

Add the data file that you need and view the file:

```
#read in data file
# Use import Dataset

library(readr)
ebola_wk <- read_delim("C:/project/data/disease/eboladata_
week.csv",
    delim = ";", escape_double = FALSE, trim_ws = TRUE)

#view the table
View(ebola_wk)

summary(ebola_wk)
```

Descriptive Analysis

Use various descriptive statistics to learn more about the data; when and where cases were reported. At the end of each section, click on **Knit** to execute the R code and view your updates.

```
#Summarize data
summary(ebola_wk)

#Group by data (need package dplyr)
dt <- data.frame(cnt=ebola_wk$cases, group=ebola_wk$week)
grp <- group_by(dt, group)

summarise(grp, sum=sum(cnt))
```

```
df<- ebola_wk
df.summary <- df %>%
  group_by(week) %>%
  summarise(
  sd = sd(cases, na.rm = TRUE),
  mean_cases = mean(cases),
  sum_cases = sum(cases)
)
df.summary
```

Plot the data

```
#View the data

plot(ebola_wk)

#boxplot
boxplot(ebola_wk ~ ebola_wk$country)
boxplot(df$cases ~ df$week , col=terrain.colors(5) )

ggplot(ebola_wk, aes(x=country, y=cases,
fill=country))+geom_violin()

#histograms
hist(ebola_wk$cases)

p<-ggplot(df, aes(x = week, y = cases)) +
geom_bar(
stat = "identity", position = position_stack()
)+
scale_color_manual(values = c("#0073C2FF", "#EFC000FF"))+
scale_fill_manual(values = c("#0073C2FF", "#EFC000FF"))
p
```

References

An, R. (2015) Health care expenses in relation to obesity and smoking among U.S. adults by gender, race/ethnicity, and age group: 1998–2011. *Public Health,* 129, 29–36.

Asokan, A. & J. Anitha (2019) Change detection techniques for remote sensing applications: A survey. *Earth Science Informatics,* 12, 143–160.

Blanford, J. I., J. Bernhardt, A. Savelyev, G. Wong-Parodi, A. M. Carleton, D. W. Titley & A. M. MacEachren (2014) Tweeting and Tornadoes. In *ISCRAM*. Pennsylvania, USA: State College.

Blanford, J. I. & A. Jolly (2021) Public Health Needs GiScience (Like Now!). In *AGILE Conference*. Greece: AGILE.

Blaschke, T. (2005) Towards a framework for change detection based on image objects. *Göttinger Geographische Abhandlungen*, 113, 1–9.

Blaschke, T. (2010) Object based image analysis for remote sensing. *ISPRS Journal of Photogrammetry and Remote Sensing*, 65, 2–16.

Bonita, R., R. Beaglehole & T. Kjellstrom (2006) *Basic Epidemiology*. Geneva, Switzerland: World Health Organization (WHO). https://apps.who.int/iris/bitstream/10665/43541/1/9241547073_eng.pdf (last accessed April 26 2021).

CDC Epi Info User Guide. https://www.cdc.gov/epiinfo/support/userguide.html (last accessed April 27 2024).

——(2012) *Principles of Epidemiology in Public Health Practice*. U.S. Department of Health and Human Services, Centers for Disease Control and Prevention. https://stacks.cdc.gov/view/cdc/6914 (last accessed April 27 2024).

Cheng, G., Y. Huang, X. Li, S. Lyu, Z. Xu, Q. Zhao & S. Xiang (2023) Change Detection Methods for Remote Sensing in the Last Decade: A Comprehensive Review. *arXiv preprint arXiv:2305.05813*.

Chetty, R., M. Stepner, B. A. Abraham, S. Lin, B. Scuderi, N. Turner, A. Bergeron & D. Cutler (2016) The association between income and life expectancy in the United States. *The Journal of the American Medical Association*, 315, 1750–1766.

Childs, C. (2004) Interpolating surfaces in ArcGIS spatial analyst. *ArcUser*, July–September, 32–35.

Comber, A. J., C. Brunsdon & R. Radburn (2011) A spatial analysis of variations in health access: Linking geography, socio-economic status and access perceptions. *International Journal of Health Geographics*, 10, 44.

Cromley, E. K. & S. L. McLafferty (2012) *GIS and Public Health*. New York: Guilford Press.

ESRI (2014) Regression Analysis Basics. https://webhelp.esri.com/arcgisdesktop/9.3/index.cfm?id=2242&pid=2239&topicname=Regression_analysis_basics (last accessed April 27 2024).

—— (2021) Classification Trees of the Interpolation Methods Offered in Geostatistical | Analyst. https://pro.arcgis.com/en/pro-app/latest/help/analysis/geostatistical-analyst/classification-trees-of-the-interpolation-methods-offered-in-geostatistical-analyst.htm (last accessed April 27 2024).

Felmlee, D., J. I. Blanford, S. Matthews & A. M. MacEachren (2020) The geography of sentiment towards the women's march of 2017. *Plos One*, 15, e0233994.

Goodwin, G. & S. Ryu (2023) Understanding the odds: Statistics in public health. *Frontiers for Young Minds*, 11, 926624.

Hakre, S., P. Masuoka, E. Vanzie & D. R. Roberts (2004) Spatial correlations of mapped malaria rates with environmental factors in Belize, Central America. *International Journal of Health Geographics*, 3, 6.

Hay, S. I., C. A. Guerra, P. W. Gething, A. P. Patil, A. J. Tatem, A. M. Noor, C. W. Kabaria, B. H. Manh, I. R. Elyazar, S. Brooker, D. L. Smith, R. A. Moyeed & R. W. Snow (2009) A world malaria map: Plasmodium falciparum endemicity in 2007. *PLoS Medicine*, 6, e1000048.

Hu, Y. (2018) Geo-text data and data-driven geospatial semantics. *Geography Compass*, 12, e12404.

Hussain, M., D. Chen, A. Cheng, H. Wei & D. Stanley (2013) Change detection from remotely sensed images: From pixel-based to object-based approaches. *ISPRS Journal of Photogrammetry and Remote Sensing*, 80, 91–106.

Liu, S. & J. Liu (2021) Public attitudes toward COVID-19 vaccines on English-language twitter: A sentiment analysis. *Vaccine*, 39, 5499–5505.

Lu, D., P. Mausel, E. Brondizio & E. Moran (2004) Change detection techniques. *International Journal of Remote Sensing*, 25, 2365–2401.

Maroko, A. R., J. A. Maantay, N. L. Sohler, K. L. Grady & P. S. Arno (2009) The complexities of measuring access to parks and physical activity sites in New York City: A quantitative and qualitative approach. *International Journal of Health Geographics*, 8, 34.

O'Sullivan, D. & D. Unwin (2010) *Geographic Information Analysis*. New York: John Wiley & Sons.

Openshaw, S. (1984) *The Modifiable Areal Unit Problem*. Norwich: Geo Books.

Schneider, A., G. Hommel & M. Blettner (2010) Linear regression analysis: Part 14 of a series on evaluation of scientific publications. *Deutsches Ärzteblatt international*, 107, 776–782.

Scott, L. (2009) *Answering Why Questions. An Introduction to Using Regression Analysis with Spatial Data*. Redlands, CA: ArcUser, Spring.

Shantikumar, S., H. Barratt & M. Kirwan (2018) Health Knowledge. https://www.healthknowledge.org.uk/public-health-textbook/research-methods/1a-epidemiology/ (last accessed 30 Oct 2023).

Szumilas, M. (2010) Explaining odds ratios. *Journal of the Canadian Academy of Child and Adolescent Psychiatry*, 19, 227.

Tewkesbury, A. P., A. J. Comber, N. J. Tate, A. Lamb & P. F. Fisher (2015) A critical synthesis of remotely sensed optical image change detection techniques. *Remote Sensing of Environment*, 160, 1–14.

Tobler, W. R. (1970) A computer movie simulating urban growth in the detroit region. *Economic Geography*, 46, 234–240.

Tomaszewski, B., J. I. Blanford, K. Ross, S. Pezanowski & A. M. MacEachren (2011) Supporting geographically-aware webdocument foraging and sensemaking. *Computers, Environment and Urban Systems*, 35, 192–207.

Vinnakota, S. & N. S. Lam (2006) Socioeconomic inequality of cancer mortality in the United States: A spatial data mining approach. *International Journal of Health Geographics*, 5, 9.

Wee, C. C., R. S. Phillips, A. T. Legedza, R. B. Davis, J. R. Soukup, G. A. Colditz & M. B. Hamel (2005) Health care expenditures associated with overweight and obesity among US adults: Importance of age and race. *The American Journal of Public Health*, 95, 159–65.

WHO (2013) WHO Methods and Data Sources for Global Burden of Disease Estimates 2000–2011. *Global Health Estimates Technical Paper WHO/HIS/HSI/GHE/2013.4*.

Ye, J., J. Hai, Z. Wang, C. Wei & J. Song (2023) Leveraging natural language processing and geospatial time series model to analyze COVID-19 vaccination sentiment dynamics on Tweets. *JAMIA Open*, 6, ooad023.

Appendix

R the Essentials

For those of you who are new to R. Here are some basics to get you started.

The '>' is your prompt to type something in the console to make things happen. Try this:

This creates a data object in R, known as a vector with 5 items, with values 1, 2, 3, and so on.

```
> x = c(1,2,3,4,5)
```

Here's another data item:

```
> y = seq(1, 100)
```

and another - a vector of 1,000 random numbers between 0 and 1:

```
> z = runif(1,000)
```

By typing the name of any data object, you can inspect its contents:

```
> x
[1] 1 2 3 4 5
```

Try examining the others. You'll see that y is a sequence of values from 1 to 100, while z is a collection of 1,000 random numbers between 0 and 1.

The power of R lies in its ability to statistically manipulate such collections of data. Using the examples above where you have created some data now try these:

```
> mean(y)
[1] 50.5
> hist(z)
> boxplot(z)
```

Now create a vector of 1,000 normally distributed random numbers:

```
> zz = rnorm(1,000)
and try examining histograms and boxplots.
```

R recognizes both $x=$seq(1,100), and $x <-$ seq(1,100) as equivalent ways to assign values to objects. As a beginner, you're likely to feel more comfortable with the '=' sign, but you should be aware of the other assignment operator as you will likely see it in the help materials. For more information go to: **http://cran.r-project.org/doc/manuals/R-intro.pdf**

Set Working Directory

Get the working directory

>getwd()

Set the working directory

>setwd("c://documents//lessondir//")

Working with Data Files

List Files	> list.files() to show hidden files > list.files(all.files=TRUE)
Reading in files	There are many different things you can do here. Here are some of the ways you can do this: > newfile <- read.table("filename",header=TRUE, sep=",") where sep=can be set to ":", "," or any other special character that might be used. If the file doesn't have a header then leave out the header or set to FALSE. OR > newfile <- read.csv("filename",header=TRUE)
Writing files and saving outputs	We are not really going to be writing any files this week but if you do need to write to a file you may be interested in the following: > outfile <-write.csv(x, file="filename", row.names=FALSE) > outfile <-write.table(x, file="filename")
Saving plots and graphs	To write your plots to a file. For a list of available functions use help(Devices) > savePlot(filename="filename.*ext*", type=*"type"*) where type=png or jpeg. OR > png("plot.png",width=x,height=y) OR Specify the height and width of the plot file you would like to create. > png("plot.png",width=650,height=400) > plot(x,y,main="Scatterplot Example") > dev.off()
Converting files	Convert DBF to CSV file library("foreign") data=read.dbf("path/to/file.dbf") write.csv(data, "path/to/file.csv", row.names=F)

Statistics and Statistical Tests

A variety of tests include:

Statistic	R code	Additional information

Descriptive Statistics	`> summary(x)`
	 `> sum(x)`
	 `> mean(x)`
	 `> median(x)`
	 where x is a column heading in the file newfile
	
	 Measures of variability
	 `> sd(x)`
	 `> IQR(x)`
		Median absolution deviation from median. This measure is less sensitive to outliers than the variance and standard deviation
		 `>mad(x)`
		 In package matrixStats
		 `> weighted.mean(x,w=x2)`
		 `> weightedMedian(x,w=x2)`
		
		 Trimmed mean drops outliers at either end of the data. In the example below where 10% is dropped from each end (lower end and higher end)
		 `> mean(x, trim=0.1)`

Converting data to z-score	`> (y-mean(x)) / sd(x)`	

Testing the mean of a sample: *t*-test	`> t.test(x, mu=m)`
	 where mu=mean of the population
	
	 Confidence interval for the mean
	 `> t.test(x, conf. level=0.95)`
		$p < 0.05$ indicates the population is not likely to be *m*;
		 $p > 0.05$ indicates no evidence.
		 If the sample size is small ($n < 30$) then the population must be normally distributed in order to derive meaningful results from the *t*-test.

Confidence interval for a median	`> wilcox.test(x, conf. int=TRUE)`	
Testing a sample proportion	`> prop.test(x,n,p)`	$p < 0.05$ indicates the proportion is not likely to be *p* $p > 0.05$ indicates no evidence.

Comparing the means of two samples	`> t.test(x,y)`
	 if paired
	 `> t.test(x,y, paired=TRUE)`
		$p < 0.05$ indicates the means are likely different;
		 $p > 0.05$ indicates no evidence.
		 $n < 20$, the data must be normally distributed;
		 if two populations have the same variance, specify var.equal=TRUE to obtain a less conservative test.

Nonparametric test for comparing two samples	`> wilcox.test(x,y)`
	 if paired
	 `> wilcox.test(x,y, paired=TRUE)`
		compare two populations but don't know the distribution but the shapes are similar

(Continued)

(Continued)

Statistic	R code	Additional information
Correlation between two variables	Pearson's method is used for normally distributed populations `> cor.test(x,y)`	$p < 0.05$ indicates the correlations are likely significant; $p > 0.05$ indicates correlations are not significant
	The Spearman's is used method for non-normal populations `> cor.test(x,y, method = "Spearman")`	
Regression	Linear regression `> lm(y ~ x, data = dataframe)`	To store results for different regression models set the outputs to an object `> lm1 <- lm(y ~ x, data = dataframe)`

	`glm`	to recall the results and formula and see the residual plots and fit
		`> plot(lm1)`
		`> summary(lm1)`

		Non-normal data and specify the distribution type (e.g. poisson, binomial)

Testing for Normality

Many statistical tests require the data to be normally distributed.

To test whether the data are normally distributed run a statistical test and visualize the data.

Statistical Test	Visualizing the Data Distribution

`>shapiro.test(x)`
$p < 0.05$ indicates the population is not likely normally distributed
$p > 0.05$ indicates no evidence so the population is likely normally distributed.
	Histogram
	`>hist(x)`
	Boxplot
	`>boxplot(x)`
	Q-Q plot (Quantile-Quantile plot)
	`>qqnorm(x)`
	`>qqline(x)`
	If the data were normally distributed the points would fall on the diagonal line.
	If the data is not normally distributed you may need to transform the data. If there are too many points above the line, the data is skewed to the left. In such cases you may want to transform the data using a logarthmic transformation. To see if a log transformation would make the data normally distributed you can apply the following:
	`>qqnorm(log(x))`
	`>qqline(log(x))`
	If a transformation doesn't prove useful then use a non-parametric test.

Grouping Data

In R there are several different ways data can be grouped.

Creating a group so that you can summarize by groups.	`> split(x,f)`
	x=variable/header/vector and *f*=factor/groupToGroupby
	`> lapply(x,f)`
	x=variable/header/vector, *f*=function; returns results in a list
	`> sapply(x,f)`
	x=variable/header/vector, *f*=function; returns results in a vector or matrix
	function can be: length, mean, sum, range, median, sd, *t*.test

Applying a function to every row or column:	`> apply(matrix,row,function)` where matrix=matrix, row (1=row by row; 2=column by column), function is the function
Applying a function to groups of data	`> tapply (variable/header/dataframe,group,function)`
Applying a function to groups of rows	`> by(variable/header/dataframe,group,function)`

Exploring Data Using Visualizations

Graphs and Plots

To display several graphs or plots together you will need to create a grid to fill.

To divide the plot window into $N \times M$ matrix

```
> par(mfrow=c(N,M))
```

After creating a matrix, call plot *n* times until each quadrant has been filled. For example if you have a 2 × 2 matrix then call plot 4 times to draw the figure in each quadrant. The plots will go from left to right starting at the top and move down.

To fill in the quadrants by column then us mfcol instead of mfrow

```
> par(mfcol=c(N,M))
```

To get you started see the list of graphs and plots.

Graph	```> plot(x, main ="Title of Graph", xlab ="Label X", ylab ="Label y")```
	Plotting a line from *x* and *y* points
	```> plot(x,y, type ="1"")```
	```> plot(dataframe, type ="1")```
	Adding a vertical or horizontal line
	draw a vertical line at *x*
	```> abline(v = x)```
	draw a horizontal line at *y*
	```> abline(h = y)```
	```> curve(x)```
Scatterplot	```> plot(x,y, main ="Title of Graph", xlab ="Label X", ylab ="Label y")```
	where *x* is one variable and *y* is another variable
Bar Chart	```> barplot2(x)```
	With confidence intervals
	```> barplot2(x, plot.ci =TRUE, ci.l =lower, ci.u =upper)```
	With error bars
	```> data1 <- read.table("filename", header =T)```
	```> attach(data1)```
	```> names(data1)```
	Calculate the heights of the bars
	```> means <- tapply(field1, field2, mean)```
	```> barplot (means,xlab ="x label", ylab ="y label", col ="green")```
Histogram	```> hist(x)```
	```> hist(x,freq =FALSE)```
	Density estimate which can be plotted over the histogram
	```> lines(density(x),lwd =3,col ="blue")```
Boxplots	```> boxplot(x)```
	Boxplot for multiple variables/vectors comparisons
	```> boxplot(x ~ f)```
	```> boxplot (x ~f, data =dataset, main ="Title", xlab ="x label", ylab ="y label")```
	where *x*=numeric valriable and *f*= the factor
	OR try
	```> plot(response~factpr(fact), notch =TRUE)```
	Order the factor levels
	```> index <- order(tapply(response,fact,mean))```
	```> ordered <- factor(rep(index,rep(20,8)))```
	```> boxplot (response~ordered, notch =T, names = as.charcter(index))```
	```> ggplot(datafile, aes(x =field, y =field)) + geom_box-plot()```
	```> ggplot(datafile, aes(x =reorder(field, field2, FUN = median), y =field2)) +geom_boxplot()```

(Continued)

(Continued)

Q-Q plot (Quantile-Quantile plot)	`> qqnorm(x)` `> qqline(x)`
Violin plots	`boxplot(ebola_weekcountrywk21_week15$cases ~ ebola_week-countrywk21_week15$country)`
	`ggplot(ebola_weekcountrywk21_week15, aes(x = country, y = cases,fill = country)) + geom_violin()`

4

Disaster Epidemiology. Health Emergencies and Hazard Considerations: Surveillance to Communication

Overview

Over the past 10 years, a number of outbreaks and pandemics ranging from respiratory diseases (COVID-19, MERS-CoV and SARS) and influenza (H1N1) to haemorrhagic fevers such as Marsburg and Ebola have taken place. Between 2012 and 2017, the WHO recorded more than 1,200 outbreaks in 168 countries, with a further 352 infectious disease events in 2018. Some of these have been localized, while others have spread globally in a short period of time (Figure 4.1).

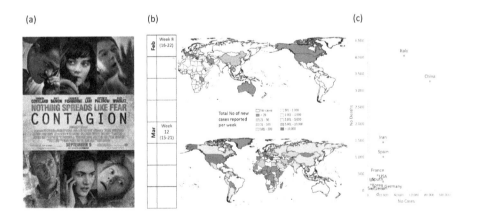

FIGURE 4.1
Movies such as (a) contagion dramatize outbreaks and how they can spread quickly. (b) In 2020, COVID-19 spread quickly around the world with (c) a high number of cases and deaths concentrated in a few countries.

Responding to Health Hazards and Disasters

Health can also be affected by many different types of hazardous events. In the last decade, more than 2.6 billion people have been affected by disasters such as earthquakes, tsunamis, landslides, cyclones, heat waves, floods or severe cold weather (WHO, 2019). Common hazardous events include transportation crashes, floods, cyclones/windstorms, outbreaks, industrial accidents, earthquakes (Coppola, 2015) and conflicts (UN, 2015). Many of which affect over 350 million people each year (Coppola, 2015). Between 2012 and 2017, the WHO recorded more than 1,200 outbreaks in 168 countries, with a further 352 infectious disease events in 2018, including Middle East respiratory syndrome coronavirus (MERS-CoV) and Ebola virus disease (EVD) (WHO, 2013). Between 2020 and 2023 over 6.8 million deaths were reported during the COVID-19 pandemic (JHU, 2023).

With climate change, as temperatures rise, in the past few years, the risk of heat and smoke caused by wildfires has also impacted the health of populations around the world (see Ebi et al., 2021; Clark and Sheehan, 2023; Heaney et al., 2022). The year 2023 was recorded as the hottest year on record with global surface temperatures close to 1.5°C above pre-industrial levels. (ECMWF, 2024).

Hazards and health emergencies can affect health in many ways (WHO, 2019). These include:

- Increase in morbidity, mortality and disabilities;
- Severe disruptions of the health system – interfere with health service delivery through damage and destruction of health facilities, interruption of health programmes, loss of health staff and overburdening of clinical services;
- Set back development gains in public health and other sectors by decades;
- Large financial costs to individuals as well as nationally; and
- May result in long-term health effects.

Most countries are likely to experience a large-scale emergency approximately once every 5 years (UNISDR, 2017), and many are prone to seasonal hazards such as monsoonal floods, cyclones and disease outbreaks. Although most international attention focuses on high-consequence disasters, hundreds of smaller-scale emergencies and other hazardous events occur locally each year, such as outbreaks, floods, fires and transportation crashes (WHO, 2019). Cumulatively, these can become costly and account for a high number of deaths, injuries, illnesses and disabilities and other compounding effects (WHO, 2019).

The health effects caused by any of the events listed in Figure 4.2 can take place at any phase of an emergency. Effects can be immediate and occur in

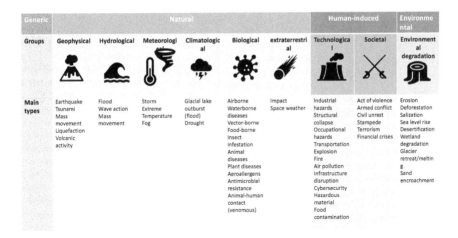

Generic	Natural						Human-induced		Environmental
Groups	Geophysical	Hydrological	Meteorological	Climatological	Biological	extraterrestrial	Technological	Societal	Environmental degradation
Main types	Earthquake Tsunami Mass movement Liquefaction Volcanic activity	Flood Wave action Mass movement	Storm Extreme Temperature Fog	Glacial lake outburst (flood) Drought	Airborne Waterborne diseases Vector-borne Food-borne Insect infestation Animal diseases Plant diseases Aeroallergens Antimicrobial resistance Animal-human contact (venomous)	Impact Space weather	Industrial hazards Structural collapse Occupational hazards Transportation Explosion Fire Air pollution Infrastructure disruption Cybersecurity Hazardous material Food contamination	Act of violence Armed conflict Civil unrest Stampede Terrorism Financial crises	Erosion Deforestation Sea level rise Salization Desertification Wetland degradation Glacier retreat/melting Sand encroachment

FIGURE 4.2
WHO classification of hazards. Adapted from WHO (2019).

the short term or have longer-term effects with lasting consequences. Thus, disaster epidemiology aims to prevent or reduce the number of deaths, illnesses, and injuries caused by a disaster event, including pandemics. However, to do so requires timely and accurate health information so decision-makers and practitioners can make assessments, figure out preventive measures, decide how to respond and determine what recovery strategies may be needed as the disaster unfolds but also after the disaster has ceased. Each type of disaster will have different health effects.

Pandemics

Sixty percent of all infectious diseases in humans are transmitted from animals, and 75% of emerging infectious diseases are caused by microbes of animal origin. These microbes 'spill over' due to increasing contact among wildlife, livestock and people driven by exponentially increasing anthropogenic changes caused by human activities and the impacts of these activities on the environment (e.g. land-use change, agricultural expansion and intensification, increased mobility and wildlife trade and consumption). An estimated 1.7 million currently undiscovered viruses are thought to exist in mammal and avian hosts. Of these, 631,000–827,000 could have the ability to infect humans. The risk of pandemics is increasing rapidly, with more than five new diseases emerging in people every year, any one of which has the potential to spread and become a pandemic (Daszak et al., 2020).

Epidemic Theory

An important measure used to understand the transmission potential of a disease is what is known as R_0 (R-naught) – the **Basic reproduction number (R_0)**. Essentially, it is the number of secondary infections caused by an infected case in a population that is entirely susceptible. It is measured by counting the number of secondary cases, as shown in Figure 4.3.

The basic reproductive number is affected by several factors:

- The rate of contact in the host population
- The probability of an infection being transmitted during contact, and
- The duration of infectiousness

By knowing R_0, for a disease (Table 4.1) it is possible to determine what level of immunity is required in a population to prevent an outbreak. For example, for measles, $R_0=11–18$ (Table 4.1). If the R_0 for measles in a population is 15, then we would expect measles to spread rapidly through the population since each new case of measles would produce 15 new secondary cases.

For individuals that cannot be vaccinated (e.g. they are too young, have immune-compromised systems, etc.), protection of these individuals is achieved through herd immunity. Herd immunity, also known as 'population immunity' is the indirect protection of a population from an infectious disease when a population is immune either through vaccination or immunity developed as a result of previous infections. Herd immunity can be achieved by vaccinating a certain percentage of the population. The percentage required varies with each disease. For measles, between 91% and 94% of the population will need to be immunized to prevent the disease from passing through a population.

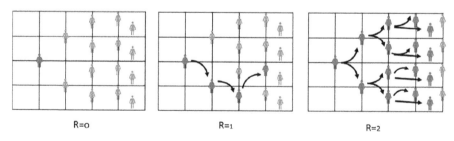

R=0   R=1   R=2

FIGURE 4.3

The reproduction number R_0. The number shows how many people are on average infected by someone who is infected. The number of transmissions remains more or less constant when the reproduction number is 1, declines when it is less than 1 and increases when the number is above 1 (R=2). In general, for an epidemic to occur in a susceptible population, R_0 must be >1, so the number of cases is increasing otherwise the disease is decreasing when $R_0<1$.

TABLE 4.1

Infectious Disease, Average R_0, Herd Immunity Thresholds (HIT) for Vaccine Preventable Diseases, Where Applicable and Known

Disease	Average R_0	Herd Immunity Threshold (Vaccine Preventable Diseases)
Measles	11–18	91%–94%
Mumps	7–14	86%–93%
Poliomyelitis	5–7	80%–86%
Rubella	6–14	83%–94%
Pertussis	10–18	90%–94%
Diphtheria	4–5	75%–80%
Varicella	7–11	86%–91%
Smallpox	4–7	83%–85%
SARS	3.5	
COVID-19 (Delta Variant)	5.8 (3.2–8)	
MERS-COV	0.7–5	
Tuberculosis	2–6	
Ebola	1.5	
Zika	2	
Influenza	1.3–4	
H1N1	1.3	
Spanish Flu 1918	1.5–3.8	

Source: Gani and Leach (2001); Luke and Rodrigue (2008); Plans-Rubio (2012); van den Driessche (2017); Liu and Rocklov (2021); Delamater et al. (2019); Liu et al. (2020).

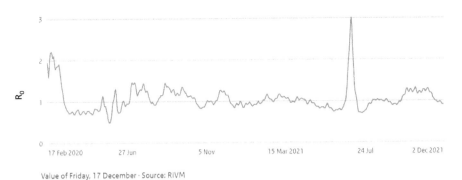

FIGURE 4.4

Changing R_0 of COVID-19 in the Netherlands. *Source:* RIVM (2021).

R_0 can be affected by many factors (Delamater et al., 2019), which is why there are a range of values and why R_0 can change over time. For example, during COVID-19, R_0 changed over time in Netherlands, as seen in Figure 4.4. Some of the complexities of R_0 are captured in the study by Delamater et al. (2019).

Epidemic Curve

In Chapter 3, you were introduced to epidemic curves to monitor case numbers. Here, they are used for passive environmental monitoring.

Monitoring of Waste Water

In the Netherlands, as COVID-19 cases subsided, monitoring of potential cases altered, resulting in more passive surveillance by monitoring concentrations of the virus in waste water (Figure 4.5).

Changing the Shape of the Epi Curve

During an outbreak or pandemic, one of the goals is to reduce the number of infections by flattening the epidemic curve through the use of control measures, as illustrated in Figure 4.6. For the COVID-19 pandemic, a variety of measures were used to control the transmission of the virus that included minimizing contact between people, wearing of masks and the use of vaccines (see Blanford et al. (2022) for full list of measures used during travel in Europe to prevent infections).

Virus particles over time

This graph shows the average number of virus particles per 100,000 inhabitants over time.

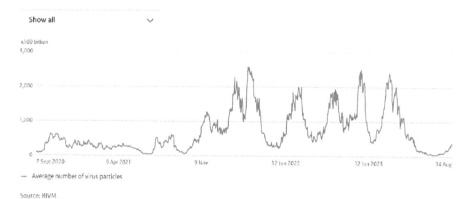

Source: RIVM

FIGURE 4.5

Graph showing the average number of COVID-19 virus particles per 100,000 inhabitants in waste water over time. *Source*: RIVM, https://coronadashboard.government.nl/landelijk/rioolwater. A map of the distribution of virus particles by municipality can be viewed in Chapter 1.

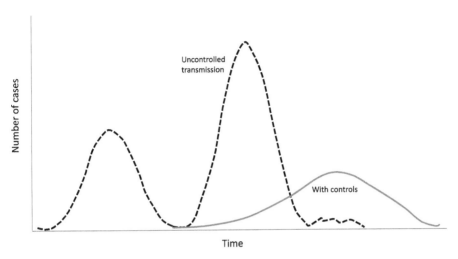

FIGURE 4.6
Flattening the epidemic curve (blue line). Keeping the curve flat will help slow the spread of the disease and reduce the burden on required services such as hospitals. Adapted from Boumans (2021).

Disaster Epidemiology

Dealing with the health effects associated with different types of disasters, both in the short- and long-term are needed. Depending on the type of hazard event (see Figure 4.2), the duration of an event or the intensity of an event, the epidemiological effects will vary. In general, there are several phases associated with a disaster that include Phase 1 (impact phase), Phase 2 (post-impact phase), Phase 3 (recovery phase) and Phase 4 (post-recovery phase) (Figure 4.7).

Phase 1 is characterized as the impact phase and generally takes place within 4 days of the event. The main health effects are physical effects due to injuries or exposures.

Phase 2 is characterized as the post-impact phase and takes place after phase 1 up to about 4 weeks. Initial contagious ailments may develop due to exposure and behaviour related risks. Psychiatric disorders and other psychological effects may start to develop due to changes in emotional states (see Figure 4.8).

Phase 3 is characterized as the recovery phase and getting back to normal. During this time, fatalities from infectious diseases with long incubation periods or latent-type illnesses may become clinically apparent and may lead to outbreaks or epidemics.

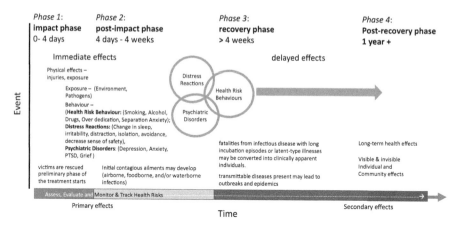

FIGURE 4.7

Summary of the potential health effects associated with each of the four phases post disaster. Compiled from variety of sources Morganstein and Ursano (2020); Brown and Murray (2013); Clayton et al. (2014); Sandifer and Walker (2018); Haelle (2021); Kola et al. (2021); Raker et al. (2020); Clark and Sheehan (2023).

Phase 4 is characterized as the post-recovery phase. During this phase, long-term health and community effects may become apparent and visible. After disasters, there may be significant economic, physical and social consequences and mental effects that may affect the victims (Safarpour et al., 2022). *"Special attention should be paid to the psychologically vulnerable, such as the injured, bereaved, and displaced. Ongoing surveillance is needed, not only for people who are directly affected, but also for those who may be affected by the economic consequences of the disaster (e.g., people who have lost their jobs)."* (Kõlves, Kõlves and De Leo, 2013). Several studies have found that the rate of developing mental disorders, fear, anxiety, stress and depression increase after disasters (see references in Safarpour et al., 2022). When not managed appropriately, individuals can develop mental disorders as a result of psychological tensions due to deaths in the family, physical disease, organ dysfunction, a lack of accommodation and unmet basic needs. The rate of developing mental disorders, fear, anxiety, stress and depression increases after a disaster (see references in Safarpour et al., 2022).

During each of the phases, there are also emotional highs and lows that can affect the mental health and well-being of the population, as depicted in Figure 4.8.

Health Effects Following Flooding Events

Flooding has a wide range of health effects that range from drowning and injuries to a range of diseases such as gastroenteritis, respiratory infections,

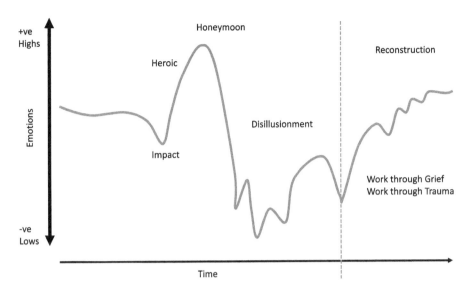

FIGURE 4.8
Emotional states associated with a disaster over time. Figure adapted from Morganstein and Ursano (2020); Clayton et al. (2014).

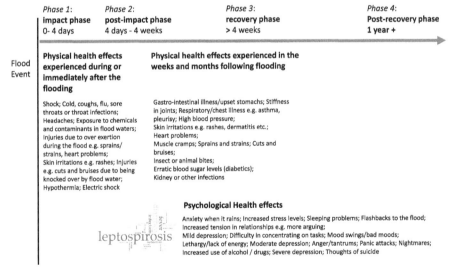

FIGURE 4.9
Summary of potential health effects for each of the four phases of a disaster following a flood event. Word cloud summarizes infectious disease transmission recorded from different flood events around the world. Data compiled from Brown and Murray (2013); Tunstall et al. (2006).

communicable diseases, epidemic diseases (e.g. cholera, diarrhoea and dengue fever), skin diseases, mental and psychological health effects, poisoning, among others (see Butsch, Beckers and Nilson and summaries in Figures 4.9 and 4.10).

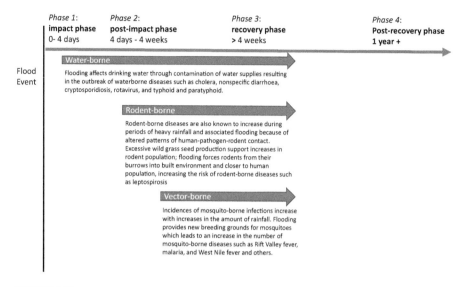

FIGURE 4.10
Summary of potential water-borne, rodent-borne and vector-borne health effects associated with different phases of a disaster following a flood event (Okaka and Odhiambo, 2018; Brown and Murray, 2013).

Flooding can interrupt clean water supply and sanitation, hinder access to and the provision of urgent medical services (UNDAC, 2018; Okaka and Odhiambo, 2018; Blanford et al., 2012) and cause population displacement, which can result in overcrowding, unhygienic conditions, and increase exposure to diseases. This in turn can lead to a reduction in productivity, an increase in demand on health services and a rise in morbidity and mortality (e.g. Figure 4.11). The diseases that are most likely to be affected by flooding are those that require vehicular transfer from host to host (water-borne, such as leptospirosis (Ifejube, et al., 2024)) or a host/vector as part of its life cycle (vector-borne (Okaka and Odhiambo, 2018; Brown and Murray, 2013)).

In light of the increased threat of flooding due to climate change, there is an increasing need for improved understanding of the association of health effects following a flood event so that preparations for post-flood effects can be improved. To achieve this:

- Surveillance of diseases are needed to provide precise knowledge of the incidence rates of infectious diseases arising from flooding across populations and geographic region (e.g. leptospirosis in India (Ifejube, et al., 2024))

- Predictions of risk effects to improve pre- and post-treatment measures. Pre- and post-treatment measures (e.g. vaccination; pre-treating mosquito habitats with insecticides; stocking of adequate drugs, medicine and rehydration fluids).

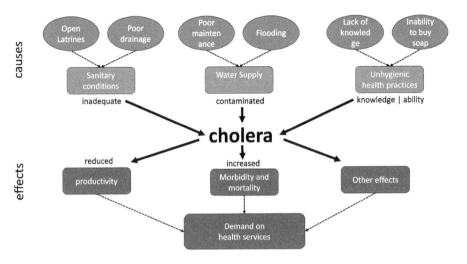

FIGURE 4.11

Example of causes and effects and how they relate to a cholera outbreak. Not all causes may occur during each event. *Source*: UNDAC (2018).

In Summary

Epidemic and pandemic-prone diseases threaten public health security and can be responsible for high levels of disease and mortality, which can have a devastating impact on the economies of the region. Therefore, when an outbreak occurs, a primary goal is to control the outbreak and prevent additional cases as quickly as possible by flattening the curve. This can be achieved through a variety of approaches that can be directed at controlling the pathogen, the source of infection, mode of transmission, portal of entry or the host. A lot of sleuthing maybe required to identify the best approach to take for the most effective control measures to implement, particularly for new pathogens. During this process it may be necessary to provide regular updates to the public and health officials. One way to achieve this is through the use of different visualizations (e.g. epi curves) and interactive map-based dashboards (e.g. JHU, WHO and other examples provided in Chapter 1). Now it is your turn to think about what information to communicate and how to communicate.

Activity - Surveillance and Effective Communication Using Different Visualizations to Understand When and Where an Outbreak is Occurring

Communication is important. What to communicate and how to communicate are vital and have also become more complex as technologies continue to evolve. Communication is about the narrative or story (see Roth, 2021 for creating the story), and the cartographic cube aids in thinking of the components – purpose, audience and interactivity (DiBiase, 1990; MacEachren and Taylor, 1994; Kraak and Ormeling, 2020; Roth, 2013) (Figure 4.12).

- **Interactivity** is the degree to which the user can manipulate the map or visualization.
- **Purpose** refers to the role of the map or visualization in providing information.
- **Audience** refers to the end-user and who will be using the information.

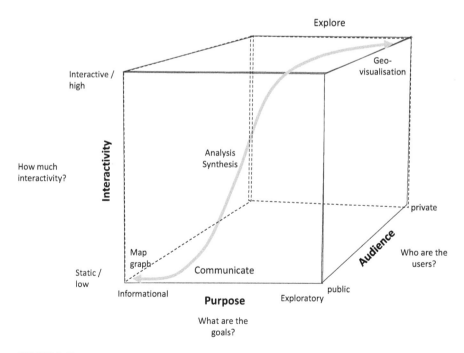

FIGURE 4.12

Communication with the cartographic cube in mind. Adapted from DiBiase (1990); MacEachren and Taylor (1994); Kraak and Ormeling (2020).

To achieve effective communication, depending on the purpose of the visualization, the data will need to be structured appropriately so that it can be mapped or graphed. Thus, the purpose of this activity is to consider the cartographic cube and evaluate

- **Structuring of the data**: Evaluate outbreak reports and assess how easy they are to map an outbreak.
- **Communicating with cartographic and data visualizations**: Evaluate different visualizations and assess their usefulness in understanding an outbreak.
- **Communicating through interactive web maps**: Explore web maps and consideration.

Evaluate Outbreak Reports and Assess How Easy It Is to Map an Outbreak

Surveillance data capturing different characteristics that may be relevant in the chain of transmission are usually collected by public health officials. These include the following:

- **Characteristics** of individuals who have been affected (e.g. age, sex, occupation, religion, ethnicity, etc.);
- **Symptoms**, date of onset and date of death (if applicable); status of case (e.g. suspect, confirmed);
- **Disease** (if known, then the disease that the person has/had);
- **Type of test used** to diagnose disease;
- **Geography** (including geographic location; e.g. where person is from, where person lives – if different);
- **Activity space** (e.g. daily travel routines, places visit);
- **Travel** by the individual (source-destination-duration) and reason for travel if any (e.g. events – religious and music festival; business; holiday; etc.).

The inclusion of geographies and in particular geographic locations can be useful for (i) mapping where the risks are, (ii) determining who may be affected, and (iii) determining the sources of transmission and exposure. For example, a person may have been exposed due to their occupation, leisure activities or food sources, or by some social interaction where they came into direct or indirect contact with an infected person.

**EXAMINE THE CONTENT OF AN OUTBREAK REPORT
(E.G. MERS-COV, SEE MIDDLE EAST RESPIRATORY
SYNDROME – CORONAVIRUS – UPDATE)**

   a. What information is contained in the report (e.g. character-istics of the person, symptoms, disease, test used, geography, activity space, travel, etc.)?

   b. Are there differences in what and how the information is reported?

   c. Do the reports change over time?

   d. Is the information reported consistently over time?

1. **Organising and mapping the data**

   a. Does the report contain sufficient geographic information that would enable you to map the information?

   b. Do you think it will be easy to map the data?

   c. What geographic information is used (place name, address, administrative boundary name, latitude and longitude coordinates, other)? (If in doubt refer to the Data Chapter)?

   d. How should the data be structured (review Tables 4.2 and 4.3)?

   e. What additional information should be included?

2. **Challenges**

   a. Are there any problems/challenges you envisage?

   b. Do you think there will be challenges mapping the data?

   c. How easy is it to organise and extract information from the report?

3. **Solutions**

   a. If you had to provide an outbreak report in the future, what changes would you make to the report (e.g. organisation of the information, additional information to include and data fields to include for capture information)?

TABLE 4.2

Summary of Surveillance Data That May Be Collected to Provide Health Officials with a Description of Whom the Case-Patients Are and Who May Be at Risk

Case No	Onset Date	Date of Death	Age	Sex	Race	Occupation	Patient Outcome	Location	Comments	Diagnostic Test	Source
1	12/05/2013	20/05/2013	56	M				Eastern region, not Al-Ahsa	Underlying medical conditions	MERS-CoV	WHO DON 29/05/2013
2	17/05/2013		85	F			currently in critical condition	Eastern region, not Al-Ahsa	Underlying medical conditions	MERS-CoV	WHO DON 29/05/2013
3	19/05/2013		76	F			Discharged from hospital 27/05/2013	Eastern region, not Al-Ahsa	Underlying medical conditions	MERS-CoV	WHO DON 29/05/2013
4	19/05/2013	26/05/2013	77	M				Eastern region, not Al-Ahsa	Underlying medical conditions	MERS-CoV	WHO DON 29/05/2013
5	18/05/2013	26/05/2013	73	M				Eastern region, not Al-Ahsa	Underlying medical conditions	MERS-CoV	WHO DON 29/05/2013
6	20/05/2013	31/05/2013	61	M				Al-Ahsa		MERS-CoV	WHO DON 31/05/2013

Source: Data extracted from WHO DON report.

TABLE 4.3
MERS-COV 2013 Ongoing Summary

Date	Country	Source of Infection	Lab-Confirmed Cases	Death	Pathogen	Comment	Source
29 May 2013	France	UAE		1	MERS-CoV	First laboratory-confirmed case in the country, with recent travel from the UAE	WHO DON 29/05/2013
29 May 2013	Saudi Arabia			1	MERS-CoV	Patient earlier reported from Al-Ahsa, an 81-year-old woman has died	WHO DON 29/05/2013
Sep 2012–29 May 2013	Global		49	27	MERS-CoV	Laboratory-confirmed cases originating in the following countries in the Middle East to date: Jordan, Qatar, Saudi Arabia, and the UAE. France, Germany, Tunisia and the United Kingdom also reported laboratory-confirmed cases; they were either transferred for care of the disease or returned from the Middle East and subsequently became ill. In France, Tunisia and the United Kingdom, there has been limited local transmission among patients who had not been to the Middle East	WHO DON 29/05/2013
Sep 2012–31 May 2013	Global		50	30	MERS-CoV		WHO DON 31/05/2013

Source: Data extracted from WHO DON report.

MIDDLE EAST RESPIRATORY
SYNDROME – CORONAVIRUS – UPDATE

29 May 2013: The Ministry of Health in Saudi Arabia has notified WHO of an additional five laboratory-confirmed cases of the Middle East respiratory syndrome coronavirus (MERS-CoV).

All five patients are from the eastern region of the country, but not from Al-Ahsa, where an outbreak began in a health care facility in April 2013. The patients had underlying medical conditions that required multiple hospital visits. The government is conducting investigations into the likely source of infection in both health care and community settings.

The first patient is a 56-year-old man with underlying medical conditions who became ill on 12 May 2013 and died on 20 May 2013.

The second patient is an 85-year-old woman with underling medical conditions who became ill on 17 May and is currently in critical condition.

The third patient is a 76-year-old woman with underlying medical conditions who became ill on 24 May 2013 and was discharged from the hospital on 27 May 2013.

The fourth patient is a 77-year-old man with underlying medical conditions who became ill on 19 May and died on 26 May 2013.

The fifth patient is a 73-year-old man with underlying medical conditions who became ill on 18 May and died on 26 May 2013.

Additionally, a patient earlier reported from Al-Ahsa, an 81-year-old woman, has died. The government is continuing to investigate the outbreaks in the country.

In France, the first laboratory-confirmed case in the country, with recent travel from the United Arab Emirates (UAE), has died.

Globally, from September 2012 to date, WHO has been informed of a total of 49 laboratory-confirmed cases of infection with MERS-CoV, including 27 deaths.

WHO has received reports of laboratory-confirmed cases originating in the following countries in the Middle East to date: Jordan, Qatar, Saudi Arabia and the UAE. France, Germany, Tunisia and the United Kingdom also reported laboratory-confirmed cases; they were either transferred for care of the disease or returned from the Middle East and subsequently became ill. In France, Tunisia and the United Kingdom, there has been limited local transmission among patients who have not been to the Middle East but have been in close contact with laboratory-confirmed or probable cases.

Based on the current situation and available information, WHO encourages all Member States to continue their surveillance for severe acute respiratory infections (SARI) and to carefully review any unusual patterns.

Health care providers are advised to maintain vigilance. Recent travellers returning from the Middle East who develop SARI should be tested for MERS-CoV, as advised in the current surveillance recommendations. Specimens from patients' lower respiratory tracts should be obtained for diagnosis where possible. Clinicians are reminded that MERS-CoV infection should be considered even with atypical signs and symptoms, such as diarrhoea, in patients who are immunocompromised. Health care facilities are reminded of the importance of systematic implementation of infection prevention and control (IPC). Health care facilities that provide care for patients suspected or confirmed to have MERS-CoV infection should take appropriate measures to decrease the risk of transmission of the virus to other patients, health care workers and visitors.

All Member States are reminded to promptly assess and notify WHO of any new case of infection with MERS-CoV, along with information about potential exposures that may have resulted in infection and a description of the clinical course. An investigation into the source of exposure should be promptly initiated to identify the mode of exposure, so that further transmission of the virus can be prevented.

Communicating with Data Visualizations

Assessment of an outbreak by place not only provides information on the geographic extent of a problem but may also demonstrate clusters or patterns that provide important aetiologic clues about the source of infection, modes of transmission and who is affected. Different visualizations of the data can be useful for evaluating cases by place and time; establishing the existence of an outbreak; and defining the extent of the outbreak and where it is located.

Epi curves are useful for showing the distribution of cases over time, however, if interventions are to be implemented in response to an outbreak, we need to know **where** these should be placed to be most effective.

Evaluate Different Visualizations and Assess Their Usefulness in Understanding the Outbreak

Earlier in Chapter 3, you were introduced to a variety of statistics and visualizations. These included **epi curves**, violin plots, **heatmaps, dot maps** and **choropleth maps** (3.8–3.10, 3.12–3.17), as well as different (spatial) descriptive statistics such as **measures of central tendency and measures of variability, central feature, mean centre** and **standard deviational ellipse.** I have created a variety of maps using different visualizations Figures 4.13–4.17 and assess their usefulness in understanding the Ebola outbreak during 2014 and 2016.

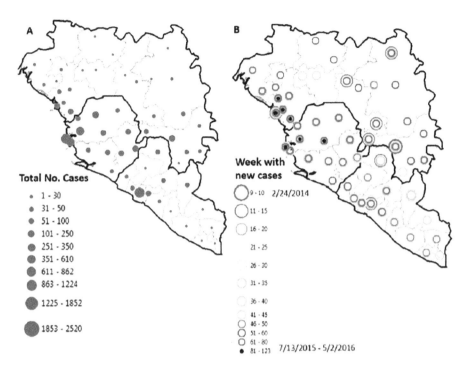

FIGURE 4.13

(a) Dot map showing the total number of cases reported within each district. Each dot represents the centroid of each district and is displayed using graduated symbols where the smallest dot represents a low number of total reported cases and a larger dot represents a higher number of total reported cases. (b) Dot map showing the duration of the outbreak at each location where each week is represented by a different colour. A circle is placed at each location where a confirmed cases was reported.

Examine each of the visualizations (Figures 4.13–4.17) and assess their usefulness in providing insights into the distribution and diffusion of the Ebola outbreak. How informative are these visualizations for communicating risk? Were you able to

- Establish the existence of an outbreak,
- Define the location of the outbreak and the cases, and
- Describe the spread of the outbreak.

If not, what would you change? Is one visualization better than the other? Or would you use several of these? What additional visualizations would you include? What combination of visualizations would you use (Figure 3.17 vs 4.16)? What limitations or challenges did you identify? Would animating the maps be useful? Would creating an interactive map be useful?

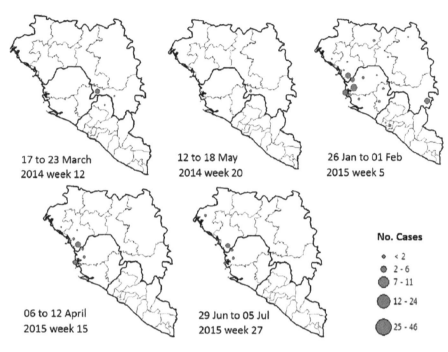

FIGURE 4.14

Dot maps showing the total number of weekly cases reported during different weeks in years 2014 and 2015.

Epi Curves

The **epidemic curve**s provide a simple visual display of the outbreak's magnitude and time trend. Review the epi curves for the Ebola outbreak of 2014 in Chapter 3 (Figures 3.8, 3.9, 3.10, and 3.12).

Dot Maps

In this example (Figure 4.13), different dot maps were created.

(Figure 4.13a) Dot map showing the **total number of cases** reported within each district. Each dot represents the centroid of each district and is displayed using graduated symbols, where the smallest dot represents a low number of total reported cases and a larger dot represents a higher number of total reported cases. (Figure 4.13b) Dot map showing the **duration of the outbreak** at each location, where each week is represented by a different colour. A circle is placed at each location, where a confirmed case has been reported.

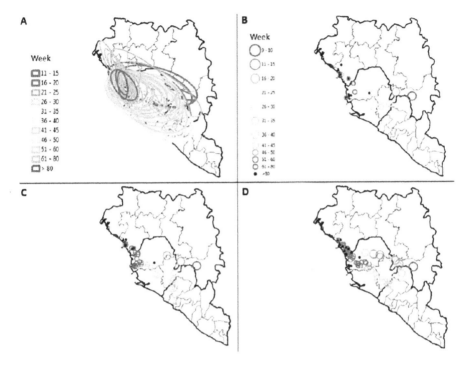

FIGURE 4.15
Visualizations capturing the spatial distribution of Ebola during each week between 2014 and 2016. (A) Directional distribution (Standard Deviational Ellipse); (B) Central feature; (C) Median Centre and; (D) Mean Centre of the Ebola outbreak.

Time Series Dot Maps

Dot maps showing the total number of weekly cases reported during different weeks in years 2014 and 2015 (Figure 4.14).

Geographic Distribution of Ebola

The directional distribution, most central feature, median and mean centre were calculated for each week and used to evaluate the distribution of the outbreak. Each analysis was weighted (weight field=count) and assessed on a weekly basis (case field=week). To use these tools, navigate to **Spatial Statistical Toolbox – Measuring Geographic Distribution**.

Weekly spatial distribution of the Ebola outbreak (Figure 4.15) is captured using different spatial descriptive statistical measures. **(A) Directional Distribution (Standard Deviational Ellipse); (B) Central Feature; (C) Median Centre** and; **(D) Mean Centre**.

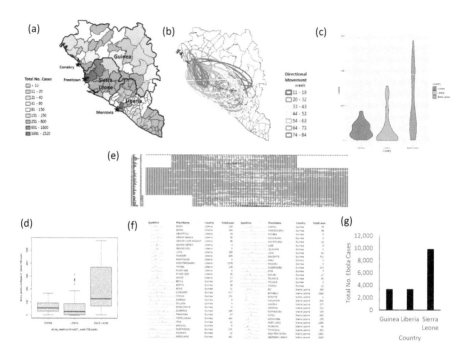

FIGURE 4.16

Multiple visualizations are used to highlight **where** the outbreak occurred. (a) Choropleth of total number of cases by district; **when** (b) the directional distribution of cases by week; (c) the intensity and duration of the distribution by country (d) captures the distribution of Ebola between countries; (e) **Intensity of outbreak**: number of cases reported by district for each week (dark red=highest; green=no cases) and (f) **peak and duration**: individual epi curves capturing the peak and duration of the outbreak in each location; and (g) **Summary of cases**: the total number of cases by reported for each country.

Combining Visualizations

Each visualization conveys different pieces of information. Some things are better shown on a map, while others are not and may require multiple types of analysis and visualizations to understand the location, distribution, mode of transmission and diffusion of an outbreak (e.g. Figure 4.16).

Communicating through Interactive Web Maps

Communication will likely take many forms, including providing regular updates by public health agencies or other local authorities. One form of communication may be to provide updates through interactive web maps. There are an ever increasing number of tools available for making interactive maps. A few examples include **ArcGIS Online**, **CartoDB**, **Mapbox** or if you have some scripting skills you can create your own webmap using **ShinyApp in R**.

Keep in mind that during an outbreak, you will need to be able to provide updates regularly, quickly and efficiently. Some design considerations that might be relevant include:

- **Data**: Storage and updates. How will the data be stored once it is acquired? The frequency of updates is likely to vary from daily, several times a day to weekly. Review Tables 4.2 and 4.3 and consider how you would structure the data to provide daily updates through maps for example. What geographic information should be captured (refer to Chapter 5 Figure 5.2)?
- **Data to map**: What data and attributes should be included in the web map? Does the data need to be pre-processed before it can be provided to the public? Will any analysis be required? If so, how and where will this be completed? What additional data is required? What background map should be used to provide context? Are there any ethical considerations?
- **Symbolization**: How will the data be symbolized?
- **Software**: What software will be used?

Here is an example of a map created to illustrate outbreaks of haemorrhagic fevers (Ebola and Marburg) since 1975 using graduated symbols. Since this is an interactive map, users can click on the points and view additional information in the table below the map. The data for this was obtained from various sources (CDC, WHO and Quammen) and placenames geocoded using geonames and googlemaps. Each point either represents the centroid of an administrative boundary (e.g. county or district (see Figure 5.2)) or the location of a town or village (Figure 4.17). This of course will affect the accuracy

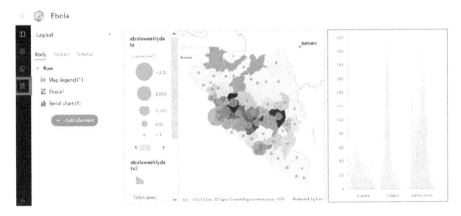

FIGURE 4.17
Dashboard of Ebola cases for Liberia, Guinea, and Siera Leone between 2014 and 2016. *Source*: Blanford (2023).

of where the actual outbreak is taking place and should be taken into consideration.

Conclusion: What did you learn about the Ebola outbreak? Were the visualizations useful in building an understanding about the outbreak and where risks were highest during the outbreak? Were these useful in understanding the movement and diffusion of the disease?

References

Blanford, J. I., S. Kumar, W. Luo & A. M. MacEachren (2012) It's a long, long walk: accessibility to hospitals, maternity and integrated health centers in Niger. *International Journal of Health Geographics*, 11, 24.

Blanford, J. I., N. Beerlage-de Jong, S. Schouten, A. Friedrich & V. Araujo-Soares (2022) Navigating travel in Europe during the pandemic: From mobile apps, certificates, quarantine to traffic light system. *Journal of Travel Medicine*, 29(3), taac006.

Boumans, M. (2021) Flattening the curve is flattening the complexity of covid-19. *History and Philosophy of the Life Sciences*, 43, 18.

Brown, L. & V. Murray (2013) Examining the relationship between infectious diseases and flooding in Europe. *Disaster Health*, 1, 117–127.

Butsch, C., L. M. Beckers & E. Nilson (2023) Health impacts of extreme weather events-Cascading risks in a changing climate. *Journal of Health Monitoring*, 8, 33–56.

Clark, K. A. & M. Sheehan (2023) Wildfires and the COVID-19 pandemic: A systematized literature review of converging health crises. *Environmental Research: Health*, 1, 022002.

Clayton, S., C. Manning & C. Hodge (2014) *Beyond Storms & Droughts: The Psychological Impacts of Climate Change*. Washington, DC: American Psychological Association and ecoAmerica.

Coppola, D. (2015) *Introduction to International Disaster Management*. Oxford, England: Butterworth-Heinemann.

Daszak, P., C. das Neves, J. Amuasi, D. Hayman, T. Kuiken, B. Roche, C. Zambrana-Torrelio, P. Buss, H. Dundarova, Y. Feferholtz, G. Foldvari, E. Igbinosa, S. Junglen, Q. Liu, G. Suzan, M. Uhart, C. Wannous, K. Woolaston, P. Mosig Reidl, K. O'Brien, U. Pascual, P. Stoett, H. Li & H. T. Ngo (2020) IPBES (2020) Workshop Report on Biodiversity and Pandemics of the Intergovernmental Platform on Biodiversity and Ecosystem Services. https://ipbes.net/sites/default/files/2020-12/IPBES%20Workshop%20on%20Biodiversity%20and%20Pandemics%20Report_0.pdf (last accessed Dec 18 2021).

Delamater, P. L., E. J. Street, T. F. Leslie, Y. Yang & K. H. Jacobsen (2019) Complexity of the basic reproduction number (R0). *Emerging Infectious Diseases*, 25, 1–4.

DiBiase, D. (1990) Visualization in earth sciences. *Bulletin of the College of Earth and Mineral Sciences*, 59, 13–18.

Ebi, K. L., A. Capon, P. Berry, C. Broderick, R. de Dear, G. Havenith, Y. Honda, R. S. Kovats, W. Ma & A. Malik (2021) Hot weather and heat extremes: Health risks. *The Lancet*, 398, 698–708.

ECMWF (2024) Global Climate Highlights 2023. Copernicus: 2023 is the hottest year on record, with global temperatures close to the 1.5°C limit. https://climate.copernicus.eu/copernicus-2023-hottest-year-record. Accessed 30 April 2024.

Gani, R., & Leach, S. (2001). Transmission potential of smallpox in contemporary populations. *Nature*, 414(6865), 748–751.

Haelle, T. (2021) Health Effects of 9/11 Still Plague Responders and Survivors. *Scientific American*, 15 p.

Heaney, A., J. D. Stowell, J. C. Liu, R. Basu, M. Marlier & P. Kinney (2022) Impacts of fine particulate matter from wildfire smoke on respiratory and cardiovascular health in California. *GeoHealth*, 6, e2021GH000578.

Ifejube, J., Kuriakosa, S.L., Anish,T.S., van Westen, C., Blanford, J.I. (2024) Analysing the outbreaks of leptospirosis after floods in Kerala, India. *International Journal of Health Geographics*. **23**, 11. *https://doi.org/10.1186/s12942-024-00372-9*

JHU (2023) Johns Hopkins University Coronavirus Resource Center. https://coronavirus.jhu.edu/map.html. Accessed 30 April, 2024

Kola, L., B. A. Kohrt, C. Hanlon, J. A. Naslund, S. Sikander, M. Balaji, C. Benjet, E. Y. L. Cheung, J. Eaton & P. Gonsalves (2021) COVID-19 mental health impact and responses in low-income and middle-income countries: Reimagining global mental health. *The Lancet Psychiatry*, 8, 535–550.

Kõlves, K., K. E. Kõlves & D. De Leo (2013) Natural disasters and suicidal behaviours: A systematic literature review. *Journal of Affective Disorders*, 146, 1–14.

Kraak, M.-J. & F. Ormeling (2020) *Cartography Visualization of Geospatial Data*, 4th Edition. Boca Raton, FL: CRC Press.

Liu, J., W. Xie, Y. Wang, Y. Xiong, S. Chen, J. Han & Q. Wu (2020) A comparative overview of COVID-19, MERS and SARS: Review article. *The International Journal of Surgery*, 81, 1–8.

Liu, Y. & J. Rocklov (2021) The reproductive number of the Delta variant of SARS-CoV-2 is far higher compared to the ancestral SARS-CoV-2 virus. *Journal of Travel Medicine*, 28, taab124.

Luke, T. C. & J.-P. Rodrigue (2008) Protecting public health and global freight transportation systems during an influenza pandemic *American Journal of Disaster Medicine*, 3, 99–107.

MacEachren, A. M. & D. R. F. Taylor (1994) *Visualization in Modern Cartography*. Oxford, England: Elsevier.

Morganstein, J. C. & R. J. Ursano (2020) Ecological disasters and mental health: Causes, consequences, and interventions. *Frontiers in Psychiatry*, 11, 1.

Okaka, F. O. & B. D. O. Odhiambo (2018) Relationship between flooding and out break of infectious diseases in Kenya: A review of the literature. *Journal of Environmental and Public Health*, 2018, 5452938.

Plans-Rubio, P. (2012) Evaluation of the establishment of herd immunity in the population by means of serological surveys and vaccination coverage. *Human Vaccines & Immunotherapeutics*, 8, 184–188.

Raker, E. J., M. Zacher & S. R. Lowe (2020) Lessons from Hurricane Katrina for predicting the indirect health consequences of the COVID-19 pandemic. *Proceedings of the National Academy of Sciences*, 117, 12595–12597.

RIVM (2021) Reproduction Number. *Coronavirus Dashboard*. https://coronadashboard.government.nl/landelijk/reproductiegetal (last accessed Dec 18 2021).

Roth, R. E. (2013) Interactive maps: What we know and what we need to know. *Journal of Spatial Information Science*, 6, 59–115.

— (2021) Cartographic design as visual storytelling: Synthesis and review of map-based narratives, genres, and tropes. *The Cartographic Journal*, 58, 83–114.

Safarpour, H., S. Sohrabizadeh, L. Malekyan, M. Safi-Keykaleh, D. Pirani, S. Daliri & J. Bazyar (2022) Suicide death rate after disasters: A meta-analysis study. *Archives of Suicide Research*, 26, 14–27.

Sandifer, P. A. & A. H. Walker (2018) Enhancing disaster resilience by reducing stress-associated health impacts. *Frontiers in Public Health*, 6, 373.

Tunstall, S., S. Tapsell, C. Green, P. Floyd & C. George (2006) The health effects of flooding: Social research results from England and Wales. *Journal of Water and Health*, 4, 365–380.

UN (2015) Recommendations for the Transport of Dangerous Goods. https://www. unece.org/fileadmin/DAM/trans/danger/publi/unrec/rev19/Rev19e_Vol_I. pdf (last accessed 27 April 2024).

UNDAC (2018) United Nations Disaster Assessment and Coordination. UNDAC Field Handbook, 137 p. https://reliefweb.int/report/world/un-disaster-assessment-and-coordination-undac-field-handbook-7th-edition-2018 (last accessed 22 Oct 2023).

UNISDR (2017) *Technical Guidance for Monitoring and Reporting on Progress in Achieving the Global Targets of the Sendai Framework for Disaster Risk Reduction.* Geneva, Switzerland: United Nations Office for Disaster Risk Reduction. https://www. unisdr.org/we/inform/publications/54970.

van den Driessche, P. (2017) Reproduction numbers of infectious disease models. *Infectious Disease Modelling*, 2, 288–303.

WHO (2013) IHR Core Capacity and Monitoring Framework: Checklist and indicators for monitoring progress in the development of IHR core capacities in states parties. https://apps.who.int/iris/bitstream/10665/84933/1/WHO_HSE_GCR_ 2013.2_eng.pdf (last accessed 27 April 2024).

— (2019) Health Emergency and Disaster Risk Management Framework. https://iris. who.int/bitstream/handle/10665/326106/9789241516181-eng.pdf?sequence=1 (last accessed 27 April 2024).

5

Data in a Nutshell: Geospatial Data, Structuring Data, Managing Data and Ethics

Overview

Ideally, a health data system contains data that is **centralized** so that health practitioners and researchers are able to (i) **access** and obtain the data efficiently; (ii) **contribute** data and (iii) **retrieve** information easily for further analysis (Figure 5.1).

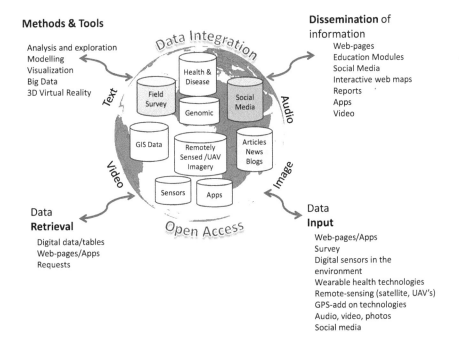

FIGURE 5.1
A centralized data infrastructure system enabling data protection, integration and sharing.
Source: Blanford (2023).

Digitalization of Health Data

As our health systems become increasingly digitalized, more and more health data is stored in private databases and, as such, can potentially be mapped and integrated with other data sources, provided there is some form of geographic information included. To ensure the quality and integrity of the data, it is important to be conscious of the many elements associated with data. These range from the **collection, storage, organization, ownership** and **use** of the data. Before starting any analysis, it is necessary to obtain data. This can be health data, disease data, environmental data or other geographic information.

Getting the Data You Need

Data may be collected in a variety of ways, at different intervals and provided in many different formats. Dealing with data is a critical part of any analysis, and the time required to organize, transform or combine (engineer) the data is often underestimated. No doubt many of you have heard of the term data wrangling; it is an essential part of dealing with data and involves **data organizing, data cleaning, data remediation** or **data munging.** It really refers to the processes of transforming the raw data and making it fit-for-purpose. In other words, it is about making the data ready for operational use. How this is done will depend on the data, the project requirements, the type of analysis you will be performing and the temporal scale utilised. Often the biggest challenge is getting the spatial data that you need in a timely manner, at the scale or resolution that you need and in the format that you need it in. However, since we are dealing with data across different domains, we have to think about the data collected in the health or disease realm as well as in the geospatial realm

Data Infrastructure

Ideally, a health data system contains data that is **centralized** as illustrated in Figure 5.1) at the start of this chapter. Having the data central is crucial for providing quality-assured information in a timely manner. This allows researchers/practitioners to (i) **access** and obtain the data they need efficiently; (ii) **contribute** data (new and update existing data) that are *standardized* and *quality* checked prior to being accepted into the central data warehouse/registry/database; (iii) have access to and use the same most-up-to-date data. The benefits of this are multifold. The data are protected and maintained over time and enable the efficient dissemination of information through different outlets and portals (e.g. intranet, internet, apps, maps, ftp, interactive web mapping applications, etc.). Although there have been many efforts aimed at spatial data infrastructures, since the COVID-19 pandemic, the importance of this has again been realized. A very nice example has been the evolution of the Johns Hopkins COVID-19 dashboard. The data is stored in github and is accessible to anyone.

Once centralized, data can be maintained and protected; efficiently retrieved and used by multiple stakeholders; and accessed through a variety of tools and technologies that will change over time. When conceptualizing a framework for monitoring public health, there are several components that need to be taken into consideration, including system design, data management and collection, analysis, interpretation, dissemination and applicability to public health programs (see below, Savel and Foldy, 2012).

- *Planning and system design* – Identifying information and sources that best address a surveillance goal; identifying who will access information, by what methods and under what conditions while considering ethical requirements; and improving analysis or action by improving the surveillance system interaction with other information systems.
- *Data collection* – Identifying potential bias associated with different collection methods (e.g., telephone use or cultural attitudes toward technology); identifying appropriate use of structured data compared with free text, most useful vocabulary, and data standards; and recommending technologies (e.g., global positioning systems and radio-frequency identification) to support easier, faster, and higher-quality data entry in the field.
- *Data management and collation* – Identifying ways to share data across different computing/technology platforms; linking new data with data from legacy systems; and identifying and remedying data-quality problems while ensuring data privacy and security.
- *Analysis* – Identifying appropriate statistical and visualization applications; generating algorithms to alert users to aberrations in health events; and leveraging high-performance computational resources for large data sets or complex analyses.
- *Interpretation* – Determining usefulness of comparing information from one surveillance program with other data sets (related by time, place, person, or condition) for new perspectives and combining data of other sources and quality to provide a context for interpretation.
- *Dissemination* – Recommending appropriate displays of information for users and the best methods to reach the intended audience; facilitating information finding; and identifying benefits for data providers.
- *Application to public health programs* – Assessing the utility of having surveillance data directly flow into information systems that support public health interventions and information elements or standards that facilitate this linkage of surveillance to action and improving access to and use of information produced by a surveillance system for workers in the field and health-care providers.

Spatial Data/Geographic Information

Spatial data "result from observation and measurement of earth phenomena" referenced to their locations on the earth's surface (Tomlinson, 1987, p. 203). That is any piece of data with a geographic component. This can be coordinates, a place name (country, county, neighbourhood, etc.), address (street name, zip code/postal code and IP address). The minute geographic information is added to the information, it can be mapped! This can be done through a variety of processes, depending on what data is available. For example, if you have an address, you will be able to obtain latitude and longitude information for that address using a process known as **geocoding**. If the geographic information references a county or some other form of administrative boundary, the information can be geocoded or joined with a spatial dataset containing the administrative boundaries.

Geography – What Geography to Include and Collect?

When collecting data, also consider including some form of geography. What geographic scale to collect will depend on not only the need but also how the data is being collected?

Geographic information can be obtained from a variety of sources (Figure 5.2). These include:

- Latitude and longitude coordinates (e.g. obtained using a device with Global Position Systems (GPS) capabilities (e.g. mobile phone); extracted from maps)
- Postal address or geographic place name (e.g. London)
- Administrative division boundary level, which is represented by an area (e.g. country, district, province, county, municipality, urban area, neighbourhood, township, village, postcode, etc.) (see Figure 5.2 for different administrative boundary levels)
- Feature or point of interest. This can be a place (e.g. a city, village, historical monument, park, etc.

If the data is to be shared, then what geographies should be included so that it can be mapped (e.g. data is connected to a geographic dataset using a common key or attribute; e.g. municipality name or some other unique identifier such as a standardized code).

The terminology used for administrative boundary levels 1–4 varies from country to country as illustrated in Table 5.1 where district, for example may

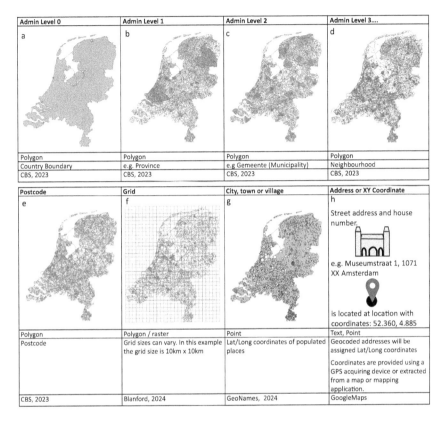

Admin Level 0	Admin Level 1	Admin Level 2	Admin Level 3....
a	b	c	d
Polygon	Polygon	Polygon	Polygon
Country Boundary	e.g. Province	e.g Gemeente (Municipality)	Neighbourhood
CBS, 2023	CBS, 2023	CBS, 2023	CBS, 2023

Postcode	Grid	City, town or village	Address or XY Coordinate
e	f	g	h — Street address and house number. e.g. Museumstraat 1, 1071 XX Amsterdam is located at location with coordinates: 52.360, 4.885
Polygon	Polygon / raster	Point	Text, Point
Postcode	Grid sizes can vary. In this example the grid size is 10km x 10km	Lat/Long coordinates of populated places	Geocoded addresses will be assigned Lat/Long coordinates. Coordinates are provided using a GPS acquiring device or extracted from a map or mapping application.
CBS, 2023	Blanford, 2024	GeoNames, 2024	GoogleMaps

FIGURE 5.2
Examples of different geographies that could be considered when collecting geographic information (administrative areas, latitude/longitude, address or place name). An example using data for the Netherlands. Data obtained from CBS, GeoNames, Google Maps or created using ArcGIS Pro.

represent level 1 in one country and level 2 admin boundaries in another country. For a full list of the different administrative boundary levels and the names of these for different countries, see list of administrative divisions by country on Wikipedia.

The main objective of the administrative division is for administration purposes. The administrative divisions have been arranged in a certain order. In general, the largest administrative subdivision of a country is named as the First-Order Administrative Division, followed by the smaller administrative subdivision of a country, which is named as the Second-Order Administrative Division. The number of orders in administrative divisions may be further divided into the Third-Order and the smaller Fourth-Order Administrative Division as captured in Table 5.1 and illustrated in Figure 5.2.

TABLE 5.1

Overview of Administrative Boundary Levels and Associated Names. Note Names Can Vary by Country

Admin Level 0	Admin Level 1	Admin Level 2	Admin Level 3+
Country Boundary	Province, State, Districts Regions Divisions Prefectures	Districts Departments Counties Municipalities Sub-prefectures Cantons	Communes, Municipalities Parish Villages Wards Enumeration Districts Townships Borough Neighbourhoods

Careful when using collecting geographic information such as place names as problems can arise. These include

1. Spelling of place names. Spelling may be incorrect or use an alternative spelling.
2. Multiple places with the same name. The same place name may be used in multiple geographic locations (e.g. London, England vs London, Ontario, Canada vs London, California, USA).
3. Use of local informal names about a place.

Whenever possible, use a GPS tracking device when surveying and collecting information to minimize possible errors or ambiguities that may arise. If this is not possible then ensure that additional information are included (e.g. place name, admin level 2, admin level 1, admin level 0, etc.).

Data Types and Formats

Geographic information can come in a variety of formats that can be vector, raster, tabular or text (Figure 5.3).

- Vector (points, polygons and lines)
- Raster (square cells assigned real world entities – imagery)
- Tabular (spreadsheet and tables)
- Text (reports, blogs, social media, etc.)

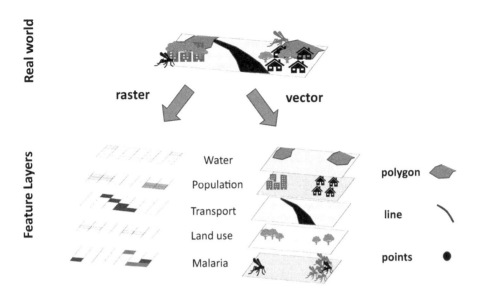

FIGURE 5.3

Illustration of different types of geographic data broken down as raster and vector. Image created by Blanford, 2023.

Data Files

There are numerous geographic information data formats! With over 60 formats as of December 7th, 2023 (GISGeography, 2023), navigating these formats is not easy for new users, let alone experienced users and requires some getting used to. To help get you familiar with the different file types that are used to store geographic information, I have compiled a list of the most common file types used to hold geographic information (Table 5.2).

Data Transformation and Processing

An important component of any analysis is preparing the data so that you can do the analysis. This could require geocoding places so that you can map each unique location and converting and integrating data using a variety of different functions, some of which are highlighted in Table 5.3.

Geocoding – Lat/Long Coordinates of Places, Place Names and Addresses

Finding latitude/longitude coordinates for addresses and places (e.g. place names) can be accomplished using a variety of geocoding tools, gazetteers or through visual inspections via an online map (e.g. Google Maps). In cases where a place cannot be located in one database, you may need to use a

TABLE 5.2

List of Common Geographic Information Data Files

Name	Extension	Description	Data Type	Source
Vector Data				
SHAPEFILES	File.shp– contains the feature geometry, file.shx – stores the shape index, file.dbf – stores attribute data, file.prj – contains projection information	Consist of a number of different files that contain information about the geography, typology, attributes, projections etc. Shapefiles consist of a number of files that **MUST** be kept together	Vector	esri
GEODATABASE	.gdb	Personal geodatabase format. Similar to an MS Access database. Stores the geographic information, attributes and ancillary information in a series of tables within a geodatabase.	Vector	esri
GEOJSON	.geojson;.json	GeoJSON is a format for encoding a variety of geographic data structures	Vector	https:// geojson.org/ javascript
GPS DATA FILES	.gpx	XML GPS data file, usually coming from a GPS device.	Vector	Garmen and other GPS devices
GOOGLE	.kml.kmz	XML plaintext file that may contain geometry, data, a web service Z files are compressed KML files	Vector	Google
SQLITE/ SPATIALITE DATABASE	.sqlite	Spatialite is an extension to sqlite to enable it spatially. Format often used with QGIS.	Vector	QGIS
Raster/Image Data				
TIFF AND TIFF WORLD FILE, GEOTIFF	.tiff	Image data when associated with a tfw ("world file") of the same filename is a georeferenced image.	Raster	
JPEG	.jpg	Image data when associated with a jpw (world file) is a georeferenced image.jpx can contain additional metadata.	Raster	
ASCII GRID	.asc	Plaintext (ASCII) format	Raster	
GRID FILES	Grid.bnd Grid.hdr Grid.sta	Compilation of files. bnd – boundary/extent metadata for the grid.hdr – metadata such as grid cell size.sta – grid cell statistics.vat – grid attribute table. Integers only	Raster	Esri

(*Continued*)

TABLE 5.2 (*Continued*)

List of Common Geographic Information Data Files

Name	Extension	Description	Data Type	Source
Tabular Data				
COMMA DELIMITED DATA FILES	.csv	Data stored in comma delimited format	Table	
TEXT DATA	.txt	can be.csv or tab delimited or just plain text	Table	
TAB-DELIMITED DATA	.tab or.txt	Data stored in tab delimited format	Table	
STANDARD DATABASE FILE	.dbf	Data stored in database file format	Table	dBASE
SPREADSHEET	.xls.ods	Data stored in spreadsheets by MS Excel or Open spreadsheets	Table	MS Google
Other				
NETCDF	.nc	Network Common Data Form (netCDF) allows for data to be created, accessed, and shared as array-oriented. Used for large datasets	Array	e.g. Satellite Imagery
HDF	.hdf	**Hierarchical Data Format** (HDF) is designed to store and organize large amounts of data.		The HDF Group e.g. Satellite Imagery
.GRIB	.grib	GRIB is used to store weather data. Defined in 1985 by the WMO. GRIB files were designed to exchange and store large volumes of gridded data and are used in Numerical Prediction models such as Weather Research Forecast (WRF) model, in reanalysis data (ERA5 and CFSR), and in forecast data, such as the NOOA Global Forecast System (GFS) or the ECMWF Integrated Forecasting System (IFS).		World Meteorological Organization (WMO).

These will change over time but this should give you some idea of the diversity of filetypes around.

combination of these services. There are now many services available. Here are a few examples.

- **Geocoding tools**: a variety of tools are available (e.g. ESRI has a built-in geocoder, Bing, Google, http://www.gpsvisualizer.com/, etc.)
- **Gazetteer**: is a database with geographical place names and features. Several databases exist such as GeoNames: http://www.geonames.org/ or the Digital Chart of World (DCW) or Open Street Map (OSM).

TABLE 5.3

Spatial Data Collection and Wrangling. Some Preprocessing Operations

Function Class	Function
Data collection	Scanning, digitizing, geocoding (address-matching), joining health data to geographical areas.
Data conversion	Importing/exporting, edgematching, clipping, raster/vector conversion
Geometric transformation	Translation, rotation, map projection, rubbersheeting
Generalization	Line thinning, line smoothing
Coordinate system	Coordinate systems, projection vs non-projected (latitude and longitude)

Source: Cromley and McLafferty (2012).

Coordinate Systems and Projections

A map projection is a systematic transformation of the latitude and longitude of a location into locations on a plane. All map projections distort the surface of the earth in one way or another. However, depending on the purpose of the map or analysis, some distortions are more acceptable while others are not; therefore, different map projections exist in order to preserve some properties of the sphere-like body at the expense of other properties. To help you decide what projections to use, a summary of the different projections, their characteristics and their uses is provided in Table 5.4. For more in depth information, view the book *Working with Map Projections: A Guide to Their Selection* (Kessler and Battersby, 2019) and the Map Projections poster created by the USGS (USGS, 2019)

GPS information is usually provided in latitude/longitude. Check the owner's manuals for additional details.

TABLE 5.4

Some Key Characteristics about Map Projections, Properties and Common

Projection Category	Properties	Common Uses
Conformal	Preserves local shapes and angles	Topographic maps, navigation charts, weather maps, weather maps
Equal area	Preserves areas	Dot density maps (e.g. population density), thematic maps (land use); quantitative information by area.
Equidistant	Preserves distance from one or two specified points to all other points on the map	Maps of airline distances, seismic maps showing distances from an earthquake epicentre Costs or charges based on straight-line distances.
Azimuthal	All directions are true from a single specified point (usually the centre) to all other points on the map	Navigation and route planning maps

Source: USGS 1993; Jenny et al., (2017).

An example:

Position Formats Supported by Garmin Outdoor Devices – Garmin GPS devices are set by factory default to hddd°mm.mmm' or hddd°mm. mmmm'. This means it is set to latitude and longitude in degrees and minutes (or lat/lon DM), with decimal minutes. There are other Position Formats available to choose from on your Garmin GPS, especially if you are using coordinates from a map or other source that are not using the same Position Format as the factory default on your Garmin GPS. To change the position format on your Garmin GPS device, refer to your device's owner's manual.

Source: https://support.garmin.com/en-US/?faq=lvWzTYlPsx6BvUDTyKfqC8

Defining Variables: Field Variables and Characteristics

Choosing the correct data type to store information in the table connected to the geographic information will enable you to store and manage your data and facilitate the retrieval of and analysis of information efficiently. There are a number of different data types as captured in Table 5.5.

Fusing and Integrating Data

A large part of any analysis is structuring data and fusing or integrating multiple data sources, some of which may be at different scales or resolutions. Take this into consideration as it could affect the results of the analysis.

TABLE 5.5
Summary of Different Data Types Used to Store Data

Data Type	Uses
Short integer	Numeric values without fractional values within a specific range; coded values
Long integer	Numeric values without fractional values within a specific range
Float (single-precision floating-point number)	Numeric values with fractional values within a specific range
Double (double-precision floating-point number)	Numeric values with fractional values within a specific range
Text	A text field represents a series of alphanumeric symbols ArcGIS Pro uses Unicode to encode characters
Dates	The date data type can store dates, times or dates and times. The default format in which the information is presented is mm/dd/yyyy hh:mm:ss and a specification of AM or PM.
ObjectID	ObjectID field is a unique system-defined ID by ArcGIS Pro that guarantees a unique identifier for each row in the table. However, it is useful to create your own unique identifier.

Joining Information in Tables to a Geographic Dataset

Non-spatial data: attribute information can be mapped and attached to spatial data through the use of a unique identifier (e.g. county boundary, zip code/postal code, etc.). This can be accomplished by performing a join or a relate through a common unique identifier, as summarized below.

Term	Functionality
Join	Is used to append fields from one table to those of another through an attribute or field common to both tables. This is commonly used to combine information and append non-spatial information to a geographic dataset. For example, add the number of votes by municipality. Depending on the software, this can be a temporary or permanent connection.
Relate	Used to create a temporary link between records in two tables using a key common to both (e.g. attribute value of a field).

Keep in mind that when using joins and relates.

- **Relates** can be one-to-many relationships or many-to-one relationships.
- **Joins**, if joining, the joins may take the first record in a table for attributes that have a one-to-many relationship. This could lead to incorrect joining of information. For example, you have some data on the distribution of a measles outbreak in the United Kingdom (UK) stored in an excel spreadsheet with geographic location information. One of the locations where measles was recorded was in London, England. Using this information, you decided to view the data and had a geographic dataset with places around the world. You decided to join the information in the excel spreadsheet with the geographic dataset using city name (e.g. London). The first record in the geographic dataset where London appears will be joined to the information in the excel spreadsheet. However, this can result in an incorrect join if the first record is London, Ontario in Canada. Double check the joins to ensure the correct information was joined. This is an easy mistake to make.

Data Collection, What to Collect and How Often?

- In this section, I have included a number of subsections about data that include data management, digital device vs paper, Sampling, Active vs passive data collection, and primary versus secondary and citizen science data sources

Data Management

For data to be useful, it should be accompanied by metadata. The metadata should not be too arduous to put together; otherwise, it will never get done. The data should have just enough descriptors so that you know the quality of the data, its currency and its usability. Today, more than ever, it is getting harder and harder to just find a dataset and use it. Many publishers require permission from the original owner unless the data is available through data hubs where the original owner or creator has shared the information. Often, datasets are now accompanied with a license and a way to cite it.

Reference datasets in the same way that you do papers. A lot of work goes into creating, maintaining and managing data. You may need to revisit a source again, so make it easy for yourself.

If you are creating or collecting data, be clear on how the data will be stored, who has access to the data and if it can be reused, how it can be reused. HDX (Humanitarian Data Exchange) provides an excellent example of metadata in action (Figure 5.4) as it provides the essential information

Before starting any data collection ensure you create a data management plan where you describe what data will be collected, how the data will be stored and managed, and what will happen to the data on completion and if it can be shared, how will the data be shared and contain any restrictions in reuse.

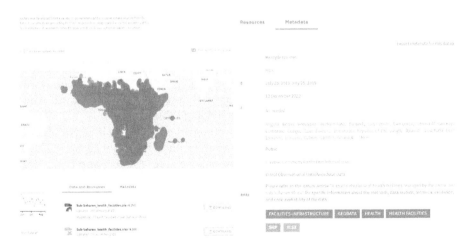

FIGURE 5.4
Example of health facilities dataset for Sub-Saharan Africa and associated metadata available from the Humanitarian Data Exchange https://data.humdata.org/dataset/health-facilities-in-sub-saharan-africa.

Collecting Data – Digital Tools, Surveys and Extraction

Data can be obtained in many different ways and from many different sources, either directly (primary data sources) or indirectly (via secondary data sources), using a variety of tools and techniques.

Much of what we do and how we collect information is done through digital devices and technologies (see the Technologies for Surveillance and Monitoring section for an overview) with varying accuracy.

During the COVID-19 pandemic, pretty much every country created an app, and in some cases, multiple apps, as shown in Sharma and Bashir (2020); Blanford et al., 2021). Regardless of whether you use digital or paper tools to collect information, ALL surveys involving human participants NEED ethical approval. This can take time, depending on the information being collected and the different stakeholders involved in the research. Do not delay submitting an approval, as it can delay the data collection process.

Digital tools for data collection. A variety of tools are available for conducting surveys and capturing a mix of survey and geographic information. In the EU, all tools need to be GDPR compliant, so if you are going to use them for collecting sensitive information, ensure that they are GDPR-compliant. Ensure that sensitive data is encrypted and stored securely.

Technologies for Surveillance and Monitoring

Today, much of our environment (and us) can be surveyed remotely through a variety of sensors and technologies that range from imagery, video, audio, text, movement/gestures and a digital footprint (e.g. payment, IP address, GPS locations (latitude (lat) and longitude (lon) coordinates), etc.). These data can provide information about the health of our environment as well as about us. Since much of this information contains some form of geographic information (lat/lon coordinates; place names), it can be easily mapped. Below are just a few examples of different technologies that collect a variety of information that can be used for collecting health data (Table 5.6).

Since all devices today have GPS technologies embedded, it is possible to map just about everything since the data can include X, Y coordinates. There are potentially many ethical concerns associated with the use and collection of data using these different types of tools and technologies. More on this later in the Ethics section at the end of the chapter.

TABLE 5.6

Types of Technologies and How They Can Be Used for Monitoring, Evaluating and Managing Health and Disease

Type	Description	Examples
Wearable technologies	There are a wide variety of health technologies now available ranging from Fitbit, running watches and a suite of elderly devices and an increasing number of health apps for mobile phone/tablets that capture information either through manual inputs or through sensors	Fitbit; Running watches; mobile phone wrist watches Health devices: pacemakers; diabetes devices Garmin running watches used to track exercise activities are also useful for monitoring your well-being over time and have been used for monitoring recovery of patients
Add-on technologies	Devices that can be added on to existing devices or technologies. Examples include adding GPS units to medical devices or adding devices that connect with an app Also see lab-on-a-chip medical devices.	GPS units added to inhalers (e.g. propellerhealth (https://propellerhealth.com/). Each time a person uses their inhaler their coordinates are recorded. e.g. microscopy, spectrometer, fluorescence spectroscopy (Smith et al., 2011) Diabetes monitoring and management (Shan et al., 2019)
SMART technologies	Interactive technologies that can listen and respond to user inputs. These continue to be embedded into technologies and combined with AI are becoming more and more sophisticated.	Examples include baby monitors; ring doorbell, TVs and various home appliances; interactive kids' toys.
Personalized physicians	This brings us to this next category which is really a mix of the aforementioned examples. In remote areas, mobile technologies are providing clinical health care solutions through telemedicine. As smart technologies mature and robots become more sophisticated and 'cuddly' as depicted in the film Big Hero 6, these technologies can also be used to collect information as well as provide personalized healthcare. Thus, adding a new dimension to health care solutions and health surveillance.	These range from telemedicine, e-health to personalized physicians.

(Continued)

TABLE 5.6 (*Continued*)

Types of Technologies and How They Can Be Used for Monitoring, Evaluating and Managing Health and Disease

Type	Description	Examples
Sensors in the environment	Today, our environments are rich with sensors. Some of these are mobile and others stationary. Regardless wireless sensors are used to collect and transmit an array of information that helps monitor the health of our environment. These can measure light, wind speed, rainfall, temperature, humidity, barometric pressure, water quality, soil moisture as well as noise and pollution and be useful for examining the relationship of health and diseases due to exposure effects (e.g. see Reis et al., 2015).	Here are a few examples: • Remote sensing imagery; UASs • Unmanned aerial systems (UAS): There has been an increase in the use of UAS to capture information remotely. Many of these systems can be programmed to fly along a specific flight path and capture imagery using a mounted camera. • Environmental monitoring: Weather stations; gauging stations; pollution sensors monitoring air quality; noise sensors monitoring noise levels • Surveillance: CCTV cameras; speed cameras • Home monitoring systems: ring doorbell; digital thermostats
Satellite data	Although satellites are unable to detect disease, they are well suited to monitor the environment. Satellite instruments make frequent, regular measurements over long time periods and cover (large parts of) the globe. Many satellite data sets are freely available.	Weather data: wind, temperature, clouds, precipitation, humidity and many more Climate data: long, homogenized time series of the above-mentioned variables, concentrations of greenhouse gases, ozone Pollution: amounts of particles, polluting gases in the atmosphere And many more land, vegetation, marine, aquatic and other variables and images (GIS)
Virtual reality and immersive technologies	The ability to create virtual environments has become more viable with the reduction in costs in technology and increasing computing speed. To highlight a few examples of how this technology is and can be used read the links provided	

(*Continued*)

TABLE 5.6 (*Continued*)

Types of Technologies and How They Can Be Used for Monitoring, Evaluating and Managing Health and Disease

Type	Description	Examples
Apps	Apps play an increasing role to collect information. These can range from the collection of informal information and behaviours (e.g. text information via social media platforms such as Twitter; Grindr (hook-up app or other dating apps; payment apps (e.g. Venmo)) to more formal information collected for a specific purpose (ZOE -COVID-19 symptom collector; women fertility apps) Surveys	See the ethics section for examples Mental disorder treatment (https://link.springer.com/article/10.1007/s00779-014-0829-5); Digital psychologist (Marshall et al., 2019) ZOE Citizen Science Project
Lab-on-a-chip (LOC) devices and medical apps	Medical Apps. This is a fast growing field where technologies are being developed that can help control our heart and many other functions. https://www.degruyter.com/document/doi/10.1515/ejnm-2014-0004/html https://www.science.org/doi/10.1126/scirobotics.abb5589	• GE Healthcare The Vscan portable ultrasound device (Jung et al., 2021) • Glucose Tracking Apps in Apple App Store (Martinez et al., 2017) • Urine Test • Blood Oxygen Monitoring • Breath Analyser • Diagnosis of infectious disease (Wood et al., 2019; Robertson et al., 2010; Wang et al., 2017; Wang et al., 2021)
AI and machine learning sentiment	Facial recognition software and language processing algorithms are useful for gaining insights into behaviours, opinions and sentiment or emotions.	https://www.getalfi.com/advertising/facial-recognition-advertising-future-is-here/ http://www.retail-innovation.com/japanese-digital-vending-with-facial-recognition

Sampling and Surveys–

As already mentioned, data can be collected in many different ways and using a variety of technologies for different purposes that range from mass surveillance and monitoring to small local based research-based studies that answer specific questions.

Geographic features extracted or digitized from imagery or maps represent a geographic object, place or thing and can be enriched with supplemental descriptive information that can be collected through participatory and survey methods, including Knowledge, Attitude, Practice (KAP)-based surveys. When geographic information is also captured with this information it can be mapped and used in further analysis.

Sampling

When collecting data different sampling strategies may be used (Figure 5.5–5.6). These range from random point/line or area based sampling to more stratified sampling strategies (Figure 5.6).

Random point/area sampling: A grid is drawn over a map of the study area as shown in Figure 5.6, each assigned a number from 1 to *n*. Random numbers are selected (e.g. using the RAND function in Excel) for the sampling locations. Sampling takes place within the grid. When sampling, the person collecting the information can be wearing a GPS tracker or using a mapping app to navigate to the required locations.

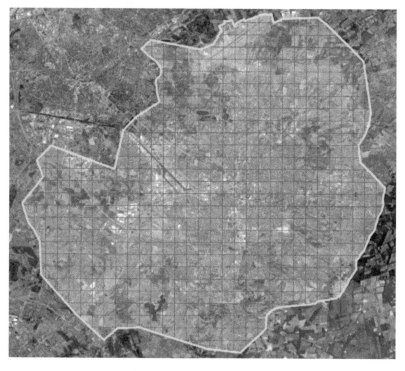

FIGURE 5.5
Integrating data sources and designing sampling. Background image available through ESRIs background maps. Grid created in ArcGIS Pro using fishnet.

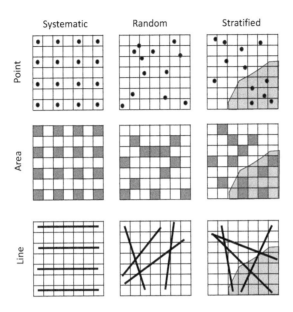

FIGURE 5.6
Sampling

Systematic point sampling: A grid can be used to create a systematic sampling approach (Figure 5.6). For example, sample every 2 metres, every 10th square or every 3rd intersection. This can be done by defining points along a transect/grid lines, at the intersections of the grid lines or in the middle of each grid square. Sampling is conducted at the nearest feasible place. Along a transect line, sampling points for vegetation/pebble data collection could be identified systematically, for example every 2 metres or every 10th pebbles.

Stratified sampling: This method is used when the parent population or sampling frame is made up of sub-sets of known size. These sub-sets make up different proportions of the total, and therefore, sampling should be stratified to ensure that results are proportional and representative of the whole (Figure 5.6).

- For example, the population can be divided into known groups, and each group is sampled using a systematic approach. The number sampled in each group should be in proportion to its known size in the parent population.
- For example, the make-up of different social groups in the population of a town can be obtained, and then the number of questionnaires carried out in different parts of the town can be stratified in line with this information.

Knowledge, Attitude, Practice (KAP) Survey

KAP surveys have been widely used for gathering information about knowledge, attitudes and practice in a population. For example, KAP have been used for planning and designing of interventions (Krentel et al., 2006; Higuera-Mendieta et al., 2016), evaluating the effectiveness of public health interventions (Sarraf-Zadegan et al., 2003), treating tropical neglected diseases (Richards Jr et al., 2016) and for assessing healthcare services (Tabash et al., 2016). KAP's can be useful for capturing opinions, understanding misconceptions and misunderstandings that may hinder the implementation of interventions or preventing behavioural changes needed. KAP surveys are useful for enhancing knowledge, attitudes and practices about specific topics, identifying what is known, and identifying knowledge gaps or variations in local cultural factors. They may also be useful for establishing a baseline reference from which assessments can be made to monitor the effectiveness of interventions or changes.

Citizen Science and Volunteered Geographic Information

Similar to KAPs, an examination of the subject matter and what people talk about, or "sentiment" analysis of the meaning of text contained in electronic messages may also provide insights into how people may respond to a situation or a topic (e.g. MacEachren et al., 2011; Blanford et al., 2014; Felmlee et al., 2020). Social media and microblogging data provide an additional layer of information and insights into the human dimension and sentiment of local communities through bursts of "conversations". Thus, visually-enabled sensemaking of place-time-attribute information can help us improve our understanding of events or topics (e.g. MacEachren et al., 2011; Tomaszewski et al., 2011).

Data Sources and Considerations

Many data sets come with metadata, which contains essential information about the data. Key information may include availability of the data (where and how to obtain the data); fitness of the data (how the data can be used); transferability of the data (how the data can be shared and technical specifications on how to handle the data); and what the attributes represent and the definition of the values (when coded). Not all data comes with metadata and not all data has been quality-assured, so it is important to assess the quality of the data.

Data quality can be variable and should be checked for

- **Currency**: age of the data. When was the data collected? How up-to-date is it? How often is it updated?
- **Accuracy and quality**: how accurate is the data spatially? How accurate is the attribute information? Was it quality-controlled?

- **Projection and scale**: determine the map projection of the data and the scale at which it was originally created or is available. For raster-based data the resolution of the cells may be of importance.

Creation: how was the data created? Was it captured through digital inputs/ sensors, digitized, transformed or derived from modelled outputs? This can be important if through any transformations or conversions there is data loss. **Source**: authoritative vs non-authoritative data sources. Who created the data. Determine if the data comes from a reliable source.

Data needs when embarking on a Geospatial Project.
A variety of data sets will likely be needed to conduct an analysis. To get you thinking, I have provided some ideas of the data sets that you will likely include in Table 5.7. These include some base information such as administrative boundaries, base infrastructure (e.g. roads) and environmental

TABLE 5.7

Variety of Data That May Be Needed When Developing a Geospatial Health Project or conducting an Analysis

Health and Disease	Environment	Transport	Socio-economic	Boundaries	Behaviour and Sentiment
Authoritative data sources (public health agencies and government agency), national statistics Non-authoritative data sources (news articles, blogs, social media (e.g. X (formerly Twitter), Meta (formerly Facebook)) Empirical data, collected by you: household survey, self-reported health status, questionnaire, in-depth interviews, group discussions, field sensors, spot checks and observations; GPS; etc.	Elevation Rivers Lakes Climate (temperature, rainfall, solar radiation, wind and humidity) Land use (satellite imagery and UAV/UAS imagery) Climate change predictions Sound Air quality	Road Rail Water Bike Air (airline connectivity) Rentals (bike, scooter and car)	Demography (census data and age-structure) Neighbourhood characteristics and lifestyle, income and level of formal education	Country Admin level 2 Admin level 3 Admin level 4 Urban boundaries Neighbourhood boundaries Public health boundaries	News reports, blogs and social media. Survey Mobile phone Payment data Exercise data Citizen Science data

information, followed by socio-economic data sources (e.g. census data). For the list determine what base data sets are needed. A good starting point is Open Street Map as it provides a range of base information.

Authoritative vs non-authoritative data sources.
A wealth of information is available. Data can come from many different sources, some of which are **authoritative** (i.e., from an authoritative source such as a national government or mapping agency (e.g. Census Bureau, National Weather Service, Statistics, Center for Disease (CDC) and European Center for Disease (ECDC))) and **non-authoritative** (i.e. from sources that haven't been collected through an agency (e.g. blogs, volunteered contributions and social media (e.g. Twitter))) Open Street Map is an excellent example of a volunteered geographic information source. We have provided a list of data sources that are by no means exhaustive and contain a mix of authoritative and non-authoritative data sources).

Today, obtaining base information to provide context is much easier. Thanks to the internet and web-map services provided by different mapping agencies and companies, it is now possible to include background maps in different software packages. For example, Open Street Map (OSM) can be integrated as a base map in R. ESRI provides users with a variety of background maps that include streets, imagery, topography and grey scales. Next time, you use a map on the web and look at the source of the information for the base map.

Data hubs, portals and repositories are growing; some are free and open, while others require you to create an account or use a subscription model and may have use restrictions. A list of data sources and repositories are provided in the Appendix: Table A.1–A.3.

Earth Observation Data

A wealth of Earth observation (EO) data is available today collected by an array of satellites (Figure 5.7). Earth observation (EO) data (satellite and in-situ) provide information that is useful for monitoring much of the environment and are useful for addressing many of the environmental and challenges we face today (e.g. pollution; Holloway et al., 2021). There are many benefits to EO data that include their characteristics of spatial consistency, accessibility, repeatability and global coverage (Anderson et al., 2017; Yin et al., 2012; Voigt et al., 2007). A variety of information about the earth is being captured (Zarnetske et al. (2019) that range from soils and glaciers to climate.

Since EO data can help with many aspects of health and planetary health, I have included a short overview. As I was putting this book together, I realized that a lot of this information is scattered, so I have done my best to pull the varying pieces together to provide an overview of what satellites and data are available (Appendix Table A.2) and how this data can be used based on the information that is captured (Table 5.6; Appendix Table A.3).

Keep in mind that I am not a remote sensing specialist, and leave this to the experts in the field. However, for different health-related outcomes, many data products about the environment are available (e.g. land use, pollution, Natural Difference Vegetation Index (NDVI), land surface temperature and rainfall).

To give you a sense of the different types of satellites capturing different aspects of our world, I have included the ESA-developed EO overview (Figure 5.7). There are many more sensors that are developed and managed by NASA (e.g. https://svs.gsfc.nasa.gov/30065#section_credits) and many other countries.

Different sensors will capture information in different wavelength ranges, as illustrated in Figure 5.8. For more details on what satellites capture what information, see the Remote Sensing text by Sabins Jr and Ellis (2020). For example, multispectral imagery captures information across a number of bands (Figure 5.8; Table A.3).

Since satellite imagery is useful for tracking and measuring anthropogenic influences across the Earth, I thought it might help to provide some information on the satellite imagery that might be of interest (Table A.3).

Spatial resolution refers to the size of the pixel in a digital image and the area represented by the pixel. A high resolution will mean that the image appears crisper, while a lower resolution image will mean that the image appears more blocky, making it more difficult to distinguish features or objects. Research and observational needs will help determine what imagery is suitable or not.

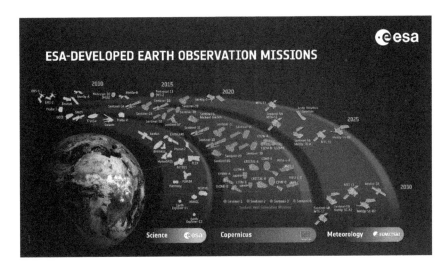

FIGURE 5.7

"The many faces of Earth." An array of different sensors capturing different aspects of our earth and our environment. *Source*: ESA https://www.esa.int/Applications/Observing_the_Earth/Earth_observing_missions ESA, CC BY-SA 3.0 IGO.

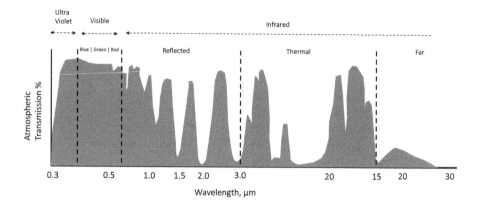

FIGURE 5.8

A generalized overview of the transmission of electromagnetic energy through the atmosphere and their associated wavelengths. Not drawn to scale. For more precise information, refer to NASA, ESA and Remote Sensing Principles, interpretation and applications (Sabins Jr and Ellis, 2020).

Spectral resolution is the amount of spectral detail in a band. Low spectral resolution means that the band covers more of the spectrum and has broader bands, while high spectral resolution means the band covers a narrower range of the spectrum.

Fake Imagery

As AI technologies continue to improve, the ability to manipulate and create fake imagery will also become easier and more feasible (e.g. Zhao et al., 2021). How to safeguard the validity of imagery is still to be determined, so ensure you check the data and sources.

Health Information – What Health Information and Characteristics are Necessary?

Health information may be collected for different purposes – routine or for surveillance during an outbreak to understand the epidemiology of a disease. A variety of data may be collected during an outbreak that may include:

- **Characteristics about the individual** – Information about the individual(s) who have been affected (e.g. age, sex, occupation, religion, ethnicity, lifestyle, etc.);
- **Symptom(s)** of an individual and date of onset and date of death (if applicable);
- **Status** of case (suspected, confirmed);
- **Disease** (if known, and that status of the person following the disease (susceptible, recovered, dead);
- **Test** used to diagnose a disease (e.g. test type, result of test (positive, negative, indeterminate);
- **Geography** – may include geographic location (e.g. where the person is from, where the person lives, if different); where the person was infected (if different);
- **Travel** by the individual (source-destination-duration) and reason for travel if any (e.g. events (religious, music festival), business, holiday, etc.);
- **Activity space** can include information places visited (e.g. coffee shop, work place, grocery store, sports activity, family or friends, etc.) and daily travel routines;
- **Additional information** – any additional information that may be relevant (e.g. underlying health conditions).

The inclusion of geography can help with viewing where infected cases are, identify potential places of transmission or exposure (e.g. occupation, leisure activities, food sources, etc.) as well as identify whether cases may overlap or be related (e.g. revisit Figure 1.10a and b, Chapter 1). Additional insights may be gained through the use of spatial analysis (review Figure 2.5 and the steps used in investigating an outbreak, Chapter 2).

Privacy and Securing Health Information

When working with health data, be mindful of privacy and maintaining anonymity, particularly when mapping. For example, in the United States the *Health Insurance Portability and Accountability Act* (HIPAA) was created to ensure the privacy and security of an individuals' health-related information – also known as Protected Health Information (PHI). Key information includes (extracted from Cromley and McLafferty (2012)):

- Name
- Residential information (anything more detailed than the state) All geographic subdivisions smaller than a state, including: street address, city, county, precinct, zip code and equivalent geocodes
  - Except for: three-digits of a zip code if, according to publicly available data from the Census Bureau; a geographic unit formed by combining all zip codes with the same three initial digits contains more than 20,000 people; and the initial three digits of a zip code for all geographic units containing 20,000 or fewer people is changed to 000.
- All elements of dates (except year) for dates directly related to an individual, including birth date, admission date, discharge date and date of death; and all ages over 89 and all elements of dates (including year) indicative of such age, except that such ages and elements may be aggregated into a single category of age 90 or older.
- Telephone, fax, e-mail and social security numbers.
- Medical record, health plan beneficiary and account numbers; website addresses.
- Certificate/license numbers, vehicle identifiers and serial numbers, including license plate numbers.
- Device identifiers and serial numbers.
- Web universal resource locators (URLs) and Internet protocol (IP) address numbers.
- Biometric identifiers, including finger and voice prints, full-face photographic images and any comparable images.
- Any other unique identifying number, characteristics or code unless permitted.

If using any information with these characteristics, be sure to anonymize and aggregate the data.

Unlike HIPAA, which is tailored to health information, the General Data Protection Regulation (GDPR) has a much broader scope and covers a wider range of personal data across all European Union member countries (EU). Personally Identifiable Information (PII) protected under GDPR includes:

- Name
- Location Information
- Personal information such as email addresses, gender, ethnicity, religious beliefs, political affiliation
- Biometric data
- Web cookies
- Any pseudonymous data that makes it easy to identify someone from it.

Health Data

An important component of health and disease monitoring is the collection of data through surveillance, where **surveillance** is the continuous monitoring of the occurrence of a disease in a population and consists of ongoing, systematic collection of data, data analysis, interpretation of data, dissemination of the information and linking health data to public health practice. Data collected for health-related purposes typically come from three sources: individual persons, the environment and health-care providers and facilities (Bonita et al., 2006; CDC 2012) (Table 5.8).

Surveillance mechanisms include compulsory notification regarding specific diseases (see Table A.4–A.9 for a list of diseases), specific disease registries (population-based or hospital-based), continuous or repeated population surveys and aggregate data.

Common sources of surveillance data include the following:

- morbidity and mortality reports for local and state health departments
- hospital records
- laboratory diagnoses
- outbreak reports
- vaccine utilization
- sickness absence records
- biological changes in agent, vectors or reservoirs
- blood banks
- citizen science and syndromic

TABLE 5.8

Criteria for Selecting and Prioritizing Health Challenges for Surveillance

Public Health Importance of the Problem	Ability to Prevent, Control or Treat the Health Problem	Capacity of Health System to Implement Control Measures For The Health Problem
• Incidence, prevalence • Severity, sequela, disabilities • Mortality caused by the problem • Socioeconomic impact • Communicability • Potential for an outbreak • Public perception and concern • International requirements	• Preventability • Control measures and treatment	• Speed of response • Economics • Availability of resources • Surveillance required

Source: CDC (2021).

Notifiable Disease Surveillance

Surprisingly, there is no global, notifiable disease surveillance (NDS) database. However, many countries have their own NDS system in place (Tables A.4–A.6 for the USA; Tables A.7–A.8 for the Netherlands; Table A.9 for the UK). As you can see from each of the examples, USA, Netherlands and the UK, the organisation of the information and list of diseases varies.

European NDS

For Europe, a variety of diseases are listed as notifiable. Many of these are reported by each country annually. What these diseases are and who is reporting them is available in a spreadsheet by year. To make this information more accessible, an interactive map was created, ENDIG – European Disease Surveillance Systems in Europe (Tjaden and Blanford, 2023), to show which countries were reporting what disease and for how many years (Figure 5.9).

Other sources of health or disease information may include the following:

- **Animal Surveillance (domestic and wildlife).**
  Given the role of animal populations in zoonotic and vector-borne diseases, surveillance of animal host populations is important (Cromley and McLafferty, 2012). These can include the collection of ticks from deer killed by hunters; rabies spread through the collection of human contact with wild animals; livestock surveillance for rift valley fever; and canine surveillance for Lyme borreliosis.
- **Vector surveillance.**
  Surveillance of vectors can be active or passive (Cromley and McLafferty, 2012).
  - **Active surveillance** occurs when a sampling design is developed and data is collected regularly over time to assess the presence and abundance of a vector (e.g. Mosquito Surveillance in Pennsylvania, USA; Taber et al., 2017).
  - **Passive surveillance** occurs when samples are collected by individuals on an ad hoc basis and is similar to data collected through citizen science.
- **Syndromic surveillance**
  Syndromic surveillance can provide an early indication of an increase in illnesses based on an assemblage of symptoms that are grouped into syndrome categories. For example, the category of "respiratory" includes cough, shortness of breath, difficulty breathing,

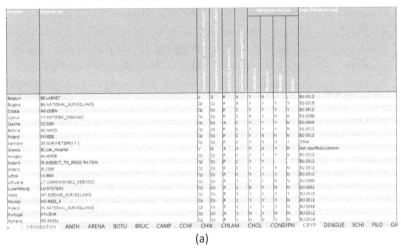

(a)

Disease surveillance systems in Europe

(b)

FIGURE 5.9
Illustration of notifiable disease spreadsheet and ENDIG an Interactive map for disease surveillance in Europe.

etc. This can be used to facilitate early intervention strategies (e.g. the use of real-time symptom tracking to predict COVID-19 cases; Menni et al., 2020) or to minimize risk and transmission (e.g. Dickens et al., 2020; Johansson et al., 2021). Syndromic surveillance, self-reported by affected individuals, focuses on symptoms instead of diagnoses, and as a result, it is less specific and more likely to identify multiple persons, without the disease of interest. As a result, syndromic surveillance relies on using large datasets and computer-based methods to look for deviations from a baseline (e.g. Valdivia et al., 2010). These can include space-time clusters and data-mining methods.

- Syndromic surveillance is used when timeliness is key for:
  - Trying to detect outbreaks when diagnosis is difficult and/or time-consuming (e.g. a new, emerging, or rare pathogen or absence of trained teams for medical surveillance in remote areas);
  - Defining the scope of an outbreak (e.g. geographic clustering) and the characteristics of the populations affected.
  - Defining the effects associated with an outbreak (e.g. COVID-19), citizen science and self-reporting were used to better understand how COVID-19 was affecting different populations.
- **Citizen Science**

    "Citizen science broadly refers to the active engagement of the general public in scientific research tasks. Citizen science is a growing practice in which scientists and citizens collaborate to produce new knowledge for science and society" (Vohland et al., 2021). Non-authoritative data sources such as Volunteered Geographic Information (VGI) and citizen science data can play an important role in obtaining information. For example, symptoms of COVID-19 were collected (e.g. https://health-study.zoe.com/) and used to identify on the range of symptoms experienced by COVID-19 infected persons. Through this process six clusters of symptoms were identified. These are summarized in Table 5.9.

TABLE 5.9

Six clusters of COVID-19 Symptoms Identified from Syndromic Surveillance Information during the COVID-19 Pandemic 2020-2023

1. **"Flu-like" with no fever**: headache, loss of smell, muscle pains, cough, sore throat, chest pain, no fever
2. **"Flu-like" with fever**: headache, loss of smell, cough, sore throat, hoarseness, fever, loss of appetite
3. **Gastrointestinal**: headache, loss of smell, loss of appetite, diarrhoea, sore throat, chest pain, no cough
4. **Severe level one, fatigue**: headache, loss of smell, cough, fever, hoarseness, chest pain, fatigue
5. **Severe level two, confusion**: headache, loss of smell, loss of appetite, cough, fever, hoarseness, sore throat, chest pain, fatigue, confusion, muscle pain
6. **Severe level three, abdominal and respiratory**: headache, loss of smell, loss of appetite, cough, fever, hoarseness, sore throat, chest pain, fatigue, confusion, muscle pain, shortness of breath, diarrhoea, abdominal pain.

Source: Wise (2020); Visconti et al. (2022); Guemes et al. (2021a); Guemes et al. (2021b).

Ethical Considerations When Working with Health and Spatial Data

Not only do we have to take into consideration the ethics of the data that is being used and mapped, but also, using apps, the security of the data collected through these devices. Here are some examples of apps collecting a lot of personal information, much of which does end up public (Table 5.10).

TABLE 5.10

Examples of Apps Collecting Personal Information and Examples of Data Breaches

App	Description	Source
Venmo	How the payment app exposes our private lives	https://www.theguardian.com/world/2018/jul/17/venmo-payments-app-default-privacy-settings-public-information
Strava	Fitness tracking app Strava gives away location of secret US army bases	https://www.theguardian.com/world/2018/jan/28/fitness-tracking-app-gives-away-location-of-secret-us-army-bases
Grindr	Grindr was a safe space for gay men. Its HIV status leak betrayed us	https://www.theguardian.com/commentisfree/2018/apr/04/grindr-gay-men-hiv-status-leak-app
Tinder	I asked Tinder for my data. It sent me 800 pages of my deepest, darkest, secrets	https://www.theguardian.com/technology/2017/sep/26/tinder-personal-data-dating-app-messages-hacked-sold
Period tracking apps Maya and Period Tracker MIA Fem: Ovulation Calculator	If you're using a period-tracking app, there's a decent chance Facebook knows all about your sex life.	https://www.theguardian.com/world/commentisfree/2019/sep/14/your-period-tracking-app-could-be-sharing-intimate-details-with-all-of-facebook

(Continued)

TABLE 5.10 *(Continued)*

Examples of Apps Collecting Personal Information and Examples of Data Breaches

App	Description	Source
	A study from Privacy International, a UK-based charity, has found that some menstruation apps have been sharing their users' intimate details with the social network – including the last time you had unprotected intercourse. Apps include Maya (> 5 million downloads) and Period Tracker MIA Fem: Ovulation Calculator (> 2 million users)	

Activity - Citizen Science. How Useful VGI and Non-Authoritative Data are for Health and Disease Studies?

What are the benefits and drawbacks associated with these data? Would you use these types of data? To help you decide, read/skim through these papers.

Citizens as sensors: the world of volunteered geography (Goodchild, 2007)

What are the societal implications of citizen mapping and mapping citizens? (CSDGS_NRC_Ch11, 2010)

Below are some other examples that might be of interest in how citizen science information can be useful for gaining insights into health and disease dynamics.

Non-authoritative data used for health and disease studies:

Citizen science information for modelling disease (US Dollar) – **highly recommended reading**

- **Open street map**: An open mapping project started in 2004 and is today a well-used data source (Haklay and Weber, 2008).
- **Flu trends** (Olson et al., 2013) and how they correspond (Valdivia et al., 2010) with cases in Europe.

- **Vaccinations**: Radzikowski, J., A. Stefanidis, K. H. Jacobsen, A. Croitoru, A. T. Crooks & P. L. Delamater (2016) The measles vaccination narrative in twitter: A quantitative analysis. *JMIR Public Health and Surveillance*, 2(1), e1.

- **Accountability with open data**: http://www.theguardian.com/global-development-professionals-network/2013/dec/02/open-data-healthcare-accountability-africa

- **Human mobility**: Enhancing mobility maps and capturing cross-border movement using geo-located tweets (Blanford et al., 2015); and the US Dollar (Hufnagel et al., 2004; Brockmann et al., 2006).

- **Disease hotspots**: Supporting Geographically-aware WebDocument Foraging and Sensemaking (Tomaszewski et al., 2011) and VAST Paper (MacEachren et al., 2011)

- **Citizen science information during a pandemic** (Birkin et al., 2021): How the symptom tracker for COVID-19 was useful for understanding how COVID-19 was affecting the population (e.g. six classifications of symptoms; Wise, 2020) and predicting COVID-19 (Menni et al., 2020; Tan et al., 2022)

Limitations: However, there are limitations to using these data:

- David Lazer, Ryan Kennedy, Gary King, and Alessandro Vespignani. 2014. "The Parable of Google Flu: Traps in Big Data Analysis." *Science*, 14 March, 343: 1203–1205.

- On Twitter, #Antivaccination Goes Viral. *Science*, 12 Apr 2013, 340(6129): 128 http://science.sciencemag.org/content/340/6129/128.1

- Social media for large studies of behaviour. *Science*. 28 Nov 2014, 346 (6213): 1063–1064 http://science.sciencemag.org/content/346/6213/1063

- It is messy data (Dobson et al., 2020).

- Limitations of obtaining data using Apps (see Birkin et al., 2021).

Ethics – Data, Apps and Bots?

What are the ethical implications of using geographic information and geospatial technologies for health and disease? Both today and into the future?

- Accelerating ethics, empathy and equity in geographic information science (Nelson et al., 2022)

- Ethical challenges of big data in public health (Vayena et al., 2015)

- Elements of an infrastructure for big urban data (Goodchild, 2022)
- Biometrics – facial recognition for detecting emotions – ethical considerations in emotion recognition technologies: a review of the literature (Katirai, 2023)
- AI and Chatbots – https://www.ncbi.nlm.nih.gov/pmc/articles/PMC7133471/ (Luxton, 2020)
- Ethics and governance of artificial intelligence for health: WHO guidance. AI for Health (World Health, 2021)

Appendix

TABLE A.1

A List of Some Data Repositories

Description	Data Sources	URLs
Citizen science mapping	Open Street Map (OSM) **Weathermob**: Social Weather Reporting, and Local and Global Weather Reports By Weathernews Inc. mPING: crowdsourcing weather reports	http://www.openstreetmap.org/ https://itunes.apple.com/us/app/weathermob-social-weather/id463729367?mt=8 http://mping.nssl.noaa.gov/
Base data	Digital Chart of the World (DCW) A bit dated but still has good base data (Rail, Road, Water, Boundaries) (these are dated so check HDX and ESRI data hubs for more up-to-date boundaries) Geonames (World placename gazetteer) OSM GoogleMaps ESRI data sets WorldPop	https://www.openstreetmap.org http://www.geonames.org/ http://www.diva-gis.org/gdata http://www.gadm.org/ http://www.esri.com/data/data-maps http://www.esri.com/software/landsat-imagery http://opendata.arcgis.com http://edcommunity.esri.com/software-and-data/Data
Data Exchange Hubs	HDX – Humanitarian Data eXchange (HDX) ESRI • Lifestyle:	https://data.humdata.org/ http://www.esri.com/landing-pages/tapestry http://www.esri.com/data/data-maps

(Continued)

TABLE A.1 (*Continued*)

A List of Some Data Repositories

Description	Data Sources	URLs
	FAO (Food and Agricultural Organization) GIS Network	http://www.esri.com/software/landsat-imagery http://opendata.arcgis.com http://edcommunity.esri.com/software-and-data/Data Global Land Cover – SHARE (GLC-SHARE) | Land and Water | Food and Agriculture Organization of the United Nations | Land and Water | Food and Agriculture Organization of the United Nations (fao.org) http://www.fao.org/geonetwork/srv/en/main.home Databases and Software | Land and Water | Food and Agriculture Organization of the United Nations | Land and Water | Food and Agriculture Organization of the United Nations (fao.org)
Environment	World climate (Historic and Future) is available from weather stations some of which are put into formats that are useful for a variety of analyses. CHIRTS Lots of environmental information that has been captured by satellites are available • NOAA • NASA • Copernicus (Europe's eyes on Earth) • Landuse change • Pollution – Tropomi • CHIRTS Daily	http://worldclim.org/ https://www.ncdc.noaa.gov/cdo-web/ https://www.ncdc.noaa.gov/cdo-web/datasets https://gis.ncdc.noaa.gov/map/viewer/#app=cdo http://www.ncdc.noaa.gov/oa/ncdc.html https://www.earthdata.nasa.gov/ https://www.copernicus.eu/en/copernicus-services/atmosphere https://www.copernicus.eu/en/copernicus-services/land https://www.copernicus.eu/en/copernicus-services/climate-change https://www.copernicus.eu/en/copernicus-services/emergency https://neo.gsfc.nasa.gov/ https://www.chc.ucsb.edu/data/chirtsdaily

(*Continued*)

TABLE A.1 (*Continued*)

A List of Some Data Repositories

Description	Data Sources	URLs
Pollution	https://ourworldindata.org/air-pollution https://www.who.int/data/gho/data/ themes/air-pollution/ambient-air-pollution	
Population	World Population http://www.worldpop. org.uk/ Census (USA) • http://www.census.gov/ • http://www.census.gov/geo/ maps-data/data/tiger.html • http://factfinder2.census.gov/faces/ nav/jsf/pages/searchresults. xhtml?refresh=t • IPUMS https://www.ipums.org/	

Health and disease data	• WHO (World Health Organization): http://www.who.int/research/en/
	• PAHO (Pan American health Organization) http://www.paho.org/hq/
	• ProMED (http://www.promedmail. org/) – an Internet-based reporting system dedicated to rapid global dissemination of information on outbreaks of infectious diseases and acute exposures to toxins.
	• DHS Data (Demographic and health surveys)
	  • http://www.dhsprogram.com/
	  • http://spatialdata.dhsprogram.com/ data/#/
	• COVID-19 Incidence: https:// coronavirus.jhu.edu/map.html (https:// github.com/CSSEGISandData/ COVID-19)
	• COVID-19 Vaccination: https://github. com/OxCGRT/covid-policy-tracker/ tree/master/images; https:// ourworldindata.org/ policy-responses-covid
	• Netherlands: COVID-19 in NL https:// data.rivm.nl/covid-19/
	• Phylogenetics: https://nextstrain.org/
	• Health Facilities – OSM and HDX, WHO
	• CDC data repository	

Health Statistics	• Lots of health statistics and information is recorded and published in documents. Unfortunately, to use this data you will need to extract the information.	Check your local Census and Statistics Bureau.

The URLs will change over time.

TABLE A.2

List of Earth Observation Hubs and Repositories Where You Can Explore and obtain EO Data,

Data Hub	Description	Data Access
ESA	Access to a variety of datasets	Sentinel-2 Toolbox – Sentinel Online (esa.int) https://land.copernicus.vgt.vito.be
USGS	Explore available imagery	EarthExplorer (usgs.gov)
ESA	Explore the Copernicus Data Space Ecosystem	Copernicus Data Space Ecosystem | Europe's eyes on Earth
NASA's Earth data	Provides a searchable library of different NASA data products.	Earthdata Search | Earthdata Search (nasa.gov) https://lpdaac.usgs.gov/tools/data-pool/
NASA's Worldview	Good for exploring different data sets	EOSDIS Worldview (nasa.gov)
Digital Earth Africa	Good for accessing and using imagery data with a focus on Africa.	https://www.digitalearthafrica.org/
Maxar Open Data Programme	Maxar releases open data for major crisis events	Open Data Program | Disaster Response Geospatial Analytics (maxar.com)
Airbus	Geo-Airbus Defense also has a collection of sample imagery. a commercial vendor.SPOT, Pleiades, and RapidEye	Sample Imagery (intelligence-airbusds.com)
Google	A platform for Earth science data and analysis	Google Earth Engine
NOAA	NOAA's Satellite and Information Service	https://www.class.noaa.gov
INPE	National Institute for Space Research, Brazil has developed an Image catalogue	http://www.dgi.inpe.br/catalogo/explore
URSC	U R Rao Satellite Centre, formerly ISRO Satellite Centre is an ISRO centre for the design, development, and construction of Indian satellites	https://www.ursc.gov.in/
ESRI	Living Atlas, Multispectral Landsat service	https://www.arcgis.com/home/item.html? id=d9b466d6a9e647ce8d1dd5fe12eb434b

TABLE A.3

Summary of EO Data and Their Use for Planetary Health Collated

Satellite Data	Description	Use	Source
NDVI	The Normalized Difference Vegetation Index (NDVI) is an indicator of the greenness of the biomes and is widely used for ecosystems monitoring.	Changes in the environment	https://land.copernicus.eu/global/products/ndvi https://neo.gsfc.nasa.gov/view.php?datasetId=MOD_NDVI_M
Sentinel 1	Monitor oceans, ice and land and to aid emergency response. The mission ended for Sentinel-1B in 2022		ESA (2019a, 2019b)
Sentinel 2	High-resolution multispectral imagery to monitor land and vegetation cover		ESA (2019a, 2019b)
Sentinel 3	Monitor oceans and lands		ESA (2019a, 2019b)
Sentinel 4	Spectrometer carried on the Meteosat Third-Generation Sounder satellites	Monitoring air quality over Europe	ESA (2019a, 2019b)
Sentinel 5	A spectrometer, primarily to monitor global air pollution and atmospheric pollution	**TROPOMI**: the TROPOspheric Monitoring Instrument aboard ESA's Sentinel-5 Precursor satellite, part of the Copernicus programme	ESA (2019a, 2019b); ESA
Sentinel 6	Radar altimeter to measure global sea-surface height for operational oceanography and for climate studies		ESA (2019a, 2019b)

(Continued)

TABLE A.3 (*Continued*)

Summary of EO Data and Their Use for Planetary Health Collated

Satellite Data	Description	Use	Source
Landsat 8	The OLI measures in the visible, near infrared, and shortwave infrared portions (VNIR, NIR, and SWIR) of the spectrum. The TIRS measures land surface temperature in two thermal bands with a new technology that applies quantum physics to detect heat. Landsat 8 images have 15-meter panchromatic and 30-meter multi-spectral spatial resolutions along a 185 km (115 mi) swath	• Band 1- Coastal and aerosol studies • Band 2- Bathymetric mapping, distinguishing soil from vegetation, and deciduous from coniferous vegetation • Band 3- Emphasizes peak vegetation, which is useful for assessing plant vigour • Band 4- Discriminates vegetation slopes • Band 5 (NIR) – Emphasizes biomass content and shorelines • Band 6 (SWIR 1) – Discriminates moisture content of soil and vegetation; penetrates thin clouds • Band 7 (SWIR 2) – Improved ability to track moisture content of soil and vegetation and thin cloud penetration • Band 8 (Pan) -15 meter resolution, sharper image definition • Band 9 (Cirrus) – Improved detection of cirrus cloud contamination • Band 10 (TIRS 1) 100 meter resolution, thermal mapping and estimated soil moisture • Band 11 (TIRS 2) 100 meter resolution, thermal mapping and estimated soil moisture	https://www.usgs.gov/landsat-missions/landsat-8 https://eos.com/find-satellite/landsat-8/ see ESRI Multispectral Landsat service
MODIS MAIAC	**MODIS MAIAC** *Multi-Angle Implementation of Atmospheric Correction*: Algorithm that combines and corrects data from both MODIS satellites, Terra and Aqua satellites	$PM_{2.5}$ data with fine spatial and temporal resolution	https://dx.doi.org/10.5067/MODIS/MCD19A2.006

TABLE A.4

List of Nationally Notifiable Diseases for 2021 in the USA: Infectious Diseases

- Anthrax
- Arboviral diseases, neuroinvasive and non-neuroinvasive
  - California serogroup virus diseases
  - Chikungunya virus disease
  - Eastern equine encephalitis virus disease
  - Powassan virus disease
  - St. Louis encephalitis virus disease
  - West Nile virus disease
  - Western equine encephalitis virus disease
- Babesiosis
- Botulism
  - Botulism, foodborne
  - Botulism, infant
  - Botulism, other
  - Botulism, wound
- Brucellosis
- Campylobacteriosis
- Carbapenemase Producing Carbapenem-Resistant Enterobacteriaceae (CP-CRE)
  - CP-CRE, *Enterobacter* spp.
  - CP-CRE, *Escherichia coli* (*E. coli*)
  - CP-CRE, *Klebsiella* spp.
- Chancroid
- *Chlamydia trachomatis* infection
- Cholera
- Coccidioidomycosis
- Congenital syphilis
  - Syphilitic stillbirth
- Coronavirus Disease 2019 (COVID-19)
- Cryptosporidiosis
- Cyclosporiasis
- Dengue virus infections
  - Dengue
  - Dengue-like illness
  - Severe dengue
- Diphtheria
- Ehrlichiosis and anaplasmosis
  - *Anaplasma phagocytophilum* infection

- Hansen's disease
- Hantavirus infection, non-Hantavirus pulmonary syndrome
- Hantavirus pulmonary syndrome
- Hemolytic uremic syndrome, post-diarrhoeal
- Hepatitis A, acute
- Hepatitis B, acute
- Hepatitis B, chronic
- Hepatitis B, perinatal virus infection
- Hepatitis C, acute
- Hepatitis C, chronic
- Hepatitis C, perinatal infection
- HIV infection (AIDS has been reclassified as HIV Stage III)
- Influenza-associated paediatric mortality
- Invasive pneumococcal disease
- Legionellosis
- Leptospirosis
- Listeriosis
- Lyme disease
- Malaria
- Measles
- Meningococcal disease
- Mumps
- Novel influenza A virus infections
- Pertussis
- Plague
- Poliomyelitis, paralytic
- Poliovirus infection, nonparalytic
- Psittacosis
- Q fever
  - Q fever, acute
  - Q fever, chronic
- Rabies, animal
- Rabies, human
- Rubella
- Rubella, congenital syndrome

- Severe acute respiratory syndrome-associated coronavirus disease
- Shiga toxin-producing *Escherichia coli*
- Shigellosis
- Smallpox
- Spotted fever rickettsiosis
- Streptococcal toxic shock syndrome
  - Syphilis
  - Syphilis, early non-primary non-secondary
  - Syphilis, primary
  - Syphilis, secondary
  - Syphilis, unknown duration or late
- Tetanus
- Toxic shock syndrome (other than streptococcal)
- Trichinellosis
- Tuberculosis
- Tularemia
- Vancomycin-intermediate *Staphylococcus aureus* and Vancomycin-resistant *Staphylococcus aureus*
- Varicella
- Varicella deaths
- Vibriosis
- Viral haemorrhagic fever
  - Crimean-Congo haemorrhagic fever virus
  - Ebola virus
  - Lassa virus
  - Lujo virus
  - Marburg virus
  - New World arenavirus – Guanarito virus
  - New World arenavirus – Junin virus
  - New World arenavirus – Machupo virus
  - New World arenavirus – Sabia virus
- Yellow Fever

(Continued)

TABLE A.4 (*Continued*)

List of Nationally Notifiable Diseases for 2021 in the USA: Infectious Diseases

- *Ehrlichia chaffeensis infection*
- *Ehrlichia ewingii infection*
- Undetermined human ehrlichiosis/anaplasmosis
- Giardiasis
- Gonorrhea
- *Haemophilus influenzae,* invasive disease

- *Salmonella* Paratyphi infection (*Salmonella enterica* serotypes Paratyphi A, B [tartrate negative], and C [*S.* Paratyphi])
- *Salmonella* Typhi infection (*Salmonella enterica* serotype Typhi)
- Salmonellosis

- Zika virus disease and Zika virus infection
  - Zika virus disease, congenital
- Zika virus disease, non-congenital
- Zika virus infection, congenital
- Zika virus infection, non-congenital

Source: CDC (2021); https://ndc.services.cdc.gov/search-results-year/.

TABLE A.5

List of Nationally Notifiable Diseases for 2021 in the USA: Non-Infectious Diseases

- Cancer
- Carbon monoxide poisoning

- Lead, elevated blood levels
  - Lead, elevated blood levels, adult (≥16 years)
- Lead, elevated blood levels, children (<16 years)

- Pesticide-related illness and injury, acute
- Silicosis

Source: CDC (2021); https://ndc.services.cdc.gov/search-results-year/.

TABLE A.6

List of Nationally Notifiable Diseases for 2021 in the USA: Outbreaks

Foodborne Disease Outbreak	Waterborne Disease Outbreak

TABLE A.7

List of Nationally Notifiable Diseases for 2021 in the Netherlands

Groep A: Mogelijk wettelijke maatregelen: gedwongen opname tot isolatie of thuisisolatie, gedwongen onderzoek, gedwongen quarantaine (inclusief medisch toezicht), verbod van beroepsuitoefening. Dit geldt voor:	Groep B1: Mogelijk wettelijke maatregelen: gedwongen opname tot isolatie of thuisisolatie, gedwongen onderzoek, verbod op beroepsuitoefening. Dit geldt voor:	Groep B2: Mogelijk wettelijke maatregelen: verbod op beroepsuitoefening. Dit geldt voor:	Groep C: Dwingende maatregelen kunnen niet opgelegd worden. Maar melding en persoonsgegevens zijn nodig om de inzet van vrijwillige/te adviseren maatregelen rondom de patiënt of anderen in de gemeenschap mogelijk te maken.
COVID-19	Humane infectie met dierlijk influenzavirus	Buiktyfus (typhoid fever)	Antrax
MERS-coronavirus	Difterie	Cholera	Bof
Pokken	Pest	Hepatitis A	Botulisme
Polio	Rabiës	Hepatitis B	Brucellose
Severe acute respiratory syndrome (SARS)	Tuberculose	Hepatitis C	Chikungunya (alléén meldingsplichtig in Caribisch Nederland: Bonaire, St. Eustatius, Saba)
Virale haemorragische koorts		Kinkhoest	Carbapenemase-producerende Enterobacteriaceae (CPE)
		Mazelen	Ziekte van Creutzfeldt-Jakob (klassieke)
		Paratyfus	Ziekte van Creutzfeldt-Jakob (variant)
		Rubella	Dengue (alléén meldingsplichtig in Caribisch Nederland: Bonaire, St. Eustatius, Saba)
		Shigatoxineproducerende *Escherichia coli* / enterohemorragische *Escherichia coli*-infectie (STEC)	Gele koorts
		Shigellose	
		Invasieve groep A-streptokokkeninfectie	
		Voedselinfectie voor zover vastgesteld bij 2 of meer patiënten met een onderlinge relatie wijzend op voedsel als bron	

(Continued)

TABLE A.7 (*Continued*)
List of Nationally Notifiable Diseases for 2021 in the Netherlands

Invasieve Haemophilus influenzae type b-infectie
Hantavirusinfectie
Legionellose
Leptospirose
Listeriose
Malaria
Meningokokkenziekte
MRSA-infectie (clusters buiten het ziekenhuis)
Invasieve pneumokokkenziekte bij kinderen geboren vanaf 01-01-2006 en bij een persoon van 60 jaar of ouder
Psittacose
Q-koorts
Tetanus
Trichinose
Tularemie
West-Nilevirus
Zikavirusinfectie

Source: RIVM (2021) https://www.government.nl/topics/animal-diseases/controlling-animal-diseases; https://business.gov.nl/regulation/obligation-report-infectious-diseases/; RIVM (2021) https://www.rivm.nl/meldingsplicht-infectieziekten/welke-infectieziekten-zijn-meldingsplichtig

TABLE A.8

Notifiable Disease with Regards to Animals in the Netherlands

BSE or mad cow disease
Foot and mouth disease
Q fever
Parrot fever (psittacosis)
Swine fever
Bird flu
Schmallenberg virus (SBV)
Equine herpes virus (EHV)
Viral infection in seals

TABLE A.9

Notifiable diseases for 2023 in the UK

- Bacillus anthracis
- Bacillus cereus
- Bordetella pertussis
- Borrelia spp.
- Brucella spp.
- Burkholderia mallei
- Burkholderia pseudomallei
- Campylobacter spp.
- Carbapenemase-producing Gram-negative bacteria
- Chikungunya virus
- Chlamydophila psittaci
- Clostridium botulinum
- Clostridium perfringens
- Clostridium tetani
- Corynebacterium diphtheriae
- Corynebacterium ulcerans
- Coxiella burnetii
- Crimean-Congo haemorrhagic fever virus
- Cryptosporidium spp.
- Dengue virus
- Ebola virus
- Entamoeba histolytica
- Escherichia coli O 157
- Francisella tularensis
- Giardia lamblia
- Guanarito virus
- Haemophilus influenzae (invasive)
- Hanta virus
- Hepatitis A
- Influenza virus
- Junin virus
- Kyasanur forest disease virus
- Lassa virus
- Legionella spp.
- Leptospira interrogans
- Listeria monocytogenes
- Machupo virus
- Marburg virus
- Measles virus
- Mpox (monkeypox) virus
- Mumps virus
- Mycobacterium tuberculosis complex
- Neisseria meningitidis
- Omsk haemorrhagic fever virus
- Plasmodium falciparum
- Plasmodium knowlesi
- Plasmodium malariae
- Plasmodium ovale
- Plasmodium vivax
- Polio virus
- Rabies virus
- Rickettsia spp.
- Rift Valley fever virus
- Rubella virus
- Sabia virus
- Salmonella spp.
- SARS coronavirus
- Shigella spp.
- Streptococcus group A (invasive)

(Continued)

TABLE A.9 (*Continued*)

Notifiable diseases for 2023 in the UK

• Hepatitis B	• Streptococcus pneumoniae (invasive)
• Hepatitis C	• Varicella zoster virus
• Hepatitis D	• Variola virus
• Hepatitis E	• Vibrio cholerae
	• West Nile virus
	• Yellow fever virus
	• Yersinia pestis

Source: Laboratories in England have a statutory duty to notify the UK Health Security Agency (UKHSA) of the identification of the following causative agents. To see what has been reported for the last 52-weeks view the data at: https://www.gov.uk/government/publications/notifiable-diseases-last-52-weeks and https://www.gov.uk/government/collections/notifications-of-infectious-diseases-noids

References

Anderson, K., B. Ryan, W. Sonntag, A. Kavvada & L. Friedl (2017) Earth observation in service of the 2030 Agenda for sustainable development. *Geo-spatial Information Science*, 20, 77–96.

Birkin, L. J., E. Vasileiou & H. R. Stagg (2021) Citizen science in the time of COVID-19. *Thorax*, 76, 636–637.

Blanford, J. I., J. Bernhardt, A. Savelyev, G. Wong-Parodi, A. M. Carleton, D. W. Titley & A. M. MacEachren (2014) Tweeting and Tornadoes. In *11th International ISCRAM Conference*. Pennsylvania, USA: State College, ISCRAM.

Bonita, R., R. Beaglehole & T. Kjellstrom (2006) *Basic Epidemiology*. Geneva, Switzerland: World Health Organization (WHO). https://apps.who.int/iris/bitstream/10665/43541/1/9241547073_eng.pdf (last accessed) April 27 2024.

Brockmann, D., L. Hufnagel & T. Geisel (2006) The scaling laws of human travel. Nature, 439, 462–465.

CDC (2012) *Principles of Epidemiology in Public Health Practice*. U.S. Department of Health and Human Services, Centers for Disease Control and Prevention. https://www.cdc.gov/ophss/csels/dsepd/SS1978/SS1978.pdf (last accessed) April 10 2024).

Cromley, E. K. & S. L. McLafferty (2012) *GIS and Public Health*. New York: Guilford Press.

Dickens, B. L., J. R. Koo, J. T. Lim, H. Sun, H. E. Clapham, A. Wilder-Smith & A. R. Cook (2020) Strategies at points of entry to reduce importation risk of COVID-19 cases and reopen travel. *Journal of Travel Medicine*, 27, 1–8.

Dobson, A. D. M., E. J. Milner-Gulland, N. J. Aebischer, C. M. Beale, R. Brozovic, P. Coals, R. Critchlow, A. Dancer, M. Greve & A. Hinsley (2020) Making messy data work for conservation. *One Earth*, 2, 455–465.

ESA Sentinel 5P. https://www.esa.int/Applications/Observing_the_Earth/Copernicus/Sentinel-5P (last accessed October 10, 2023).

— (2019a) Earth Observation: Copernicus Sentinel Satellite Data-Open Access at ESA. Available online: https://open.esa.int/copernicus-sentinel-satellite-data/ (last accessed October 10, 2023

— (2019b) Sentinels: Space for Copernicus, 5 p. https://esamultimedia.esa.int/multi-media/publications/BR-319/BR319.pdf (last accessed Aug 10 2023).

Felmlee, D., J. I. Blanford, S. Matthews & A. M. MacEachren (2020) The Geography of sentiment towards the Women's March of 2017. *Plos One*, 15, e0233994.

GISGeography (2023) The Ultimate List of GIS Formats and Geospatial File Extensions. https://gisgeography.com/gis-formats/ (last accessed 22 Dec 2023)

Goodchild, M. F. (2022) Elements of an infrastructure for big urban data. *Urban Informatics*, 1, 3.

Guemes, A., S. Ray, K. Aboumerhi, M. R. Desjardins, A. Kvit, A. E. Corrigan, B. Fries, T. Shields, R. D. Stevens, F. C. Curriero & R. Etienne-Cummings (2021a) Author correction: A syndromic surveillance tool to detect anomalous clusters of COVID-19 symptoms in the United States. *Scientific Reports*, 11, 17939.

— (2021b) A syndromic surveillance tool to detect anomalous clusters of COVID-19 symptoms in the United States. *Scientific Reports*, 11, 4660.

Haklay, M. & P. Weber (2008) OpenStreetMap: User-generated street maps. *IEEE Pervasive Computing*, 7, 12–18.

Higuera-Mendieta, D. R., S. Cortés-Corrales, J. Quintero & C. González-Uribe (2016) KAP surveys and dengue control in Colombia: Disentangling the effect of sociodemographic factors using multiple correspondence analysis. *PLoS Neglected Tropical Diseases*, 10, e0005016.

Holloway, T., D. Miller, S. Anenberg, M. Diao, B. Duncan, A. M. Fiore, D. K. Henze, J. Hess, P. L. Kinney, Y. Liu, J. L. Neu, S. M. O'Neill, M. T. Odman, R. B. Pierce, A. G. Russell, D. Tong, J. J. West & M. A. Zondlo (2021) Satellite monitoring for air quality and health. *The Annual Review of Biomedical Data Science*, 4, 417–447.

Hufnagel, L., D. Brockmann & T. Geisel (2004) Forecast and control of epidemics in a globalized world. *Proceedings of the National Academy of Sciences of the United States of America*, 101, 15124–15129.

Jenny, B., B. Šavrič, N. D. Arnold, B. E. Marston & C. A. Preppernau (2017) A Guide to Selecting Map Projections for World and Hemisphere Maps. In *Choosing a Map Projection*, eds. M. Lapaine & E. L. Usery, 213–228. Cham: Springer International Publishing.

Johansson, M. A., H. Wolford, P. Paul, P. S. Diaz, T. H. Chen, C. M. Brown, M. S. Cetron & F. Alvarado-Ramy (2021) Reducing travel-related SARS-CoV-2 transmission with layered mitigation measures: Symptom monitoring, quarantine, and testing. *BMC Medicine*, 19, 94.

Jung, E. M., J. Dinkel, N. Verloh, M. Brandenstein, C. Stroszczynski, F. Jung & J. Rennert (2021) Wireless point-of-care ultrasound: First experiences with a new generation handheld device. *Clinical Hemorheology and Microcirculation*, 79, 463–474.

Kessler, F. & S. Battersby (2019) *Working with Map Projections: A Guide to Their Selection*. Boca Raton, FL: CRC Press.

Krentel, A., P. Fischer, P. Manoempil, T. Supali, G. Servais & P. Rückert (2006) Using knowledge, attitudes and practice (KAP) surveys on lymphatic filariasis to prepare a health promotion campaign for mass drug administration in Alor District, Indonesia. *Tropical Medicine & International Health*, 11, 1731–1740.

Luxton, D. D. (2020) Ethical implications of conversational agents in global public health. *The Bulletin of the World Health Organization*, 98, 285–287.

MacEachren, A. M., A. Jaiswal, A. C. Robinson, S. Pezanowski, A. Savelyev, P. Mitra, X. Zhangi & J. Blanford (2011) SensePlace2: GeoTwitter Analytics Support for Situational Awareness. In *Visual Analytics Science and Technology (VAST), IEEE Conference, Providence, RI, USA*, 181–190.

Marshall, J. M., D. A. Dunstan & W. Bartik (2019) The digital psychiatrist: In search of evidence-based apps for anxiety and depression. *Front Psychiatry*, 10, 831.

Martinez, M., S. B. Park, I. Maison, V. Mody, L. S. Soh & H. S. Parihar (2017) iOS Appstore-based phone apps for diabetes management: Potential for use in medication adherence. *JMIR Diabetes*, 2, e12.

Menni, C., A. M. Valdes, M. B. Freidin, C. H. Sudre, L. H. Nguyen, D. A. Drew, S. Ganesh, T. Varsavsky, M. J. Cardoso, J. S. El-Sayed Moustafa, A. Visconti, P. Hysi, R. C. E. Bowyer, M. Mangino, M. Falchi, J. Wolf, S. Ourselin, A. T. Chan, C. J. Steves & T. D. Spector (2020) Real-time tracking of self-reported symptoms to predict potential COVID-19. *Nature Medicine*, 26, 1037–1040.

Nelson, T. A., M. F. Goodchild & D. J. Wright (2022) Accelerating ethics, empathy, and equity in geographic information science. *Proceedings of the National Academy of Sciences of the United States of America*, 119, e2119967119.

Olson, D. R., K. J. Konty, M. Paladini, C. Viboud & L. Simonsen (2013) Reassessing google flu trends data for detection of seasonal and pandemic influenza: A comparative epidemiological study at three geographic scales. *PLOS Computational Biology*, 9, e1003256.

Reis, S., E. Seto, A. Northcross, N. W. Quinn, M. Convertino, R. L. Jones & et al. (2015) Integrating modelling and smart sensors for environmental and human health. *Environmental Modelling & Software*, 74, 238–246.

Richards Jr, F. O., R. E. Klein, O. de León, R. Mendizábal-Cabrera, A. L. Morales, V. Cama, C. G. Crovella, C. E. Díaz Espinoza, Z. Morales & M. Sauerbrey (2016) A knowledge, attitudes and practices survey conducted three years after halting ivermectin mass treatment for onchocerciasis in Guatemala. *PLoS Neglected Tropical Diseases*, 10, e0004777.

Robertson, C., K. Sawford, S. L. Daniel, T. A. Nelson & C. Stephen (2010) Mobile phone-based infectious disease surveillance system, Sri Lanka. *Emerging Infectious Diseases*, 16, 1524–1531.

Sabins Jr, F. F. & J. M. Ellis (2020) *Remote Sensing: Principles, Interpretation, and Applications*. Long Grove, IL: Waveland Press.

Sarraf-Zadegan, N., G. Sadri, H. Malek-Afzali, M. Baghaei, N. Mohammadi-Fard, S. Shahrokhi, H. Tolooie, M. Poormoghaddas, M. Sadeghi & A. Tavassoli (2003) Isfahan Healthy Heart Programme: A comprehensive integrated community-based programme for cardiovascular disease prevention and control. Design, methods and initial experience. *Acta Cardiologica*, 58, 309–320.

Savel, T. G. & S. Foldy (2012) The role of public health informatics in enhancing public health surveillance. *MMWR Surveillance Summary*, 61, 20–24.

Shan, R., S. Sarkar & S. S. Martin (2019) Digital health technology and mobile devices for the management of diabetes mellitus: State of the art. *Diabetologia*, 62, 877–887.

Sharma, T. & M. Bashir (2020) Use of apps in the COVID-19 response and the loss of privacy protection. *Nature Medicine*, 26, 1165–1167.

Smith, Z. J., K. Chu, A. R. Espenson, M. Rahimzadeh, A. Gryshuk, M. Molinaro, D. M. Dwyre, S. Lane, D. Matthews & S. Wachsmann-Hogiu (2011) Cell-phone-based platform for biomedical device development and education applications. *PLoS One*, 6, e17150.

Tabash, M. I., R. A. Hussein, A. H. Mahmoud, M. D. El-Borgy & B. A. Abu-Hamad (2016) Impact of an educational program on knowledge and practice of health care staff toward pharmaceutical waste management in Gaza, Palestine. *Journal of the Air & Waste Management Association*, 66, 429–438.

Taber, E. D., M. L. Hutchinson, E. A. H. Smithwick & J. I. Blanford (2017) A decade of colonization: The spread of the Asian tiger mosquito in Pennsylvania and implications for disease risk. *Journal of Vector Ecology*, 42, 3–12.

Tan, Y. R., A. Agrawal, M. P. Matsoso, R. Katz, S. L. M. Davis, A. S. Winkler, A. Huber, A. Joshi, A. El-Mohandes, B. Mellado, C. A. Mubaira, F. C. Canlas, G. Asiki, H. Khosa, J. V. Lazarus, M. Choisy, M. Recamonde-Mendoza, O. Keiser, P. Okwen, R. English, S. Stinckwich, S. Kiwuwa-Muyingo, T. Kutadza, T. Sethi, T. Mathaha, V. K. Nguyen, A. Gill & P. Yap (2022) A call for citizen science in pandemic preparedness and response: Beyond data collection. *BMJ Global Health*, 7, e009389.

Tjaden, N. & J. I. Blanford (2023) ENDIG: Interactive geovisualisation of notifiable diseases for disease surveillance systems in Europe. *AGILE GIScience Series*, 4, 1–6.

Tomaszewski, B., J. I. Blanford, K. Ross, S. Pezanowski & A. M. MacEachren (2011) Supporting geographically-aware webdocument foraging and sensemaking. *Computers, Environment and Urban Systems*, 35, 192–207.

Tomlinson, R. F. (1987) Current and potential uses of geographical information systems: The North American experience. *International Journal of Geographical Information Systems*, 1, 208–218.

USGS (2019) Map Projections. https://pubs.usgs.gov/gip/70047422/report.pdf. Last accessed. May 2, 2024.

Valdivia, A., J. Lopez-Alcalde, M. Vicente, M. Pichiule, M. Ruiz & M. Ordobas (2010) Monitoring influenza activity in Europe with google flu trends: Comparison with the findings of sentinel physician networks - results for 2009–10. *Eurosurveillance*, 15, 19621.

Vayena, E., M. Salathe, L. C. Madoff & J. S. Brownstein (2015) Ethical challenges of big data in public health. *PLOS Computational Biology*, 11, e1003904.

Visconti, A., B. Murray, N. Rossi, J. Wolf, S. Ourselin, T. D. Spector, E. E. Freeman, V. Bataille & M. Falchi (2022) Cutaneous manifestations of SARS-CoV-2 infection during the Delta and Omicron waves in 348 691 UK users of the UK ZOE COVID Study app. *British Journal of Dermatology*, 187, 900–908.

Vohland, K., A. Land-Zandstra, L. Ceccaroni, R. Lemmens, J. Perelló, M. Ponti, R. Samson & K. Wagenknecht (2021) *The Science of Citizen Science*. London: Springer Nature.

Voigt, S., T. Kemper, T. Riedlinger, R. Kiefl, K. Scholte & H. Mehl (2007) Satellite image analysis for disaster and crisis-management support. *IEEE Transactions on Geoscience and Remote Sensing*, 45, 1520–1528.

Wang, C., M. Liu, Z. Wang, S. Li, Y. Deng & N. He (2021) Point-of-care diagnostics for infectious diseases: From methods to devices. *Nano Today*, 37, 101092.

Wang, Y., L. Yu, X. Kong & L. Sun (2017) Application of nanodiagnostics in point-of-care tests for infectious diseases. *International Journal of Nanomedicine*, 12, 4789–4803.

Wise, J. (2020) Covid-19: Study reveals six clusters of symptoms that could be used as a clinical prediction tool. *BMJ*, 370, m2911.

Wood, C. S., M. R. Thomas, J. Budd, T. P. Mashamba-Thompson, K. Herbst, D. Pillay, R. W. Peeling, A. M. Johnson, R. A. McKendry & M. M. Stevens (2019) Taking connected mobile-health diagnostics of infectious diseases to the field. *Nature*, 566, 467–474.

World Health, O. (2021) *Ethics and Governance of Artificial Intelligence for Health: WHO Guidance.* Geneva, Switzerland: World Health Organization (WHO).

Yin, H., T. Udelhoven, R. Fensholt, D. Pflugmacher & P. Hostert (2012) How normalized difference vegetation index (NDVI) trends from advanced very high resolution radiometer (AVHRR) and système probatoire d'observation de la terre vegetation (SPOT VGT) time series differ in agricultural areas: An inner Mongolian case study. *Remote Sensing*, 4, 3364–3389.

Zarnetske, P. L., Q. D. Read, S. Record, K. D. Gaddis, S. Pau, M. L. Hobi, S. L. Malone, J. Costanza, K. M. Dahlin, A. M. Latimer, A. M. Wilson, J. M. Grady, S. V. Ollinger & A. O. Finley (2019) Towards connecting biodiversity and geodiversity across scales with satellite remote sensing. *Global Ecology and Biogeography*, 28, 548–556.

Zhao, B., S. Zhang, C. Xu, Y. Sun & C. Deng (2021) Deep fake geography? When geospatial data encounter Artificial Intelligence. *Cartography and Geogrpahic Information Science*, 48, 338–352.

6

Health and Disease in Dynamically Changing Environments: Mapping and Modelling Vector-Borne Diseases

Overview

Vector-borne diseases are illnesses caused by pathogens and parasites in human populations and account for more than 17% of all infectious diseases (WHO, 2016b). Every year, there are more than 1 billion cases and over 1 million deaths from vector-borne diseases (WHO, 2016b). The distribution of these diseases is determined by complex, dynamic environmental and social factors. Recent years have seen the emergence of new diseases and the re-emergence of old diseases in new and existing areas (e.g. swift movement of West Nile virus (WNV) across the continental US; resurgence of dengue in the Americas, with local transmission reported in Florida; arrival of chikungunya in the Americas with outbreaks in Europe; malaria in Europe and the dispersion of Zika). Vector-borne diseases are a significant cause of human morbidity and mortality and are a worldwide concern, often taking just one bite to become infected (Figure 6.1).

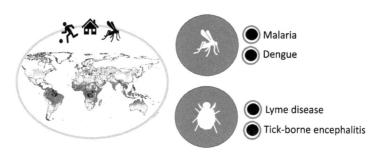

FIGURE 6.1

Infographic summarizing common vectors (mosquitoes and ticks) and the diseases that they transmit, such as malaria, dengue, Lyme and Tick-borne encephalitis (TBE). Adapted from (WHO, 2014).

DOI: 10.1201/9781003435082-6

Vectors and the Diseases They Transmit

Vectors are living organisms involved in the transmission of many diseases (Table 6.1). For some vector-borne diseases, the transmission cycle also involves an intermediate host organism in which the agent develops or multiplies and a reservoir population of organisms that, in addition to human hosts, maintain the agent. Many of these vectors are bloodsucking insects that ingest disease-producing micro-organisms during a blood meal from an infected host (human or animal) and later inject them into a new host during their next blood meal. Mosquitoes are the best known disease vector; others include ticks, flies, sandflies, fleas, triatomine bugs and some freshwater aquatic snails (Table 6.1; WHO (2021)).

TABLE 6.1

Key Vectors and The Diseases They Transmit

Vector		Disease Caused	Type of Pathogen
Mosquito	*Aedes*	Chikungunya	Virus
		Dengue	Virus
		Lymphatic filariasis	Parasite
		Rift Valley fever	Virus
		Yellow fever	Virus
		Zika	Virus
	Anopheles	Lymphatic filariasis	Parasite
		Malaria	Parasite
	Culex	Japanese encephalitis	Virus
		Lymphatic filariasis	Parasite
		West Nile fever	Virus
Aquatic snails		Schistosomiasis (bilharziasis)	Parasite
Blackflies		Onchocerciasis (river blindness)	Parasite
Fleas		Plague (transmitted from rats to humans)	Bacteria
		Tungiasis	Ectoparasite
Lice		Typhus	Bacteria
		Louse-borne relapsing fever	Bacteria
Sandflies		Leishmaniasis	Parasite
		Sandfly fever (phlebotomus fever)	Virus
Ticks		Crimean-Congo haemorrhagic fever	Virus
		Lyme disease	Bacteria
		Relapsing fever (borreliosis)	Bacteria
		Rickettsial diseases (e.g.: spotted fever and Q fever)	Bacteria Virus
		Tick-borne encephalitis	Bacteria
		Tularaemia	
Triatome bugs		Chagas disease (American trypanosomiasis)	Parasite
Tsetse flies		Sleeping sickness (African trypanosomiasis)	Parasite

Source: WHO (2021).

Vector-borne diseases may be *endemic* or permanently present even when controlled; *hyperendemic*, exhibiting high and continued incidence; *hypoendemic*, affecting only a small proportion of the population at risk, or not endemic; and *holoendemic*, when almost every person in a population is affected.

Factors Important in Vector-Borne Diseases

The epidemiological triad summarizes the factors (agent, host and environment) (Figure 2.2, Chapter 2) that influence the distribution of diseases. Since many of the vectors are small, they are influenced by environmental factors such as temperature, rainfall, wind and solar radiation. Of these factors, temperature plays an important role in various life-history traits, such as the rate of development of the insect vector and also of the disease agent inside its host. So for many vector-borne diseases, not only do you have to consider the disease agent (virus, parasite or bacteria) but also a vector that obtains and transmits the disease to a single or multiple hosts. Thus, we have to consider the ecology of the disease, the ecology of the vector and the interaction of these two in the environment. As we work our way through the epidemiological triad (What, Where and Who) (Figure 6.2), we can create a general summary of factors relevant for mapping and modelling vector-borne diseases.

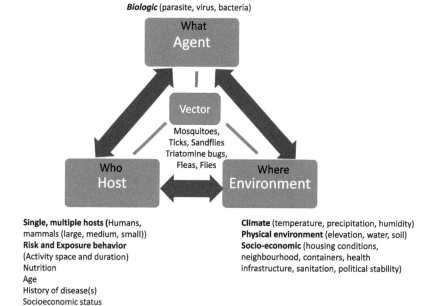

FIGURE 6.2

A generalized epidemiologic triad for vector-borne diseases. Image created by Blanford (2021).

- **What?** What are the vector(s)? What is the disease agent?
- **Where?** What environmental factors influence the distribution of the vector? What environmental factors influence the occurrence of the disease? Where are these located?
- **Who?** Who are the hosts (single/multiple)? Who is affected?

Exposure, Morbidity and Risk

The risk of becoming infected with a vector-borne disease is dependent on a number of factors, including the presence of the vector and disease and the risk of exposure (human-risk environment interactions based on the behaviour and activity space of the individual). Essentially, the time spent in an exposure environment (frequency and duration), location type (indoor vs outdoor), time of day (day vs night) and activity space composition (no. of different locations visited) (Perkins et al., 2014; Stoddard et al., 2009) (Figure 6.3).

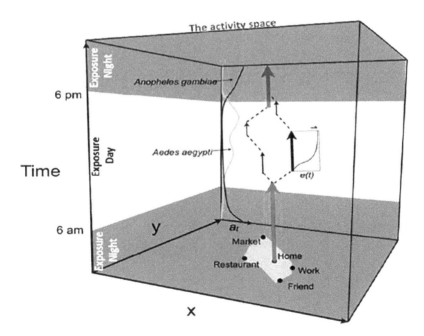

FIGURE 6.3

The activity space of an individual through time and space. As the individual moves throughout the day from home, to work, visiting a friend, shopping at the market and eating at a restaurant they are exposed to different risk environments for different durations. During daylight hours, they are exposed to one set of vectors, and during the night, they are exposed to another set of vectors. *Source*: Stoddard et al. (2009).

Think about your daily activity space. Are you exposing yourself to different vectors? If so, which ones? And where?

Dispersion and Diffusion: Diseases on the Move

The movement of humans plays an important role in the distribution and re-distribution of diseases, as illustrated in Figure 6.4. For example, an individual acquires a disease through exposure in one environment and then moves to a new geographic location, where they can introduce the disease through human-vector interactions in the environment. The vector acquires the disease and subsequently infects new individuals that pass through during a blood meal.

In many cases, travellers are often unaware they have acquired a disease when they return home, as symptoms may be delayed due to the length of the incubation period (time interval from infection to onset of symptoms) (Table 6.2).

For example, in Figure 6.5 mobility tracks for several individuals were mapped and overlayed with a malaria risk map created by the Malaria Atlas Project. The movement tracks show individuals moving in and out of

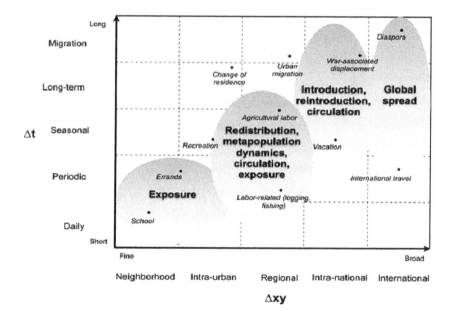

FIGURE 6.4

A framework for human mobility. Movements are characterized in terms of their spatial and temporal scale, which are defined in terms of physical displacement (Δxy) and time spent (Δt, frequency and duration). *Source*: Stoddard et al. (2009).

TABLE 6.2

Summary of Incubation Periods for Different Diseases

Disease	Incubation Period	Source
Malaria	7 (10)–15	http://www.who.int/mediacentre/factsheets/fs094/en/
West Nile virus (WNV)	3–14	http://www.who.int/mediacentre/factsheets/fs354/en/
Yellow Fever	3–6	http://www.who.int/mediacentre/factsheets/fs100/en/
Dengue	4–10	http://www.who.int/mediacentre/factsheets/fs117/en/
Rift Valley Fever	2–6	http://www.cdc.gov/vhf/rvf/symptoms/index.html
Chikungunya	3–7	http://www.cdc.gov/chikungunya/hc/clinicalevaluation.html
Zika	~ 3–12 days	http://www.who.int/mediacentre/factsheets/zika/en/ http://www.health.govt.nz/our-work/diseases-and-conditions/zika-virus
Japanese encephalitis	5–15	http://www.cdc.gov/japaneseencephalitis/symptoms/

areas with different malaria prevalence (Figure 6.5). As you can see from the figure, several individuals spend time in areas with high malaria prevalence and then move into areas with low or no malaria prevalence. Thus, illustrating the possibility of acquiring the parasite in one region and transporting it to another.

Some examples of the spread of diseases to new geographic locations include the following:

- **Chikungunya**: Antibodies of the virus were detected in German aid workers in the 1990's (Eisenhut et al., 1999), however, it was not until 2007 that the virus reached Europe (Simon et al., 2008) with outbreaks occurring in Italy (Rezza et al., 2007). In 2005 the virus moved from Kenya to India and in 2013 the virus moved to a number of islands in the Caribbean before going global (Lanciotti and Valadere, 2014);

- **Zika**: Zika, first isolated in 1947 in Uganda, had the first human case reported in Nigeria in the 1950's (Musso and Gubler, 2016). The first outbreak was reported in 2007 in the Federated States of Micronesia with outbreaks in the South Pacific in 2014 and 2015 before emerging in Brazil and the Caribbean in 2016 (Musso and Gubler, 2016; PAHO, 2016; WHO, 2016a);

- **Malaria**: Local outbreaks of malaria have been reported in California (1980s) and Houston (1990s) in the USA (Martens and Hall, 2000)

(a)   (b)

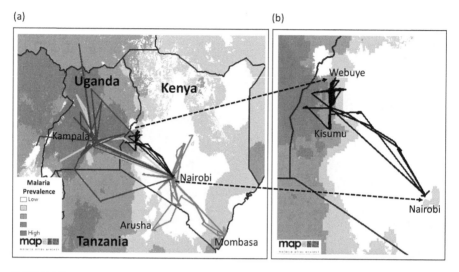

FIGURE 6.5

Activity space of four individuals over one year in and around Kenya. Individual movement obtained from geolocated tweets (e.g. Blanford et al., 2015) was integrated with malaria risk areas (obtained from the Malaria Atlas Project (http://www.map.ox.ac.uk/) (Hay et al., 2009) to show potential within country and cross-border movement of the malaria parasite between diverse geographic locations. Image created by Blanford (2021).

    with local transmission occurring in Florida and Texas (2023) (CDC, 2023c) and parts of Italy and France (Armengaud et al., 2008)

- **Dengue**: Dengue has been absent for over 60 years in the USA; however, local transmission has resulted in outbreaks in Hawaii (2001), Texas (2005) (Adalja et al., 2012; Bouri et al., 2012) and Florida (2009–2011; 2013) (Adalja et al., 2012; Radke et al., 2012). Local transmission now occurs in different parts of the USA. Absent from Japan for 70 years a dengue outbreak occurred in 2014 (Wang and Nishiura, 2021).

- **West Nile Virus (WNV)**: WNV distribution across the USA 2002–2008 (see Sugumaran et al., 2009 Figure 6.6).

Although cases of imported vector-borne diseases such as dengue, chikungunya and malaria are routinely recorded in countries where the disease is not prevalent, many of these **do not** result in local outbreaks. In the case of mosquito-vectored diseases, for the disease to be transmitted, a mosquito-vector is required. Therefore, knowing the distribution and changes in distribution of highly competent vectors of a disease is equally important. This includes the establishment of vectors in novel areas, since these too can raise public health concerns. For example, *Aedes albopictus*, the Asian tiger mosquito, a potential vector of dengue fever virus (DENV) (Mitchell, 1995), chikungunya, West Nile and La Crosse virus (Bonilauri et al., 2008; Paupy et al., 2009), has spread to new areas through the transportation

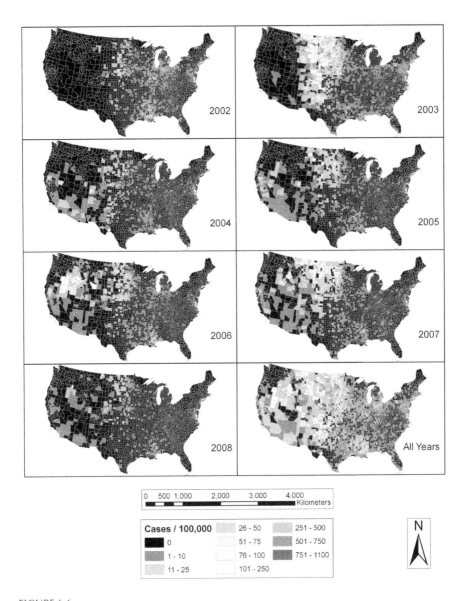

FIGURE 6.6

WNV human incidence by year. Distribution of WNV in the US from 2002–2008. Each panel shows the number of human WNV cases per 100,000 people for a single year or for all of the years combined. *Source*: Sugumaran et al. (2009).

of goods, particularly used tires (Hawley et al., 1987), and international travel (Benedict et al., 2007; Weaver and Reisen, 2010). Local populations have now become established in new locations (e.g. Pennsylvania, USA (Taber et al., 2017)).

Mosquito-Vectored Diseases

There are several economically important diseases vectored by mosquitoes many of which have emerged in new locations. If we are to map the risk of these vectored diseases knowing something about the biology of the system (e.g. host-pathogen-environment is useful).

An important driver of many mosquito-vectored diseases is temperature, as it affects many life history traits (e.g. population dynamics (rate of development) and biting rates), as well as the ***Extrinsic Incubation Period (EIP)*** – the length of time it takes for the pathogen (e.g. the malaria parasite) to complete development within the mosquito from initial acquisition via an infected blood meal to the point at which it can be transmitted to another host via a further blood meal.

West Nile Virus

The West Nile virus (WNV) was first identified in New York City in 1999 and has subsequently spread across the USA, reaching the west coast in 2004 (Figure 6.6, Sugumaran et al., 2009). The West Nile virus is now endemic in the USA.

WNV is maintained in the natural environment through a transmission cycle between mosquitoes and birds (Weaver and Barrett, 2004), http://www.cdc.gov/westnile/transmission/. Several *Culex* mosquito species have been implicated as important vectors of WNV in North America, including *Culex pipiens, Cx restuans, Cx salinarius* (mainly found in the Northeastern States) (Andreadis et al., 2004; Darsie and Hutchinson, 2009; Mackay et al., 2008), *Cx quinquefasciata* (Southern States) (Gibbs et al., 2006; Godsey et al., 2005) and *Cx tarsalis* (Western States) (Goddard et al., 2002). Although these mosquitoes predominantly feed on birds such as the American robin, they also feed on other hosts, thus acting as bridge vectors, causing periodic disease "spillovers" into humans and other "dead end" hosts such as horses (Figure 6.7).

Temperature affects virus replication. WNV can take as short as 7 days at temperatures of 30°C and as long as ~30 days at cooler temperatures of 20°C (Reisen et al., 2006) (Figure 6.9). Thus, the potential risk of WNV transmission increases significantly with increasing temperatures. Similar effects of temperature on parasite/virus incubation have also been documented with other mosquito diseases (e.g. malaria: Craig et al., 1999; Paaijmans et al., 2009; Mordecai et al., 2013; dengue: Lambrechts et al., 2011; Watts et al., 1987). Temperature also affects various aspects of mosquito life history and overall vectorial capacity via impacts on the rate of development (e.g. *Ae albopictus*: Delatte et al., 2009; *Cx restuans* and *Cx pipiens*: Gong et al., 2011; survival: Loetti et al., 2008; Reisen, 1995; Rueda et al., 1990 for *Culex*; and feeding rate/gonotrophic cycle: Eldridge, 1968; Reisen et al., 2010 for *Culex*).

In summary, the epidemiologic triad of WNV is summarized in Figure 6.8.

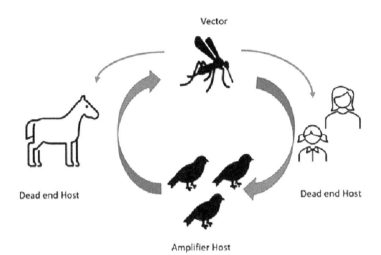

FIGURE 6.7
West Nile virus transmission cycle. Image created by Blanford (2023) adapted from (CDC).

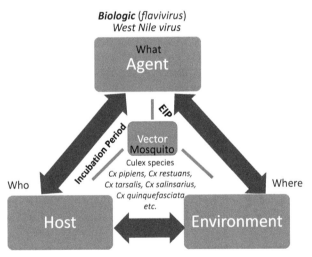

Birds
Dead-end hosts (horses)
Humans Risk and Exposure behavior (Activity space)

Temperature affects the life-history of mosquito; affects Extrinsic Incubation Period (EIP)
Rainfall, Irrigation, Water bodies for breeding sites
Land use habitats and hosts

FIGURE 6.8
The key agent, environmental factors and hosts important in the epidemiologic triad for West Nile virus in the USA.

Identify Areas of Risk through the Integration of Temperature and Surveillance Data

Now that you have an understanding of the factors important in the disease dynamics of WNV, let's see how this information can be used to identify areas of risk. To do so I draw on the mosquito-borne virus response plan used in California to minimize WNV risks in the human population (MVCA_California, 2023). In this example, areas at risk of WNV are identified through the integration of a number of different factors important in the transmission dynamics of WNV. These include temperature, mosquito populations, mosquito infection rates and hosts infected (surveillance of chickens) with the virus or killed by the virus (dead bird infections) (Figure 6.9b). Infection rates are estimates of the number of infected mosquitoes per 1,000 tested. The minimum infection rate (MIR), is calculated: ([number of positive pools / total specimens tested]×1000), with the data representing a single species or species group collected over a time period and geographic area (CDC, 2024a). The MIR uses the assumption that a positive pool contains only one infected mosquito, an assumption that may be invalid. Calculations can be done using the PooledInfRate package (https://github.com/CDCgov/PooledInfRate) (CDC, 2024a).

Data for mosquitoes and infection rates are collected through surveillance and may include mosquitoes, sentinels and dead birds, as well as the number of human cases. Surveillance data can be combined with temperature data using an Multi-Criteria Decision Analysis Method (MCDA) that captures the expert knowledge about a system. Each factor is ranked between 1 and 5, where 1=low risk and 5=high risk (Figure 6.9b). The breakdown of each risk level corresponds to the theoretical understanding of the host-pathogen interaction in the environment (Figure 6.9a vs Figure 6.9b). Thus, in areas where temperatures are not conducive to EIP (the time it takes for the mosquito to acquire the virus and then be able to transmit it), mosquito populations and MIRs are low and the risk of WNV transmission is also low. The method outlined here is used in California to determine when vector control is needed in real-time so as to minimize the risk of human infections.

GIS has been used to map and model different mosquito vectored diseases at different scales, ranging from household and village levels (e.g. Parker et al., 2019; Parker et al., 2023; Manning et al., 2021) to global (e.g. Hay et al., 2009; Gething et al., 2012), as well as examine the distribution of the vector (e.g. invasion of the *Aedes albopictus* in Pennsylvania (Taber et al., 2017) or worldwide (Mordecai et al., 2017) for *Aedes albopictus* or *Aedes aegypti*); distribution of anophelines (Kyalo et al., 2017) and malaria vectors (Kiszewski et al., 2004) . Due to on-going surveillance, Finland has just detected a new *Culex* species (potential vector of WNV) (Brown, 2023).

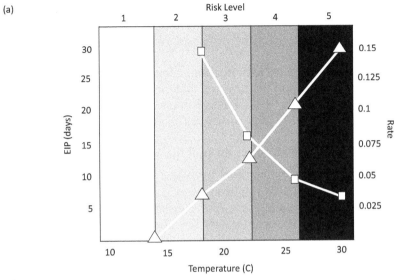

FIGURE 6.9
(a) Risk levels for WNV related to temperature. Risk levels increase as EIP decreases and the rate of virus development (1/EIP) increases (Reisen et al., 2006). (b) Summary of thresholds for individual surveillance indicators used in calculating the California Mosquito-Borne Virus Surveillance and Response Plan risk levels. *Source:* Barker et al. (2010).

Risk Level	Temperature (F) (daily average)	Abundance (adult mosquito) 5 yr average	Mosquito MIR/1,000	Chicken seroconversions	Birds (dead bird infections)	Human Cases
1	<50	<50%	0	0 in region	0 in region	
2	57-65	50-90%	0.1-1.0	>= 1 in region, 0 in agency	>= 1 in region, 0 in agency	
3	66-72	91-151%	1.1-2.0	1 flock in agency	1 in agency	>= 1 in region, 0 in agency
4	73-79	151-300%	2.1-5.0	2 flocks in agency	2-5 in agency	1 in agency
5	>79	>300%	>5.0	> 2 flocks in agency	> 5 in agency	> 1 in agency

Ticks

Tick-vectored diseases can have devasting health consequences (Mac et al., 2019) and be costly. Treatment can range between 10.1 and 20.1 billion (Euro) in humans (Davidsson, 2018) and be equally as devasting in animals, particularly in economically important livestock (e.g. in cattle in Tanzania economic losses exceed US$ 346 million (Kivaria, 2006); in dairy production systems in India estimates are US$ 787 million (Singh et al., 2022)). With increasing resistance to acaricides (Githaka, et al., 2022) the economic burden and health consequences will no doubt continue to be problematic. More work is clearly needed to not only fully grasp the health and economic consequences of tick-vectored diseases but also the distribution of ticks and the pathogens they vector.

In the meantime, we can try to better understand the spatial distribution of the different tick species and the diseases that they transmit so as to improve our understanding of where risks are (e.g. maps of bacteria infected ticks in France (Jumpertz et al., 2023) and Australia (Egan et al., 2021).

Important Tick Vectored Diseases around the World

A variety of diseases are transmitted by ticks (see Parola et al., 2001, 2013; Boulanger et al., 2019). To highlight a few: Crimean-Congo haemorrhagic fever, Lyme disease, Relapsing fever (borreliosis), Rickettsial diseases (spotted fever and Q fever), Tick-borne encephalitis and Tularaemia (WHO, 2016b). These can be viruses, bacteria and parasites, Table 6.3. It is not possible to go into the details about all of these diseases, however, the information provided over the next few pages should give you some insights into different tick-vectored disease systems around the world.

South East Asia

Knowledge of ticks in Southeast Asia is patchy. To date, 97 species have been recorded, making it one of the most diverse worldwide. Work on tick-borne diseases of stock and companion animals, as well as of humans, is in its infancy, and the medical, veterinary and socio-economic importance of these diseases is largely unknown (Tan et al., 2021; Petney et al., 2019).

Central and South America

Similar to Southeast Asia, knowledge of ticks in South and Central America is patchy. A few articles have captured some of the key species in Central America (Bermúdez et al., 2022) and South America (Guglielmone et al., 2006), but more work is needed.

Africa

In Africa, many tick species are prevalent (e.g. Table 6.4, Figure 6.10). They mainly affect livestock and other wildlife, as documented by Madder, Horak and Stoltsz (unknown) (Table 6.4, Figure 6.10). However, Ixodes ticks are also important vectors of several organisms causing diseases in humans in sub-Saharan Africa. These are *Rickettsia conori*, the cause of tick bite fever or tick typhus; *Coxiella burneti*, the cause of Qfever; and the virus causing Crimean-Congo Haemorrhagic fever. In addition, argasid ticks of the *Ornithodoros moubata* complex can transmit *Borrelia duttoni*, the cause of tick-borne relapsing fever (TBRF), to humans (Madder et al. unknown) (e.g. in Tanzania (Talbert, 2005)).

TABLE 6.3

Overview of Human Infectious Diseases Transmitted by Ticks around the World

Disease(s)	Infectious Agents	Vector Ticks	Geographical Distribution
Virus (arbovirus)			
Tick-borne encephalitis	*Flavivirus*	*I. ricinus; I. persulcatus*	Asia, Europe
Crimean-Congo hemorrhagic fever	*Nairovirus*	*Hyalomma marginatum*	Europe, Asia, Africa
Bacteria			
Q fever or coxiellosis	*Coxiella burnetii*	*Rhipicephalus* spp. *Dermacentor* spp.	Worldwide
Lyme borreliosis	*Borrelia burgdorferi* sensu lato	*Ixodes* spp.	Northern Hemisphere
Borrelioses/tick-borne recurrent fever	*Borrelia* spp.	*Ornithodoros* spp; *Ixodes* spp.	Europe, Northern America, tropical and subtropical regions
Mediterranean spotted fever	*Rickettsia conorii*	*R. sanguineus*	Africa, Asia, Europe
Tick-borne African fever	*Rickettsia africae*	*Amblyomma* spp.	Sub-Saharan Africa
Tick-borne lymphadenopathy (TIBOLA)	*Rickettsia slovaca, R. raoultii, R. rioja*	*Dermacentor* spp.	Europe
Tularemia	*Francisella tularensis*	Various genera of ticks	Worldwide
Human granulocytic anaplasmosis	*Anaplasmataceae phagocytophilum*	*I. ricinus; I. pacificus; I. scapularis*	Europe, Northern America, Russia
Parasites			
Babesioses	*Babesia divergens, B. microti, B. venatorum*	*I. ricinus*	Europe

Source: Parola et al. (2013); Boulanger et al. (2019).

TABLE 6.4

Summary of Tick Species, Diseases They Transmit in Africa

Vector	Disease	Causative Agent	Animals Affected
A. hebraeum	Heartwater	*Ehrlichia ruminantium*	Cattle, goats, sheep
	Benign bovine theileriosis	*Theileria mutans, T. velifera*	Cattle
	Foot abscesses		Goats
A. variegatum	Heartwater	*Ehrlichia ruminantium*	Cattle, goats, sheep
	Benign bovine theileriosis	*Theileria mutans, T. velifera*	Cattle
	Bovine anaplasmosis	*Anaplasma bovis*	Cattle
	Nairobi sheep disease	Nairobi sheep disease	Sheep
	Bovine dermatophilosis	virus *Dermatophilus congolensis*	Cattle, sheep, goats, horses
H. dromedarii	Oriental theileriosis	*Theileria annulata*	Cattle
H. truncatum	Sweating sickness	Toxin	Calves
	Equine piroplasmosis	*Babesia caballi*	Horses, donkeys,
	Toxicosis	Crimean-Congo	mules
	Crimean-Congo	Haemorrhagic fever	Dogs
	Haemorrhagic fever	(CCHF) virus	Humans
	Boutonneuse fever	*Rickettsia conorii*	
I. rubicundus	Spring lamb paralysis	Toxins	Lambs
R. microplus	African redwater	*Babesia bigemina*	Cattle
	(babesiosis)	*Babesia bovis*	Cattle
	Asian redwater	*Anaplasma marginale*	Cattle
	(babesiosis)	*Borrelia theileri*	Cattle, sheep,
	Galsickness		goats, horses
	(anaplasmosis)		
	Borreliosis/spirochaetosis		
R. decoloratus	African redwater	*Babesia bigemina*	Cattle
	(babesiosis)	*Anaplasma marginale*	Cattle
	Galsickness	*Borrelia theileri*	Cattle, sheep,
	(anaplasmosis)		goats, horses
	Borreliosis/spirochaetosis		
R. appendiculatus	East Coast fever, Corridor	*Theileria parva*	Cattle, African
	disease	*Theileria taurotragi*	buffalo
	Benign bovine theileriosis	*Anaplasma bovis*	Cattle, eland
	Bovine anaplasmosis	Nairobi sheep disease	Cattle
	Nairobi sheep disease	virus Toxins	Sheep, goats
	Tick toxicosis		Cattle, antelope
R. evertsi evertsi	Ovine theileriosis	*Theileria separata/T. ovis*	Sheep
	Equine babesiosis	*Babesia caballi*	Horses, donkeys,
	Equine theileriosis	*Theileria equi*	mules
	Galsickness	*Anaplasma marginale*	Horses, donkeys,
	(anaplasmosis)	*Borrelia theileri*	mules
	Borreliosis/spirochaetosis	Toxins	Cattle
	Tick paralysis		Cattle, sheep, goats, horses Lambs, dogs
R. zambeziensis	East Coast fever, Corridor	*Theileria parva*	Cattle, African
	disease	*Theileria taurotragi*	buffalo
	Benign bovine theileriosis	*Anaplasma bovis*	Cattle, eland Cattle
	Bovine anaplasmosis		

Source: Madder et al. (unknown).

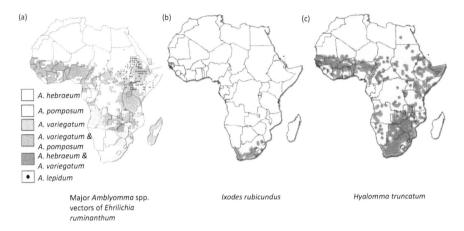

FIGURE 6.10

Tick distributions across Africa for *Amblyomma* spp. vectors, *Ixodes rubicundus and Hyalomma truncatum. Source*: Madder et al. (unknown) available through Creative Common License.

Europe

In Europe there are a number of tick-vectored diseases. These include tick-borne encephalitis and Lyme borreliosis which are vectored by *Ixodes Ricinus; Ixodes persulcatus* (Figure 6.11a, b) (ECDC 2011, 2014, 2023d,e), Mediterranean spotted fever vectored by *Rhipicephalus sanguineus* (Figure 6.11c) (Kubiak et al., 2024; ECDC 2023f) and Crimean-Congo hemorrhagic fever vectored by *Hyalomma marginatum* (Figure 6.11d) (ECDC 2023b,c).

North America

In the USA, 15 different diseases are transmitted to humans (CDC, 2016) by 7 different tick species distributed throughout the country (Figure 6.12). Of these, Lyme disease (Figure 6.13) is the most reported with over 30,000 cases reported annually (CDC, 2024b).

Between 2017 and 2021, nearly half a million people in the US were affected by a potentially life-threatening red meat allergy caused by the saliva of a lone-star tick (Thompson et al., 2023). The CDC issued a warning over alpha-gal syndrome (AGS), a red-meat allergy primarily associated with lone-star tick saliva (Thompson et al., 2023). Despite the growing number of individuals affected by AGS, many healthcare providers have low awareness of AGS (Carpenter et al., 2023), with 42% of doctors having never heard of AGS and 35% reporting to be "not too confident" in their ability to diagnose AGS.

(a) *Ixodes ricinus*

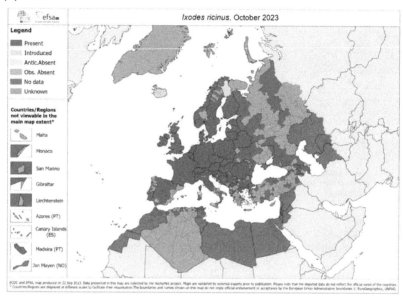

(b) *Ixodes persulcatus*

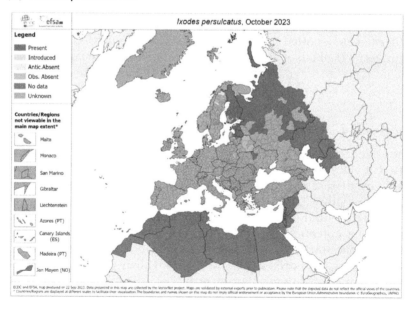

FIGURE 6.11
Distribution of a variety of tick species in Europe 2023. Maps obtained from ECDC (2023g).
(a) *Ixodes ricinus*, and (b) *Ixodes persulcatus* vectors of Lyme and tick-borne encephalitis (TBE),
(c) *Rhicephalus sanguineus* a vector of Mediterranean spotted fever, (d) *Hyalomma marginatum* a
vector of Crimean-Congo hemorrhagic fever.

(Continued)

(c) *Rhicephalus sanguineus*

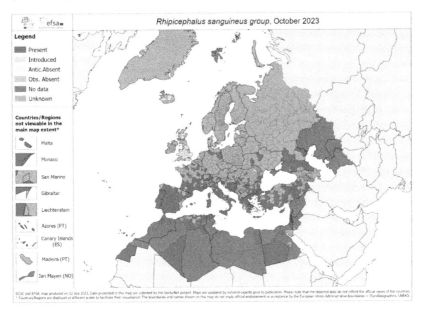

(d) *Hyalomma marginatum*

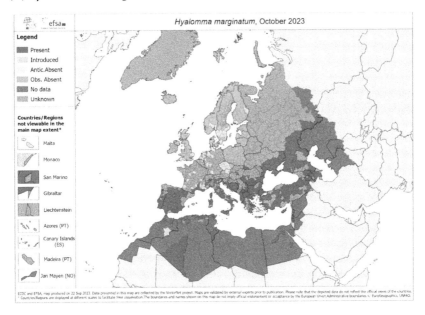

FIGURE 6.11 (*Continued*)
Distribution of a variety of tick species in Europe 2023. Maps obtained from ECDC (2023g).
(a) *Ixodes ricinius*, and (b) *Ixodes persulcatus* vectors of Lyme and tick-borne encephalitis (TBE),
(c) *Rhicephalus sanguineus* a vector of Mediterranean spotted fever, (d) *Hyalomma marginatum* a
vector of Crimean-Congo hemorrhagic fever.

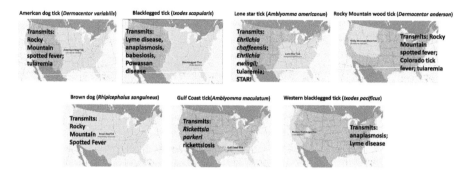

FIGURE 6.12

Geographic distribution of tick species and the diseases they may transmit in the United States of America. *Source*: Maps obtained from CDC (2023b).

FIGURE 6.13

Lyme disease reports in the USA during 2001 and 2019. *Source*: CDC.

Lyme Disease

Although Lyme disease has been linked to ticks since 1978, the causative agent remained unknown until 1982 (Stone et al., 2017). "Lyme disease" was named after the town of Lyme, Connecticut, where the disease was originally reported (Stone et al., 2017). Lyme disease, an infection caused by a corkscrew-shaped spirochete bacterium (*Borrelia burgdorferi*) (CDC, 2013). The bacteria is transmitted to humans through the bite of an infected tick belonging to the genus *Ixodes*, also known as the blacklegged tick or deer tick.

Once the tick acquires the *Borrelia* bacteria while feeding on an infected host (e.g. adult ticks feed on deer, dogs and cattle in fall and spring (Burgdorfer and Keirans, 1983); nymphs feed on squirrels; and larvae feed on mice, chipmunks, voles, etc. (Radolf et al., 2012; Werden et al., 2014)), they in turn can transmit the bacteria to other hosts, including humans (Todar, 2012).

Lyme disease is reported throughout the USA (Figure 6.12). The number of reported human cases is highest during the summer months and lowest in the winter months. This corresponds with the life cycle of the tick and also with increased outdoor activity. Lyme disease is spreading (Figure 6.13) due to migratory birds and deer, changing ecosystems (e.g. development and reforestation) (LRA, n.d.), changes in climate (Ogden et al., 2014) and range expansion of tick hosts (e.g. mice and deer) (Roy-Dufresne et al., 2013). Not only are tick-borne illnesses increasing in the USA, but there has also been an increase reported in northern Europe (Githeko et al., 2000).

Data

Now that you have an understanding of the factors influencing ticks, the next step is to find data that can be used to map and model disease risk. For Lyme disease, surveillance data and health-reported cases can be useful.

- **Surveillance information**: Different types of surveillance can be used for collecting data and for understanding risk and changes in risk. These can include surveillance of the vector, animals (canine, wildlife and livestock) and humans (e.g. Figure 6.9a,b for WNV).

- **Veterinary surveillance**: Information from pets are useful. They can also act as "sentinels" or indicator of Lyme disease prevalence in the area due to their frequent outdoor exposure (Guerra et al., 2001; Lindenmayer et al., 1991). Positive correlations between canine seropositive rates and human incidence rates have been found (Guerra et al., 2001).

- **Health reporting**: Lyme disease is a notifiable disease in the USA. Data at the county level can be obtained from the CDC and are available as reported cases of human incidence (e.g. case data by county can be obtained from CDC https://www.cdc.gov/lyme/datasurveillance/lyme-disease-maps.html)

- **Citizen science**: With the availability of apps, health officials and researchers are gaining additional insights into risk behaviours, and where bites are occurring from citizens (e.g. Tick Apps – Fernandez et al., 2019; Antonise-Kamp et al., 2017).

- **Environmental data**: Since ticks are also governed by the environment, a variety of environmental data may also be useful for creating maps delineating tick habitats or areas with suitable environmental

TABLE 6.5

Key Factors Important to Tick Survival and Habitat Suitability

Tick Species	Geographic Location	Factors	Source
A. americanum	S. Missouri, Texas	Forest, humidity, vegetation	Brown et al. (2011); Texas_A&M_AE (n.d.)
I. scapularis	N. America	Negative association: urban, wetlands, saturated soils	Glass et al. (1994)
I. scapularis	N. Eastern USA Maryland, Wisconsin, Illinois, Michigan, Massachusetts, Connecticut; Canada	Temperature, precipitation, vapour pressure, land cover, deciduous forest, leaf litter, deer abundance, small mammal richness and abundance, canopy cover	Brownstein et al. (2003); Kitron et al. (1991); Brownstein et al. (2005); Diuk-Wasser et al. (2012); Githeko et al. (2000); Glass et al. (1994); Guerra et al. (2002); Guerra et al. (2001); Kitron and Kazmierczak (1997); Lindenmayer et al. (1991); Moore et al. (2014); Ogden et al. (2014); Roy-Dufresne et al. (2013); Werden et al. (2014)
Ixodes ricinus	Europe, Finland	Vegetation, host availability, urban, temperature	Estrada-Pena (2008); Junttila et al. (1999)
B. microplus (Ixodidae)	S. America	Temperature, humidity	Estrada-Peña (1999)

Source: Williams (2014); Kopsco et al., (2022).

conditions for the survival of ticks. Some of these may include land use/land cover, the Normalized Difference Vegetation Index (NDVI) (Randolph, 2000), climate (e.g. temperature and rainfall) and soil type. Review the factors listed in Table 6.5.

Identifying Potential Risk Areas through Habitat Mapping of the Vector

Mapping of tick-vectored diseases has included creating dot maps of reported clinical cases of a disease (Figure 6.13) or mapping the distribution of tick-species known to vector particular diseases (Figure 6.10, 6.11a–d, 6.12). Key factors important to tick survival are summarized for different species in Table 6.5 and include land use and land cover, temperature and moisture (rainfall) Additional variables that have been used in studies include the NDVI, elevation, soil type, water availability, food and timber demand, urban land use (Kopsco et al., 2022).

Based on the background information provided earlier, the epidemiologic triad for Lyme disease in the USA can be summarized as follows: The bacteria is vectored by several tick species; tick populations are driven by environmental factors such as temperature, precipitation, soil moisture, vegetation type and the presence of suitable hosts (e.g. large, medium and small mammals). Humans are an accidental host and become exposed to the disease due to their behaviour (e.g. hiking, camping, dog-walking, gardening and land use change) when passing through an environment infested with ticks or when a host comes into the human's habitat.

Response (Control and Prevention, Recovery and Elimination)

Once areas of risk can be identified, different strategies can be used to reduce disease transmission. This can occur in multiple ways as follows:

- **Self-protection** through altered human behaviour by avoiding risky environments, applying repellents and/or taking prophylactics (where applicable) and being vaccinated when applicable.
- **Vector control** through manipulation of habitats (e.g. reducing favourable tick habitats), targeted spraying of insecticides and/or biocontrol agents and manipulation of host populations.
- **Education, awareness and communication** educating and communicating to the public where and when risks may occur and what to do if you do find a tick or are bitten, as highlighted in the alert by the CDC (Figure 6.14).

Recovery may result in improved access to health care facilities that test and treat the disease. Improving diagnostic testing and treatment of severe cases. Availability of necessary treatment in a timely manner. More on this a little later, when we tackle accessibility to health facilities.

Elimination is a little more difficult and no doubt will require a variety of approaches, as demonstrated by the elimination of malaria in Suriname (Hiwat et al., 2018; van Eer et al., 2018; PAHO, 2023) where four main strategies were used (Malaria_Program_Suriname):

- Improved access to diagnosis, treatment and case investigation
- Prevention in risk areas with long-lasting impregnated mosquito nets
- Awareness-building to change behaviour
- National and regional partnerships

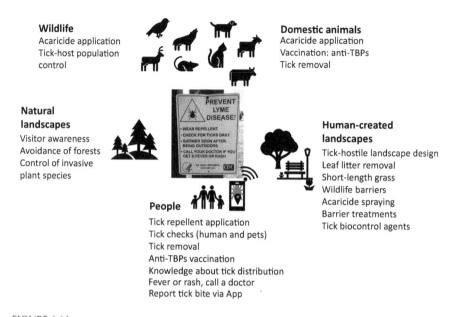

Wildlife
Acaricide application
Tick-host population
control

Domestic animals
Acaricide application
Vaccination: anti-TBPs
Tick removal

Natural landscapes
Visitor awareness
Avoidance of forests
Control of invasive
plant species

Human-created landscapes
Tick-hostile landscape design
Leaf litter removal
Short-length grass
Wildlife barriers
Acaricide spraying
Barrier treatments
Tick biocontrol agents

People
Tick repellent application
Tick checks (human and pets)
Tick removal
Anti-TBPs vaccination
Knowledge about tick distribution
Fever or rash, call a doctor
Report tick bite via App

FIGURE 6.14
Intervention strategies for managing ticks in wildlife, domestic animals, natural landscapes, human-created landscapes and through people. Compiled from Černý et al. (2020); CDC (2023a); Image of Lyme disease warning sign posted at a popular hiking trail entrance in Pennsylvania. Image credit Blanford (2016).

In Suriname, the Malaria Program is dealing with a population at risk estimated to be around 80,000 people, mostly living in and around the gold mining areas.

In the last decade, with support from the Global Fund and applying the best practices generated by the AMI-RAVREDA network (USAID-PAHO), Suriname has reached the Roll Back Malaria and the Millennium Development Goals for Malaria. The Annual Parasitic Index has dropped from circa 88 in 2004 to 1.06 in 2015. This means that currently the annual number of cases is nearing the elimination level of 1 case per 1,000 people at risk.

Malaria has practically been eliminated in the stable villages of the interior, which previously had the highest transmission rates in the Americas, as captured in Figure 6.15 which shows changing distribution of malaria in 2000, 2010, and 2020. Malaria in Suriname has decreased to less than 90 autochthonous (nationally transmitted) cases a year. Hospital admissions for malaria have decreased enormously (by 97%) from 377 in 2003 to 11 in 2015. Deaths due to malaria have also dropped from 24 in 2000 to 1 in 2013. In 2014 and 2015, no deaths were recorded.

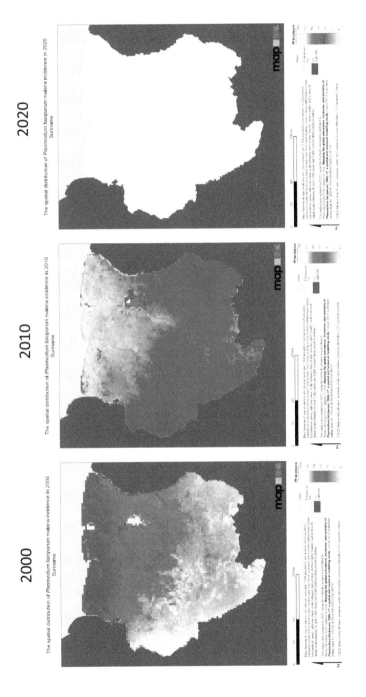

FIGURE 6.15
The spatial distribution of *Plasmodium falciparum* malaria incidence in Suriname 2000–2020. *Source:* Malaria Atlas Project.

Mapping Vectors of Disease and Ecology of Disease: Different Approaches

Now that you know a bit more about different vector-borne diseases, the next step is to map the spatial and/or temporal distribution and risk of the vectors and/or the diseases they transmit. Depending on the data that is available and the resolution and scale at which it is available will determine what approaches to take.

The approach used may be more data-driven or model-driven. A *data-driven approach* is one where information is derived from the data without a prior notion of what the theoretical framework should be (Anselin, 1989). In other words, one lets the "data speak for themselves" (Gould, 1981) and attempts to derive information on spatial pattern, spatial structure and spatial interaction without the constraints of a pre-conceived theoretical notion. A *model-driven approach* is one which will start from a theoretical specification and is subsequently confronted with the data.

Approaches may be data-driven or model-driven depending on what data is available at the time of analysis, what is already known about the system being mapped/modelled and what needs to be known. This can range from mapping *where* and *when* a vector or disease is distributed, evaluating control and response efforts, determining who is at risk or exploring what interventions are needed.

Data-Driven Approach

Mapping a disease or vector with known occurrence of the disease or vector. The information may be represented by a point (x, y) or an area (administration boundary level (county, township, etc.), zip code/postcode) or some other polygon-based feature.

Niche models, species distribution models (SDMs) or climate envelopes, are useful for mapping the spatial distribution of a species or disease using known locations of where that disease or species has been reported. These essentially act as a training dataset from which predictions are made.

A variety of tools have been developed that use different algorithms, statistical and/or "machine learning" methods. These can be classified as "profile," "regression" and "machine learning" methods (Hijmans and Elith, 2019). Profile methods only consider "presence-only" data, not absence or background data. Regression and machine learning methods use both presence and absence data.

Profile methods include the following:

- **BIOCLIM** (Booth et al., 2014) and **DOMAIN** (Carpenter et al., 1993) are classic "climate-envelope-models" that have been extensively used for species distribution modelling and are available through the "**dismo**" packages in R. These models use environmental distance and presence-only data to find areas of similarity.

- **Similarity Index** (ArcGIS 10.2 and above) – Compares the z-score of presence only data to identify areas of similarity.

Regression-based methods include the following:

- **GLM** uses general linear modelling methods or mixed-model linear models
- **Kriging** or geostatistical interpolation methods use presence only data to generate an estimated surface from a scattered set of points with z-values.

Machine learning methods include the following:

- **MaxEnt** uses maximum entropy methods with presence only data to predict locations of a disease or species (Phillips et al., 2008; Phillips et al., 2017).
- **GARP** modelling (Illoldi-Rangel et al., 2004; Stockwell, 1999)
- **Support Vector Machine (SVM)** algorithm for predicting multivariate or structured outputs

Below are some examples of how these methods have been used and continue to be used.

- **MaxEnt**: Climate Change and Range Expansion of the Asian Tiger Mosquito (*Aedes albopictus*) in Northeastern USA: Implications for Public Health Practitioners (Rochlin et al., 2013); Climate change effects on chikungunya transmission in Europe: geospatial analysis of vector's climatic suitability and virus' temperature requirements (Fischer et al., 2013; Tjaden et al., 2021); Schistosomiasis (Scholte et al., 2012). **Similarity index**: Mosquito habitat and dengue risk potential in Kenya (Attaway et al., 2014).
- **Bioclim**: A species distribution modelling package (Booth et al., 2014).
- **GARP**: A species distribution modelling method used to map current and future potential wintering distributions of eastern North American monarch butterflies (Oberhauser and Peterson, 2003).
- **SVM**: Support vector machines has been used for predicting distribution of Sudden Oak Death in California (Guo et al., 2005).

Model-Driven Approach

Model-driven approaches are based on a theoretical understanding of the disease or distribution of a vector. These can be biological and based on laboratory and/or field experiments and observations or theoretical, where model simulations are used to fit patterns of observations. Environmental data are integrated with the models to make predictions about the distribution of the vector or disease. These types of outputs are useful for capturing the distribution of the disease while also validating theoretical understanding.

Different approaches can include the following:

- **Models**. Mathematical and/or statistical models derived from laboratory and/or field observations.
- **Multi-criteria decision analysis (MCDA) method** (also known as surface overlay methods or suitability analysis). In this method, the user is able to combine different criteria, such as environmental information, socio-economic information based on expert knowledge. The way the information is combined or ranked indicates the strength and importance of membership in a set (Malczewski and Rinner, 2015; Raines et al., 2010).
- **Simulation or dynamic models**. Many systems are complex and dynamic. In such cases, geocomputational modelling approaches with different time and evaluation steps can be programmed (e.g. Blanford et al., 2013).

Summary

During this chapter, you learned about different vector-borne diseases and some of the approaches that can be used to map and model them. Now it is your turn to create a risk map for a vector-borne disease.

Activity – Mapping Malaria

Malaria remains a major public health burden with around 247 million cases reported worldwide in 2021 and 619,000 deaths (WHO, 2023). Maps showing the global distribution of malaria provide a geographical framework for monitoring malaria incidence as well as the evaluation of the impact on malaria control worldwide. In 2009, the Malaria Atlas Project (MAP) produced a map showing the global distribution of malaria endemicity and now produces regular global updates (Gething et al., 2012; Weiss et al., 2019; Hay et al., 2009).

Malaria in humans is caused by five *Plasmodium* parasite species. These include *Plasmodium falciparum, P. vivax, P. ovale, P. malariae* and *P. knowlesi* (WHO, 2023). *P. falciparum* and *P. vivax* are the biggest threats. *P. falciparum* is

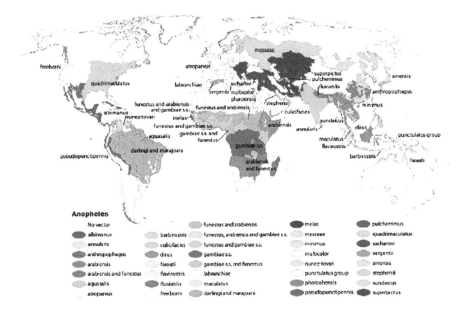

FIGURE 6.16
Distribution of malaria vectors around the world. *Source*: Kiszewski et al. (2004).

the deadliest malaria parasite and the most prevalent on the African continent, while *P. vivax* is the dominant malaria parasite in most countries outside of sub-Saharan Africa. *Plasmodium* parasites are transmitted through the bite of female *Anopheles* mosquitoes (WHO, 2023). The key vector is the *Anopheles* mosquito, of which there are many different species distributed around the world (Figure 6.16).

Temperature and rainfall play an important role in the dynamics and distribution of malaria and the mosquito vector that transmits the parasite.

Malaria – More Than Just an Annoying Mosquito?

In reality, malaria is much more than an annoying mosquito, as captured in the malaria cycle (Figure 6.17). Malaria is governed by a variety of factors that influence the vector, the parasite and the hosts. The environment and the interactions that take place within the environment influence the outcome of malaria. This involves exposure to the vector and the host. What part of the cycle is mapped depends on what data is available.

For example, A (Figure 6.17), – Mosquito-Parasite-Environment: Essentially when the female mosquito takes a blood feed she ingests the gametocytes of the parasite. The parasite develops inside of the

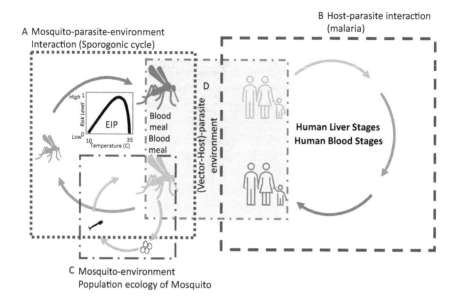

FIGURE 6.17

A simplified illustration of the malaria parasite life cycle adapted from (CDC). During a blood meal, a malaria-infected female *Anopheles* mosquito inoculates sporozoites into the human host. Image created by Blanford (2023).

mosquito (sporogonic cycle) and after 6–18 days, the parasites are found (as "sporozoites") in the mosquito's salivary glands. When the *Anopheles* mosquito takes a blood meal on another human, the sporozoites are injected with the mosquito's saliva and start another human infection when they parasitize the liver cells. The time it takes to complete the sporogonic cycle can vary and is largely driven by temperature. Thus, EIP – the time it takes for the mosquito to transmit the parasite can be as quick as 6–10 days at temperatures around 28°C–30°C or take in excess of 50 days at temperatures around 16°C with no development below 16°C (Craig et al., 1999; Paaijmans et al., 2009; Paaijmans et al., 2010; Mordecai et al., 2019; Mordecai et al., 2013).

B (Figure 6.17) – the parasite resides in the human host (e.g. malaria cases; seroprevalence).

C (Figure 6.17) – the environment suitable for mosquitoes to survive and increase in population (e.g. habitats such as waterbodies, land use, etc.)

D (Figure 6.17) – factors that capture mosquito and human interactions. These can be socio-economic related factors (e.g. house type, income levels, education, accessibility to health facilities, land use, distance to water and other breeding areas for specific mosquito species) or behaviour related.

Mapping Malaria – Mapping Malaria Risk Using Different Methods

Many different geospatial methods have been used to map malaria, ranging from suitability analysis (Figure 6.18a) and Bayesian geostatistical methods (Figure 6.18b) to geocomputational host-pathogen-environment models (Figure 6.18c). A and C capture the Mosquito-Parasite-Environment interactions (A in Figure 6.17) using biological and environmental information and B using clinical data to capture the Host-parasite interaction (B in Figure 6.17).

MARA/AMRA. The Mapping Malaria Risk in Africa (MARA) Collaboration (Atlas du Risque de la Malaria en Afrique – ARMA) was initiated in 1996 to provide an accurate atlas of malaria risk for sub-Saharan Africa. The main aim of the project was to generate malaria risk maps that could be used to implement malaria control activities across the continent (MARA, n.d.). The maps created by MARA represent the theoretical distribution of malaria based on climate suitability of the vector and parasite derived from long-term climate averages of precipitation and temperature. Thus, the maps portray the climate suitable for stable malaria transmission (e.g. Figure 6.18a) (Craig et al., 1999).

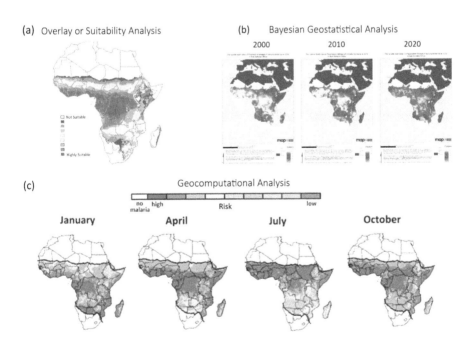

FIGURE 6.18

Different approaches for mapping malaria risk. (a) Suitability analysis (e.g. Craig et al., 1999); (b) Bayesian geostatistical methods (Weiss et al., 2019; Hay et al., 2009); (c) Geocomputational methods coupled with host-pathogen-environment models (Blanford et al., 2013).

To create the maps, a set of decision rules that capture the biological development constraints of the parasite and vector were derived by reviewing published laboratory and field studies throughout Africa. A model was developed that uses three variables: mean monthly temperatures, winter minimum temperatures and total monthly precipitation. Areas of stable transmission were determined by identifying areas where temperature and precipitation are favourable at the same time of the year and continue long enough, 5 months in length. Data were combined using a fuzzy logic model where values of 0=not suitable and 1=very suitable. This form of spatial analysis is also known as multi-criteria decision analysis, surface overlays or suitability analysis.

Response – Intervention and Recovery from Malaria

What types of interventions are necessary to reduce the risk of malaria?

Many approaches are used to control malaria, some targeting the parasite, some targeting the vector and some targeting the human, as summarized in Table 6.6 (WHO, 2022).

In 2023, a new malaria vaccination was released and has been rolled out to 12 African countries since July (Ahmed, 2023).

Overview of Mapping Malaria Risk in Africa

Now it is your turn to create a malaria risk map using a multi-criteria decision analysis, suitability analysis modelling approach or a geostatistical approach.

As with any project, there are several steps required to conduct the analysis. The steps are broken down as follows:

Step 1: **Understand** the system you are modelling. In this case, get a handle on the ecology of malaria using the epidemiologic triad.

Step 2: **Get some data**. Collect the data needed to conduct the analysis. In this case, we will use data obtained from the Malaria Atlas Project (MAP) and published scientific articles.

Step 3: **Explore and transform the data**. Examine the data and get the data into the right format for the analysis.

Step 4: **Perform the analysis**. Identify the methods to use and conduct the analysis.

Step 5: **Evaluate the results**.

Step 6: **Communicate** the findings and how to respond.

Software: ArcGISPro will be used to perform the analysis.

TABLE 6.6

Summary of Different Types of Malaria Control Approaches and Types

Vector Control Types	Products
Indoor Residual Spraying (IRS)	Clothianidin Clothianidin + Deltamethrin Chlorfenapyr New IRS active ingredient
Insecticide-treated mosquito net (ITN)	Pyrethroid + PBO Pyrethroid + repurposed active ingredient (alphacypermethrin + chlorfenapyr) Pyrethroid + IGR (alphacypermethrin+ pyriproxyfen) Next generation LLINs
Targeted Sugar Baits (TSB)	Targeted sugar baits designed to attract and kill mosquitoes (active ingredient 2)
Other	Larvicide Vector trap Eave tubes Spatial repellents Genetic control (gene drive/population suppression) Peridomestic combined repel and lure devices
Housing modifications	Screens, doors, ceiling, insecticide-impregnated materials to the eaves (gap of the wall and bottom of roof)
Personal protection	Mosquito repellents, protective clothing, mosquito nets
Malaria Diagnostics	Testing kits (self, blood)
Malaria Medicines	Anti-malarial treatment Prophylaxis Malaria treatment
Malaria vaccine candidates for *P. falciparum, P. vivax* by life cycle stage	Pre-erythrocytic stage Blood stage Sexual, sporogonic or mosquito stage (interrupting transmission) Malaria in pregnancy (targeting VAR2CSA antigens)

Source: WHO (2022).

Relevant Readings:

- Craig, M. H., et al. (1999) A climate-based distribution model of malaria transmission in Africa. *Parasitology Today*, 15(10).
- Kioko, K. & J. I. Blanford (2023) Malaria in Kenya during 2020: malaria indicator survey and suitability mapping for understanding spatial variations in prevalence, intervention and risk. *AGILE GIScience Series*. 4(31).
- Blanford, J. I., et al. (2013) Implications of temperature variation for malaria parasite development across Africa. *Scientific Reports* 3, 1300.
- Mordecai, E. A., et al. (2019) Thermal biology of mosquito-borne disease. *Ecology Letters*, 22(10), 1690–1708.

- Mordecai, E. A., et al. (2013) Optimal temperature for malaria transmission is dramatically lower than previously predicted. *Ecology Letters*, 16(1), 22–30.
- Giorgi, E., C. Fronterrè, P. M. Macharia, V. A. Alegana, R. W. Snow & P. J. Diggle (2021) Model building and assessment of the impact of covariates for disease prevalence mapping in low-resource settings: To explain and to predict. *The Journal of the Royal Society Interface*, 18, 20210104. http://doi.org/10.1098/rsif.2021.0104.

Data

Data Set	Description	Date	Source
Malaria prevalence	Prevalence of malaria across Africa. **map_pr20160503Africa**	2016	Malaria Atlas Project
Monthly temperature	Mean monthly temperature surfaces	2020	Worldclim
Monthly rainfall	Monthly rainfall surfaces	2020	Worldclim
Country boundaries	Country boundaries for all countries in Africa	2020	ESRI
Admin 2 boundaries	Admin boundary level 2 for all countries in Africa	2020	ESRI

Ecology of Malaria

Define the epidemiologic triad for malaria for Africa summarizing the key components and factors required for the prevalence of malaria. These include the Agent, Vector(s), Environment and Host.

Distribution of Malaria in Africa

Where Is Malaria Most Prevalent in Africa?

Using the information in the **map_pr20160503Africa** explore where malaria is prevalent in Africa. Create a map and change the symbology to get a better understanding of prevalence (Figure 6.19).

Assess the map or maps you created. Where is malaria prevalent in Africa? Where are the highest malaria prevalence rates? Show variation in parasite infection rates. Summarize malaria prevalence rates by country and administrative boundary.

Malaria Prevalence across Africa

What information is missed when you create a dot map? If malaria incidence occurs in the same location, then when the points are mapped they will plot

(a)   (b)

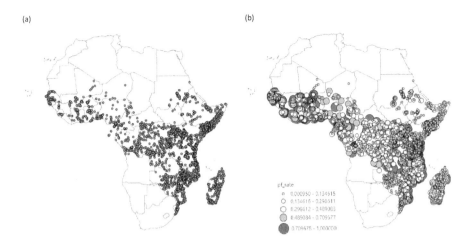

FIGURE 6.19

Malaria distribution in 2016 across Africa using different visualizations. (a) Points and (b) points symbolized using graduated symbols and colour to illustrate the different falciparum parasite (pf) prevalence rates for each point. *Data Source*: MAP. Image created in ArcGISPro.

on top of one another, which could hide information and mask what may be taking place. To get a better understanding of how malaria incidence may change across countries, there are several different analyses that can be performed. Create a density map to show where there are concentrations of cases. What did you find? Were there areas where malaria incidence was more concentrated?

Mapping Malaria: *Where:* A Data Approach

If you have additional information associated with each point, such as the number of deaths, total number of cases and/or malaria incidence rates, then use the spatial interpolation methods to create a malaria prevalence map.

Spatial Interpolation Methods – Some Background

For those who are new to interpolation methods, interpolation is a procedure used to predict the values of cells at locations that lack sampled points. There are a variety of methods available (Table 6.7). These methods are based on the principle of spatial autocorrelation or spatial dependence (e.g. Tobler's First Law of Geography), which measures the degree of relationship/dependence between near and distant objects (Childs, 2004). This correlation is used to measure (Childs, 2004):

TABLE 6.7

A Summary of Different Interpolation Methods Available (e.g. Natural Neighbours, Trend method, Inverse Distance Weighting, Spline and a Variety of Geostatistical/ Kriging Methods

Deterministic	**Inverse distance weighted (IDW)** interpolation makes the assumption that things that are close to one another are more alike than those that are further away. To predict a value for any unmeasured location, IDW uses the measured values surrounding the prediction location (number of points). The measured values closest to the prediction location have more influence on the predicted value than those farther away (e.g. set using the power function). IDW is an exact interpolator and will never predict values above the maximum measured value or below the minimum measured value.
Geostatistical Methods/Kriging	Kriging is an advanced geostatistical method that generates an estimated surface from a scattered set of points with z-values. These include • **Empirical Bayesian kriging (EBK)** is an interpolation method that accounts for the error in estimating the underlying semivariogram through repeated simulations. • **EBK Regression Prediction** is a geostatistical interpolation method that uses Empirical Bayesian Kriging with explanatory variable rasters that are known to affect the value of the data that you are interpolating.

Source: ESRI (2021); Childs (2004).

- Similarity of objects within an area
- The degree to which a spatial phenomenon is correlated to itself in space
- The level of interdependence between the variables
- Nature and strength of the interdependence

A number of different interpolation methods are available that can be used to map different phenomena, such as temperature, rainfall, air pollution, disease prevalence, minerals, etc., across space.

Mapping Malaria Risk – A Geostatistical Approach

For this, you will need to add the **Geostatistical Analyst** in ArcGIS Pro.

Add the data **map_pr20160503Africa** to your project and start creating a variety of surfaces using the different interpolation methods. Try a few different interpolation methods and examine each of the outputs. How similar and different are the results? How representative are the surfaces?

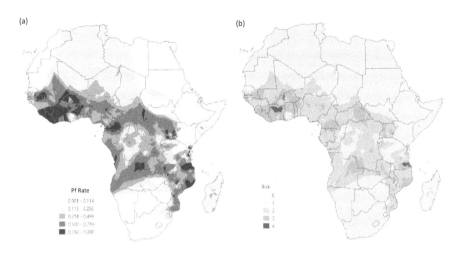

FIGURE 6.20
Maps illustrating (a) malaria prevalence rates and (b) malaria risk based on reclassified values
(Table 6.8).

Use the **Identify Tool** to interact with the data. Also, explore how to handle coincidental samples. What option works best (mean values, minimum value and maximum value). Figure 6.20a is the output of a Kriging method.

Evaluate what you find. Where were the prevalence rates highest? Did you gain additional insights compared to the malaria maps you created earlier? (Where is malaria most prevalent in Africa?)

Who Is Affected?

Now to figure out who is affected and where?

Determine what risk levels to use. Reclassify the malaria prevalence risk map you created (**Spatial Analyst Tools – Reclass – Reclassify**). To do so, define what the risk values represent by reclassifying the risk map to a range of values. Perhaps define four categories where 1=low risk and 4=very high risk, as shown below in Table 6.8 and Figure 6.20b. You will need to decide what is best and include your reasoning. To get some ideas, examine the following paper on what classifications to consider:

Giorgi, E., C. Fronterrè, P. M. Macharia, V. A. Alegana, R. W. Snow & P. J. Diggle (2021) Model building and assessment of the impact of covariates for disease prevalence mapping in low-resource settings: To explain and to predict. *The Journal of the Royal Society Interface*, 18, 20210104. http://doi.org/10.1098/rsif.2021.0104.

TABLE 6.8

Definition of Risk Levels and How These Relate
to the Malaria Prevalence Risk Values

Malaria Prevalence Risk Value Range	Risk Level (Reclassified Value Range)
≤0.25	1
0.25–0.5	2
0.5–0.75	3
≥0.75	4

Next, use zonal statistics (**Spatial Analyst Tools – Zonal – Zonal Statistics as Table**) to calculate the total population in each risk category and determine the population that is affected by malaria. Use the WorldPop dataset with the risk levels you created previously (e.g. Table 6.8).

Discussion and Conclusion

What **actions should be taken** to reduce the risk of malaria?
How might these maps be used for setting priorities and planning?
What interventions would you recommend based on your analysis?
What additional information would be useful?

Mapping Malaria: *Where and When:* A Suitability Mapping Approach

Multi-criteria decision analysis or suitability analysis, as mentioned earlier, allows for expert knowledge to be combined. To do so, different membership operators can be used to combine different types of information (e.g. OR, AND, product or sum).

Data use one of four main numbering systems: ratio, interval, ordinal or nominal scales. However, due to the potential different ranges of values (e.g. temperatures can range from –30°C to +50°C; land use 1–100; slope 0%–10%) and the different types of numbering systems used, each layer will first need to be reclassified or transformed to a common ratio scale before factors can be combined. Common scales can be predetermined using a range, for example, from 1–5 or 1–10 or 0–1, with higher values being more favourable than lower values or belonging to a specific set (e.g. 1) or not (0). Scales can also be discrete binary (0=no/absent, 1=yes/present).

One of the first malaria maps made for Africa used a fuzzy logic approach as described by Craig et al. (1999) (e.g. Figure 6.18a). The model is summarized in 4 main steps:

1. **Rescale climate data:** Based on the biological factors associated with malaria Craig et al. converted the climate data to several climate suitability maps of fractions between 0 (conditions unsuitable, U) and 1 (conditions suitable, S) using the following formula:

$$y = \cos^2\left[\frac{x-U}{S-U} \times \frac{\pi}{2}\right] \tag{6.1}$$

where y is the fuzzy suitability of climate value x. In the decreasing curve, fuzzy membership is equal to y; in the increasing curve, it is $(1-y)$. Substituting in the values,

for **rainfall**, $U=0$, $S=80$ mm per month;

for **average temperature** $U=18$, $S=22$°C for the increasing curve and $S=32$, $U=40$°C for the decreasing curve;

for **winter minimum temperature** (mean daily minimum of coldest month) $U=4$, $S=6$°C.

2. **Evaluate suitability of temperature and rainfall at each location:** Because favourable temperature and rainfall conditions have to coincide temporally for transmission to occur, the 12-monthly fuzzy rain and temperature images were overlaid month by month. The minimum suitability rating was calculated at each point, according to whichever (rain or temperature) was more limiting.

3. **Evaluate suitable conditions for a certain "time window":** Furthermore, suitable conditions have to occur for a certain "time window," constituting a transmission season, long enough for vector populations to increase and for the transmission cycle to be completed. In North Africa (>8° north), the highest value spanning any three, and in the rest of Africa, any five consecutive months were calculated.

4. **Combine data to create a malaria risk map for Africa:** Lastly, combine the outputs to create a map highlighting areas of climate suitability for malaria transmission throughout Africa.

For the purpose of this exercise, you will create a malaria risk map for only one month (July). Of course, if you have time, you can perform the analysis for each month.

1. **Rescale climate data:** First, rescale the climate information based on the following criteria: risk value=0 represents low risk while risk value=3 represents high risk. For temperature, based on the EIP of the parasite (Figure 6.21).

   The malaria host-pathogen-environment performance curve (Figure 6.21), where T_o=optimum temperature for the parasite; T_{max}=maximum temperature threshold for the parasite; T_{min}= minimum temperature threshold for the parasite; and what this means for parasite development.

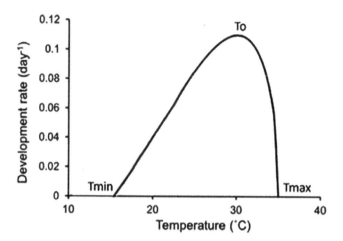

FIGURE 6.21
Thermal performance of malaria parasite development within the mosquito.

2. Evaluate suitability of temperature and rainfall at each location
Because favourable temperature and rainfall conditions have to coincide for transmission to occur and combine the rescaled climate data.

3. Evaluate suitable conditions
Evaluate the map created in (2).

Reclassify Climate Data

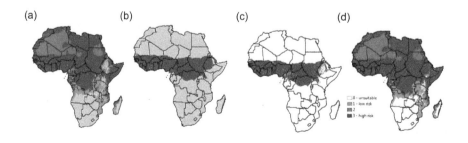

FIGURE 6.22
Reclassified (a) temperature and (b) rainfall data based on the risk categories defined in Table 6.9. The risk maps from (a) and (b) were combined using the overlay method (c) AND and (d) OR.

TABLE 6.9

Reclassification of Climate Information

Reclassified Group Value and Associated Risk	Temperature Range	Rainfall Range
1. (low risk, unsuitable)	<17, >30	<80
2.	18–21	
3.	26–30	
4. (high risk, highly suitable)	22–25	≥80

Sources: Craig et al. (1999); Paaijmans et al. (2010); Blanford et al. (2013); Mordecai et al. (2013).

Temperature Data

To reclassify the temperature layer, use the **Reclassify** tool (navigate to **Spatial Analyst Tools – Reclass – Reclassify**) and use the values highlighted in Table 6.9. Darker areas contain temperatures that are highly suitable for malaria transmission and lighter areas are less or unsuitable Figure 6.22a.

Assess the map:

- Where are the most suitable areas and where are the least suitable areas?
- Why might this be the case?

Rainfall – Create Rainfall Suitability Map

This is a little easier than the temperature maps because there is only one limitation (rainfall <80 mm for each month is unsuitable). For this, we will use a CON Statement (Conditional Statement) to reclassify the rainfall layer. Rainfall greater than or equal to 80 mm will be assigned suitable (3) and anything less will be assigned unsuitable (0) (Figure 6.22b).

Con is the **conditional statement (CON)** (navigate to **Spatial Analyst Tools – Map Algebra – Raster Calculator**) and performs a conditional if/else evaluation on each of the input cells of an input raster and works as follows:

CON (Statement to evaluate, if **TRUE** then do this, otherwise if **FALSE** then do this)

Con (for rainfall values greater or equal to 80 mm, assign them a value of 3, otherwise assign them a value of 0).

Assess the map:

- Where are the most suitable areas and where are the least suitable areas?
- Why might this be the case?

Evaluate Suitability of Temperature and Rainfall at Each Location. Combine Data to Create Malaria Risk Maps for Africa

Because favourable temperature and rainfall conditions have to coincide temporally for transmission to occur, the 12-monthly fuzzy rain and temperature images were overlaid month-by-month. You will do this for July. The minimum suitability rating was calculated at each grid cell, according to whichever (rain or temperature) was more limiting. **In other words, compare the suitability values for rainfall and temperature for each cell and assign the suitability value.**

To do this, use the Fuzzy Overlay method (navigate to **Spatial Analyst Tools – Overlay – Fuzzy Overlay**). Use the OR or AND overlay type.

Create a Suitability Map for July

For July, combine the reclassified layers you created for rainfall and temperature (e.g. Figure 6.22c (AND), Figure 6.22d (OR)).

Discussion and Conclusion

Describe how the overlay method works and what the map shows.

- Evaluate the outputs from the OR and AND overlay types. What do the maps show?
- Discuss where the key malaria transmission areas are and where they are not.
- Why is malaria transmission not in certain areas?
- How do your maps compare with the point map you created earlier?
- Discuss any problems you encountered and any limitations you may have identified.
- What **actions should be taken** to reduce the risk of malaria?
- How might these maps be used for setting priorities and planning?
- What interventions would you recommend based on your analysis?
- What additional information would be useful?

Optional: Year round map of areas suitable for malaria in Africa (optional – Figure 6.18a)

To examine year round transmission, repeat the steps 1–3 and then combine the outputs for all months to create a suitability map.

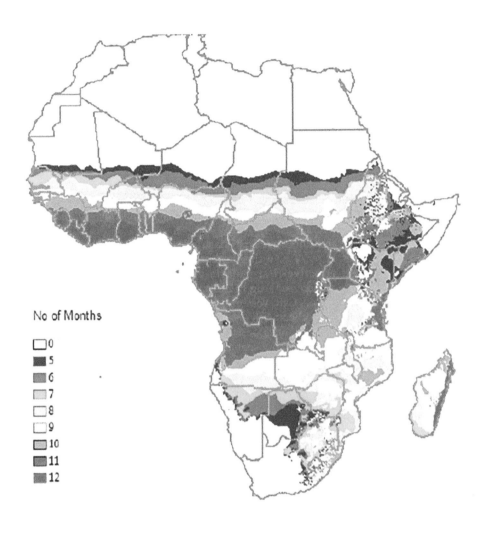

No of Months

- ☐ 0
- ■ 5
- ■ 6
- ☐ 7
- ☐ 8
- ☐ 9
- ▨ 10
- ▨ 11
- ■ 12

Who Is Affected?

Now to figure out who is affected and where.

Using the malaria map created with the suitability method for July, determine the total population at risk of malaria. Use the zonal statistic (**Spatial Analyst Tools – Zonal – Zonal Statistics as Table**) to calculate the total population in each risk category, where the categories are as follows: 1=unsuitable, 2 and 3=somewhat suitable and 4=most suitable. To view the table, click on the cylinder in contents and search for the zonal table. Right click on the table to open it and to create a chart from the data.

Determine the population that will be affected by malaria. Assess what percentage of the population will be most affected.

What actions to take: How might these maps be used to plan interventions (hint: see the elimination strategies used in Suriname and the different interventions used to control malaria)? Where interventions are most needed? What interventions would you recommend using?

References

Adalja, A. A., T. K. Sell, N. Bouri & C. Franco (2012) Lessons learned during dengue outbreaks in the United States, 2001–2011. *Emerging Infectious Diseases*, 18, 608–614.

Ahmed, K. (2023) 'Safe and Effective': First Malaria Vaccine to be Rolled Out in 12 African Countries. https://www.theguardian.com/global-development/2023/jul/06/safe-and-effective-first-malaria-vaccine-to-be-rolled-out-in-12-african-countries (last accessed Aug 13 2023).

Andreadis, T. G., J. F. Anderson, C. R. Vossbrinck & A. J. Main (2004) Epidemiology of West Nile virus in Connecticut: A five-year analysis of mosquito data 1999–2003. *Vector-Borne and Zoonotic Diseases*, 4, 360–378.

Anselin, L. (1989) What is special about spatial data? Alternative perspectives on spatial data analysis. *NCGIA Technical Reports*. https://api.semanticscholar.org/CorpusID:1861693.

Antonise-Kamp, L., D. J. M. A. Beaujean, R. Crutzen, J. E. van Steenbergen & D. Ruwaard (2017) Prevention of tick bites: An evaluation of a smartphone app. *BMC Infectious Diseases*, 17, 744.

Armengaud, A., F. Legros, E. D'Ortenzio, I. Quatresous, H. Barre, S. Houze, P. Valayer, Y. Fanton & F. Schaffner (2008) A case of autochthonous *Plasmodium vivax* malaria, Corsica, August 2006. *Travel Medicine and Infectious Disease*, 6, 36–40.

Attaway, D. F., K. H. Jacobsen, A. Falconer, G. Manca, L. R. Bennett & N. M. Waters (2014) Mosquito habitat and dengue risk potential in Kenya: Alternative methods to traditional risk mapping techniques. *Geospatial Health*, 9, 119–130.

Barker, C. M., V. L. Kramer & W. K. Reisen (2010) *Decision Support System for Mosquito and Arbovirus Control in California*. Earthzine: An IEEE publication. https://earthzine.org/decision-support-system-for-mosquito-andarbovirus-control-in-california/.

Benedict, M. Q., R. S. Levine, W. A. Hawley & L. P. Lounibos (2007) Spread of the tiger: Global risk of invasion by the mosquito Aedes albopictus. *Vector Borne Zoonotic Disease*, 7, 76–85.

Bermúdez C, S., L. Domínguez A, A. Troyo, V. M. Montenegro H & J. M. Venzal (2022) Ticks infesting humans in Central America: A review of their relevance in public health. *Current Research in Parasitology & Vector-Borne Diseases*, 2, 100065.

Blanford, J. I., S. Blanford, R. G. Crane, M. E. Mann, K. P. Paaijmans, K. V. Schreiber & M. B. Thomas (2013) Implications of temperature variation for malaria parasite development across Africa. *Scientific Reports*, 3, 1300.

Blanford, J. I., Z. Huang, A. Savelyev & A. M. MacEachren (2015) Geo-located tweets. Enhancing mobility maps and capturing cross-border movement. *PLoS One*, 10, e0129202.

Bonilauri, P., R. Bellini, M. Calzolari, R. Angelini, L. Venturi, F. Fallacara, P. Cordioli, P. Angelini, C. Venturelli, G. Merialdi & M. Dottori (2008) Chikungunya virus in *Aedes albopictus*, Italy. *Emerging Infectious Diseases*, 14, 852–854.

Booth, T. H., H. A. Nix, J. R. Busby & M. F. Hutchinson (2014) BIOCLIM: The first species distribution modelling package, its early applications and relevance to most current MAXENT studies. *Diversity and Distributions*, 20, 1–9.

Boulanger, N., P. Boyer, E. Talagrand-Reboul & Y. Hansmann (2019) Ticks and tick-borne diseases. *Médecine et maladies infectieuses*, 49(2), 87–97.

Bouri, N., T. K. Sell, C. Franco, A. A. Adalja, D. A. Henderson & N. A. Hynes (2012) Return of epidemic dengue in the United States: Implications for the public health practitioner. *Public Health Reports*, 127, 259–256.

Brown, H. E., K. F. Yates, G. Dietrich, K. MacMillan, C. B. Graham, S. M. Reese, W. S. Helterbrand, W. L. Nicholson, K. Blount, P. S. Mead, S. L. Patrick & R. J. Eisen (2011) An acarologic survey and Amblyomma americanum distribution map with implications for tularemia risk in Missouri. *American Journal of Tropical Medicine and Hygiene*, 84, 411–419.

Brown, P. (2023) Virus-Carrying Tropical Mosquitos Found in Finland as Climate Heats. https://www.theguardian.com/news/2023/aug/04/virus-carrying-tropical-mosquitos-found-in-finland-as-climate-heats (last accessed Aug 12 2023).

Brownstein, J. S., T. R. Holford & D. Fish (2003) A climate-based model predicts the spatial distribution of the Lyme disease vector Ixodes scapularis in the United States. *Environ Health Perspect*, 111, 1152–1157.

—— (2005) Effect of climate change on Lyme disease risk in North America. *Ecohealth*, 2, 38–46.

Burgdorfer, W. & J. E. Keirans (1983) Ticks and Lyme disease in the United States. *Annals of Internal Medicine*, 99, 121.

Carpenter, A., N. A. Drexler, D. W. McCormick, J. Thompson, G. Kersh, S. Commins & J. Salzer (2023) Health care provider knowledge regarding alpha-gal syndrome - United States, March-May 2022. *Morbidity and Mortality Weekly Report*, 72, 809–814.

Carpenter, G., A. N. Gillison & J. Winter (1993) Domain: A flexible modelling procedure for mapping potential distributions of plants and animals. *Biodiversity Conservation*, 2, 667–680.

CDC About Malaria. Biology. Lifecycle. https://www.cdc.gov/malaria/about/biology/index.html (last accessed Aug 12 2023).

—— West Nile Virus Transmission Cycle. https://www.cdc.gov/westnile/resources/pdfs/13_240124_west_nile_lifecycle_birds_plainlanguage_508.pdf (last accessed Oct 12 2023).

—— (2013) Lyme Disease. https://www.cdc.gov/lyme/stats/ (last accessed Aug 5 2014).

—— (2016) Tickborne Diseases of the United States. https://www.cdc.gov/ticks/diseases/index.html (last accessed Oct 12 2021).

—— (2023a) Avoiding Ticks. https://www.cdc.gov/ticks/avoid/index.html (last accessed Oct 25 2023).

—— (2023b) Geographic Distribution of Ticks that Bite Humans. https://www.cdc.gov/ticks/geographic_distribution.html (last accessed Aug 11 2023).

—— (2023c) Locally Acquired Malaria Cases Identified in the United States. https://emergency.cdc.gov/han/2023/han00494.asp (last accessed Aug 8 2023).

—— (2024a) Mosquito Surveillance Software. https://www.cdc.gov/mosquitoes/mosquito-control/professionals/MosqSurvSoft.html (last accessed May 3 2024).

—— (2024b) Lyme Disease Surveillance Data. https://www.cdc.gov/lyme/data-research/facts-stats/surveillance-data-1.html. (last accessed May 21 2024).

Černý, J., G. Lynn, J. Hrnková, M. Golovchenko, N. Rudenko & L. Grubhoffer (2020) Management options for Ixodes ricinus-associated pathogens: A review of prevention strategies. *International Journal of Environmental Research and Public Health*, 17, 1830.

Childs, C. (2004) Interpolating Surfaces in ArcGIS Spatial Analyst. *ArcUser*, July–September, pp. 32–35.

Clark, N. J., S. Tozer, C. Wood, S. M. Firestone, M. Stevenson, C. Caraguel, A. L. Chaber, J. Heller & R. J. Soares Magalhães (2020) Unravelling animal exposure profiles of human Q fever cases in Queensland, Australia, using natural language processing. *Transboundary and Emerging Diseases*, 67, 2133–2145.

Craig, M. H., R. W. Snow & D. Le Sueur (1999) A climate-based distribution model of malaria transmission in Africa. *Parasitology Today*, 15, 2133–2145.

Cutler, S. J., G. A. Paiba, J. Howells & K. L. Morgan (2002) Q fever-a forgotten disease? *The Lancet Infectious Diseases*, 2, 717–718.

Darsie, R. F. & M. L. Hutchinson (2009) The Mosquitoes of Pennsylvania. In Technical bulletin #2009-001 of the Pennsylvania Vector Control Association.

Delatte, H., G. Gimonneau, A. Triboire & D. Fontenille (2009) Influence of temperature on immature development, survival, longevity, fecundity, and gonotrophic cycles of *Aedes albopictus*, Vector of Chikungunya and Dengue in the Indian Ocean. *Journal of Medical Entomology*, 46, 33–41.

Davidsson, M. (2018) The financial implications of a well-hidden and ignored chronic lyme disease pandemic. *Healthcare*. 6(1), 16.

Delsing, C. E., B. J. Kullberg & C. P. Bleeker-Rovers (2010) Q fever in the Netherlands from 2007 to 2010. *Netherlands Journal of Medicine*, 68, 382–387.

Diuk-Wasser, M. A., A. G. Hoen, P. Cislo, R. Brinkerhoff, S. A. Hamer, M. Rowland, R. Cortinas, G. Vourc'h, F. Melton, G. J. Hickling, J. I. Tsao, J. Bunikis, A. G. Barbour, U. Kitron, J. Piesman & D. Fish (2012) Human risk of infection with Borrelia burgdorferi, the Lyme disease agent, in eastern United States. *American Journal of Tropical Medicine and Hygiene*, 86, 320–327.

ECDC (2023) Tick Maps. https://www.ecdc.europa.eu/en/disease-vectors/surveillance-and-disease-data/tick-maps (last accessed Aug 19 2023).

—— (2011) Tick-borne encephalitis in Europe. https://www.ecdc.europa.eu/sites/default/files/media/en/healthtopics/vectors/world-health-day-2014/Documents/factsheet-tick-borne-encephalitis.pdf (last accessed May 3 2024).

—— (2014) Ixodes ricinus - Factsheet for experts. https://www.ecdc.europa.eu/en/disease-vectors/facts/tick-factsheets/ixodes-ricinus (last accessed May 3 2024).

—— (2023a) Dermacentor reticulatus - current known distribution: October 2023. https://www.ecdc.europa.eu/en/publications-data/dermacentor-reticulatus-current-known-distribution-october-2023 (last accessed May 3 2024).

—— (2023b) Hyalomma marginatum - current known distribution: October 2023. https://www.ecdc.europa.eu/en/publications-data/hyalomma-marginatum-current-known-distribution-october-2023 (last accessed May 3 2024).

—— (2023c) Hyalomma marginatum - Fact sheet. https://www.ecdc.europa.eu/en/disease-vectors/facts/tick-factsheets/hyalomma-marginatum (last accessed May 3 2024).

—— (2023d) Ixodes persulcatus - current known distribution: October 2023. https://www.ecdc.europa.eu/en/publications-data/ixodes-persulcatus-current-known-distribution-october-2023 (last accessed May 3 2024).

—— (2023e) Ixodes ricinus - current known distribution: October 2023. https://www.ecdc.europa.eu/en/publications-data/ixodes-ricinus-current-known-distribution-october-2023 (last accessed May 3 2024).

—— (2023f) Rhipicephalus sanguineus - current known distribution: October 2023. https://www.ecdc.europa.eu/en/publications-data/rhipicephalus-sanguineus-current-known-distribution-october-2023 (last accessed May 3 2024).

—— (2023g) Tick maps. https://www.ecdc.europa.eu/en/disease-vectors/surveillance-and-disease-data/tick-maps (last accessed Aug 19 2023).

Egan, S. L., et al. (2021) The bacterial biome of ticks and their wildlife hosts at the urban–wildland interface. *Microbial Genomics*, 7(12), 000730.

Eisenhut, M., et al. (1999) Seroprevalence of dengue, chikungunya and Sindbis virus infections in German aid workers. *Infection*, 27(2), 82–85.

Eldridge, B. F. (1968) The effect of temperature and photoperiod on blood-feeding and ovarian development in mosquitoes of the *Culex pipiens* complex. *American Journal of Tropical Medicine & Hygiene*, 17, 133–140.

ESRI (2021) Classification Trees of the Interpolation Methods Offered in Geostatistical Analyst. https://pro.arcgis.com/en/pro-app/latest/help/analysis/geostatistical-analyst/classification-trees-of-the-interpolation-methods-offered-in--geostatistical-analyst.htm (last accessed Aug 10 2023).

Estrada-Peña, A. (1999) Geostatistics and remote sensing using NOAA-AVHRR satellite imagery as predictive tools in tick distribution and habitat suitability estimations for Boophilus microplus (Acari: Ixodidae) in South America. *Veterinary Parasitology*, 81, 73–82.

Fernandez, M. P., G. M. Bron, P. A. Kache, S. R. Larson, A. Maus, D. Gustafson Jr, J. I. Tsao, L. C. Bartholomay, S. M. Paskewitz & M. A. Diuk-Wasser (2019) Usability and feasibility of a smartphone app to assess human behavioral factors associated with tick exposure (the tick app): Quantitative and qualitative study. *JMIR Mhealth Uhealth*, 7, e14769.

Fischer, D., S. M. Thomas, J. E. Suk, B. Sudre, A. Hess, N. B. Tjaden, C. Beierkuhnlein & J. C. Semenza (2013) Climate change effects on Chikungunya transmission in Europe: Geospatial analysis of vector's climatic suitability and virus' temperature requirements. *International Journal of Health Geographics*, 12, 51.

Gething, P. W., I. R. Elyazar, C. L. Moyes, D. L. Smith, K. E. Battle, C. A. Guerra, A. P. Patil, A. J. Tatem, R. E. Howes, M. F. Myers, D. B. George, P. Horby, H. F. Wertheim, R. N. Price, I. Mueller, J. K. Baird & S. I. Hay (2012) A long neglected world malaria map: Plasmodium vivax endemicity in 2010. *PLoS Neglected Tropical Diseases*, 6, e1814.

Gibbs, S. E., M. C. Wimberly, M. Madden, J. Masour, M. J. Yabsley & D. E. Stallknecht (2006) Factors affecting the geographic distribution of West Nile virus in Georgia, USA: 2002–2004. *Vector Borne Zoonotic Disease*, 6, 73–82.

Githaka, N. W., et al. (2022) Acaricide resistance in livestock ticks infesting cattle in Africa: Current status and potential mitigation strategies. *Current Research in Parasitology & Vector-Borne Diseases*, 2, 100090.

Githeko, A. K., S. W. Lindsay, U. E. Confalonieri & J. A. Patz (2000) Climate change and vector-borne diseases: A regional analysis. *Bulletin of the World Health Organization*, C, 1136–1147.

Glass, G. E., F. P. Amerasinghe, J. M. Morgan, 3rd & T. W. Scott (1994) Predicting *Ixodes scapularis* abundance on white-tailed deer using geographic information systems. *American Journal of Tropical Medicine and Hygiene*, 51, 538–544.

Goddard, L. B., A. E. Roth, W. K. Reisen & T. W. Scott (2002) Vector competence of California mosquitoes for West Nile virus. *Emerging Infectious Diseases,* 8, 1385–1391.

Godsey, M. S., M. S. Blackmore, N. A. Panella, K. Burkhalter, K. Gottfried, L. A. Halsey, R. Rutledge, S. A. Langevin, R. Gates, K. M. Lamonte, A. Lambert, R. S. Lanciotti, C. G. M. Blackmore, T. Loyless, L. Stark, R. Oliveri, L. Conti & N. Komar (2005) West Nile virus epizootiology in the Southeastern United States, 2001. *Vector-Borne and Zoonotic Diseases,* 5, 82–89.

Gong, H. F., A. T. DeGaetano & L. C. Harrington (2011) Climate-based models for West Nile *Culex* mosquito vectors in the Northeastern US. *International Journal of Biometeorology,* 55, 435–446.

Gould, P. (1981) Letting the data speak for themselves. *Annals of the Association of American Geographers,* 71, 166–176.

Guerra, M., E. Walker, C. Jones, S. Paskewitz, M. R. Cortinas, A. Stancil, L. Beck, M. Bobo & U. Kitron (2002) Predicting the risk of Lyme disease: Habitat suitability for Ixodes scapularis in the north central United States. *Emerging Infectious Diseases,* 8, 289–297.

Guerra, M. A., E. D. Walker & U. Kitron (2001) Canine surveillance system for Lyme borreliosis in Wisconsin and northern Illinois: Geographic distribution and risk factor analysis. *American Journal of Tropical Medicine and Hygiene,* 65, 546–552.

Guglielmone, A. A., L. Beati, D. M. Barros-Battesti, M. B. Labruna, S. Nava, J. M. Venzal, A. J. Mangold, M. P. J. Szabó, J. R. Martins, D. González-Acuña & A. Estrada-Peña (2006) Ticks (Ixodidae) on humans in South America. *Experimental & Applied Acarology,* 40, 83–100.

Guo, Q., M. Kelly & C. H. Graham (2005) Support vector machines for predicting distribution of sudden Oak death in California. *Ecological Modelling,* 182, 75–90.

Hawley, W. A., P. Reiter, R. S. Copeland, C. B. Pumpuni & G. B. Craig, Jr. (1987) *Aedes albopictus* in North America: Probable introduction in used tires from northern Asia. *Science,* 236, 1114–1116.

Hay, S. I., C. A. Guerra, P. W. Gething, A. P. Patil, A. J. Tatem, A. M. Noor, C. W. Kabaria, B. H. Manh, I. R. Elyazar, S. Brooker, D. L. Smith, R. A. Moyeed & R. W. Snow (2009) A world malaria map: *Plasmodium falciparum* endemicity in 2007. *PLoS Med,* 6, e1000048.

Hijmans, R. J. & J. Elith (2019) Species Distribution Modeling. https://rspatial.org/sdm/index.html (last accessed Aug 10 2024).

Hiwat, H., B. Martínez-López, H. Cairo, L. Hardjopawiro, A. Boerleider, E. C. Duarte & Z. E. Yadon (2018) Malaria epidemiology in Suriname from 2000 to 2016: Trends, opportunities and challenges for elimination. *Malaria Journal,* 17, 1–13.

Illoldi-Rangel, P., V. Sanchez-Cordero & A. T. Peterson (2004) Predicting distributions of Mexican mammals using ecological niche modeling. *Journal of Mammalogy,* 85, 658–662.

Jumpertz, M., et al. (2023) Bacterial agents detected in 418 ticks removed from humans during 2014-2021, France. *Emerging Infectious Diseases,* 29(4), 701–710.

Junttila, J., M. Peltomaa, H. Soini, M. Marjamaki & M. K. Viljanen (1999) Prevalence of Borrelia burgdorferi in Ixodes ricinus ticks in urban recreational areas of Helsinki. *Journal of Clinical Microbiology,* 37, 1361–1365.

Kioko, K. & J. I. Blanford (2023) Malaria in Kenya during 2020: Malaria indicator survey and suitability mapping for understanding spatial variations in prevalence, intervention and risk. *AGILE GIScience Series,* 4, 31.

Kiszewski, A., A. Mellinger, A. Spielman, P. Malaney, S. E. Sachs & J. Sachs (2004) A global index representing the stability of malaria transmission. *American Journal of Tropical Medicine and Hygiene*, 70, 486–498.

Kitron, U., J. K. Bouseman & C. J. Jones (1991) Use of the ARC/INFO GIS to study the distribution of Lyme disease ticks in an Illinois county. *Preventive Veterinary Medicine*, 11, 243–248.

Kitron, U. & J. J. Kazmierczak (1997) Spatial analysis of the distribution of Lyme disease in Wisconsin. *American Journal of Epidemiology*, 145, 558–566.

Kivaria, F. M. (2006) Estimated direct economic costs associated with tick-borne diseases on cattle in Tanzania. *Tropical Animal Health and Production*, 38, 291–299.

Kopsco, H. L., R. L. Smith & S. J. Halsey (2022) A scoping review of species distribution modeling methods for tick vectors. *Frontiers in Ecology and Evolution*, 10, 893016.

Kyalo, D., P. Amratia, C. W. Mundia, C. M. Mbogo, M. Coetzee & R. W. Snow (2017) A geo-coded inventory of anophelines in the Afrotropical Region south of the Sahara: 1898–2016. *Wellcome Open Research*, 2, 57.

Lambrechts, L., K. P. Paaijmans, T. Fansiri, L. B. Carrington, L. D. Kramer, M. B. Thomas & T. W. Scott (2011) Impact of daily temperature fluctuations on dengue virus transmission by *Aedes aegypti*. *Proceedings of the National Academy of Sciences of the United States of America*, 108, 7460–7465.

Lanciotti, R. S. & A. M. Valadere (2014) Transcontinental movement of Asian genotype chikungunya virus. *Emerging Infectious Diseases*, 20, 1400–1402.

Lindenmayer, J. M., D. Marshall & A. B. Onderdonk (1991) Dogs as sentinels for Lyme disease in Massachusetts. *American Journal of Public Health*, 81, 1448–1455.

Loetti, M. V., N. E. Burroni, P. Prunella & N. Schweigmann (2008) Effect of temperature on the development time and survival of preimaginal *Culex hepperi* (Diptera: Culicidae). *Revista de la Sociedad Entomológica Argentina*, 67, 79–85.

Logan, J. J., A. G. Hoi, M. Sawada, A. Knudby, T. Ramsay, J. I. Blanford, N. H. Ogden & M. A. Kulkami (2023) Risk factors for Lyme disease resulting from residential exposure amidst emerging *Ixodes scapularis* populations: A neighbourhood-level analysis of Ottawa, Ontario. *PLoS One*, 18, e0290463.

LRA (n.d.) What Is Lyme Disease? https://www.lymeresearchalliance.org/prevention-what-is-lyme.html (last accessed Feb 5 2014).

Mac, S., S. R. da Silva & B. Sander (2019) The economic burden of Lyme disease and the cost-effectiveness of Lyme disease interventions: A scoping review. *PLoS ONE*, 14, e0210280.

Mackay, A. J., A. Roy, M. M. Yates & L. D. Foil (2008) West Nile virus detection in mosquitoes in East Baton Rouge Parish, Louisiana, from November 2002 to October 2004. *Journal of the American Mosquito Control Association*, 24, 28–35.

Madder, M., I. Horak & Stoltsz (unknown) Ticks: Tick Importance and Disease Transmission. https://www.afrivip.org/sites/default/files/importance_complete.pdf (last accessed Nov 11 2022).

Malaria_Program_Suriname Achievements. https://www.malariasuriname.com/achievements/ (last accessed Aug 13 2023).

Malczewski, J. & C. Rinner (2015) *Multicriteria Decision Analysis in Geographic Information Science*. Berlin, Germany: Springer.

Manning, J., S. Chea, D. M. Parker, J. A. Bohl, S. Lay, A. Mateja, S. Man, S. Nhek, A. Ponce & S. Sreng (2021) Humoral Immunity Against Aedes Aegypti Salivary Proteins Associated with Development of Inapparent Dengue: A Longitudinal Observational Cohort in Cambodia. https://ssrn.com/abstract=3857656.

MARA (n.d.) Mapping Malaria Risk in Africa (MARA). https://www.mara-data-base.org/login.html;jsessionid=A5F86B6BB5BDE6CA6D92B848E94E9CA1 (last accessed Aug 10 2020).

Martens, P. & L. Hall (2000) Malaria on the move: Human population movement and malaria transmission. *Emerging Infectious Diseases*, 6, 103–109.

Mitchell, C. J. (1995) Geographic spread of *Aedes albopictus* and potential for involvement in arbovirus cycles in the Mediterranean basin. *Journal Vector Ecology*, 20, 44–58.

Moore, S. M., R. J. Eisen, A. Monaghan & P. Mead (2014) Meteorological influences on the seasonality of Lyme disease in the United States. *American Journal of Tropical Medicine and Hygiene*, 90, 486–496.

Mordecai, E. A., J. M. Caldwell, M. K. Grossman, C. A. Lippi, L. R. Johnson, M. Neira, J. R. Rohr, S. J. Ryan, V. Savage & M. S. Shocket (2019) Thermal biology of mosquito-borne disease. *Ecology Letters*, 22, 1690–1708.

Mordecai, E. A., J. M. Cohen, M. V. Evans, P. Gudapati, L. R. Johnson, C. A. Lippi, K. Miazgowicz, C. C. Murdock, J. R. Rohr & S. J. Ryan (2017) Detecting the impact of temperature on transmission of Zika, dengue, and chikungunya using mechanistic models. *PLoS Neglected Tropical Diseases*, 11, e0005568.

Mordecai, E. A., K. P. Paaijmans, L. R. Johnson, C. Balzer, T. Ben-Horin, E. de Moor, A. McNally, S. Pawar, S. J. Ryan, T. C. Smith & K. D. Lafferty (2013) Optimal temperature for malaria transmission is dramatically lower than previously predicted. *Ecology Letters*, 16, 22–30.

Musso, D. & J. Gubler Duane (2016) Zika virus. *Clinical Microbiology Reviews*, 29(3), 487–524.

MVCA_California (2023) California Mosquito-borne virus surveillance and response plan.

Oberhauser, K. & A. T. Peterson (2003) Modeling current and future potential wintering distributions of eastern North American monarch butterflies. *Proceedings of the National Academy of Sciences of the United States of America*, 100, 14063–14068.

Odhiambo, J. N., C. Kalinda, P. M. Macharia, R. W. Snow & B. Sartorius (2020) Spatial and spatio-temporal methods for mapping malaria risk: A systematic review. *BMJ Global Health*, 5, e002919.

Ogden, N. H., M. Radojevic, X. Wu, V. R. Duvvuri, P. A. Leighton & J. Wu (2014) Estimated effects of projected climate change on the basic reproductive number of the Lyme disease vector Ixodes scapularis. *Environmental Health Perspectives*, 122, 631–638.

Paaijmans, K. P., S. Blanford, A. S. Bell, J. I. Blanford, A. F. Read & M. B. Thomas (2010) Influence of climate on malaria transmission depends on daily temperature variation. *Proceedings of the National Academy of Sciences of the United States of America*, 107, 15135–15139.

Paaijmans, K. P., A. F. Read & M. B. Thomas (2009) Understanding the link between malaria risk and climate. *Proceedings of the National Academy of Sciences of the United States of America*, 106, 13844–13849.

PAHO (2016) Latest Global Situation Report on Zika. https://www.paho.org/hq/index.php?option=com_content&view=article&id=11669&Itemid=41716&lang=en (last accessed May 2 2024).

— (2023) Bringing Malaria Prevention and Outreach to Apetina, Suriname: A Successful Site Visit and Collaborative Effort in Suriname's Interior. https://www.paho.org/en/news/11-4-2023-bringing-malaria-prevention-and-out-reach-apetina-suriname-successful-site-visit-and (last accessed Aug 13 2023).

Parker, D. M., S. T. T. Tun, L. J. White, L. Kajeechiwa, M. M. Thwin, J. Landier, V. Chaumeau, V. Corbel, A. M. Dondorp, L. von Seidlein, N. J. White, R. J. Maude & F. Nosten (2019) Potential herd protection against *Plasmodium falciparum* infections conferred by mass antimalarial drug administrations. *eLife*, 8, e41023.

Parker, D. M., et al. (2023) Determinants of exposure to Aedes mosquitoes: a comprehensive geospatial analysis in peri-urban Cambodia. *Acta Tropica*, 239, 106829.

Parola, P. & D. Raoult (2001) Ticks and tickborne bacterial diseases in humans: An emerging infectious threat. *Clinical Infectious Diseases*, 32(6), 897–928.

Parola, P., et al. (2013) Update on tick-borne rickettsioses around the world: A geographic approach. *Clinical Microbiology Reviews*, 26(4), 657–702.

Paupy, C., H. Delatte, L. Bagny, V. Corbel & D. Fontenille (2009) *Aedes albopictus*, an arbovirus vector: From the darkness to the light. *Microbes and Infection*, 11, 1177–1185.

Pennisi, E. (2018) A Tropical Parasitic Disease has Invaded Europe, Thanks to a Hybrid of Two Infectious Worms. Genomic Study Helps Explain How Schistosomiasis Gained a Foothold on Corsica. *Science*.

Perkins, T. A., A. J. Garcia, V. A. Paz-Soldan, S. T. Stoddard, R. C. Reiner, Jr., G. Vazquez-Prokopec, D. Bisanzio, A. C. Morrison, E. S. Halsey, T. J. Kochel, D. L. Smith, U. Kitron, T. W. Scott & A. J. Tatem (2014) Theory and data for simulating fine-scale human movement in an urban environment. *Journal of the Royal Society Interface*, 11, 20140642.

Petney, T. N., W. Saijuntha, N. Boulanger, L. Chitimia-Dobler, M. Pfeffer, C. Eamudomkarn, R. H. Andrews, M. Ahamad, N. Putthasorn & S. V. Muders (2019) Ticks (Argasidae, Ixodidae) and tick-borne diseases of continental Southeast Asia. *Zootaxa*, 4558, 1–89.

Phillips, S. J., et al. (2017) Opening the black box: An open-source release of Maxent. *Ecography*, 40(7), 887–893.

Phillips, S. J. & M. Dudík (2008) Modeling of species distributions with Maxent: New extensions and a comprehensive evaluation. *Ecography*, 31(2), 161–175.

Radke, E. G., C. J. Gregory, K. W. Kintziger, E. K. Sauber-Schatz, E. A. Hunsperger, G. R. Gallagher, J. M. Barber, B. J. Biggerstaff, D. R. Stanek, K. M. Tomashek & C. G. Blackmore (2012) Dengue outbreak in Key West, Florida, USA, 2009. *Emerging Infectious Diseases*, 18, 135–137.

Radolf, J. D., M. J. Caimano, B. Stevenson & L. T. Hu (2012) Of ticks, mice and men: Understanding the dual-host lifestyle of Lyme disease spirochaetes. *Nature Reviews Microbiology*, 10, 87–99.

Raines, G. L., D. L. Sawatzky & G. F. Bonham-Carter (2010) New fuzzy logic tools in ArcGIS 10. *ArcUser*, 2, 8–13.

Randolph, S. E. (2000) Ticks and tick-borne disease systems in space and from space. *Advances in Parasitology*, 47, 217–243.

Reisen, W. K. (1995) Effect of temperature on culex tarsalis (diptera, culicidae) from the coachella and san-joaquin valleys of california. *Journal of Medical Entomology*, 32, 636–645.

Reisen, W. K., Y. Fang & V. M. Martinez (2006) Effects of temperature on the transmission of West Nile virus by *Culex tarsalis* (Diptera: Culicidae). *Journal of Medical Entomology*, 43, 309–317.

Reisen, W. K., T. Thiemann, C. M. Barker, H. Lu, B. Carroll, Y. Fang & H. D. Lothrop (2010) Effects of warm winter temperature on the abundance and gonotrophic activity of *Culex* (Diptera: Culicidae) in California. *Journal of Medical Entomology*, 47, 230–237.

Rezza, G., L. Nicoletti, R. Angelini, R. Romi, A. C. Finarelli, M. Panning, P. Cordioli, C. Fortuna, S. Boros, F. Magurano, G. Silvi, P. Angelini, M. Dottori, M. G. Ciufolini, G. C. Majori, A. Cassone & C. S. group (2007) Infection with chikungunya virus in Italy: An outbreak in a temperate region. *Lancet*, 370, 1840–1846.

Rochlin, I., D. V. Ninivaggi, M. L. Hutchinson & A. Farajollahi (2013) Climate change and range expansion of the Asian tiger mosquito (*Aedes albopictus*) in Northeastern USA: Implications for public health practitioners. *PLoS One*, 8, e60874.

Roy-Dufresne, E., T. Logan, J. A. Simon, G. L. Chmura & V. Millien (2013) Poleward expansion of the white-footed mouse (Peromyscus leucopus) under climate change: Implications for the spread of lyme disease. *PLoS One*, 8, e80724.

Rueda, L. M., K. J. Patel, R. C. Axtell & R. E. Stinner (1990) Temperature-dependent development and survival rates of *Culex quinquefasciatus* and *Aedes aegypti* (Diptera: Culicidae). *Journal of Medical Entomology*, 27, 892–898.

Schneeberger, P. M., C. Wintenberger, W. Van der Hoek & J. P. Stahl (2014) Q fever in the Netherlands-2007–2010: What we learned from the largest outbreak ever. *Médecine et Maladies Infectieuses*, 44, 339–353.

Scholte, R. G., O. S. Carvalho, J. B. Malone, J. Utzinger & P. Vounatsou (2012) Spatial distribution of Biomphalaria spp., the intermediate host snails of Schistosoma mansoni, in Brazil. *Geospatial Health*, 6, S95–S101.

Simon, F., H. Savini & P. Parola (2008) Chikungunya: A paradigm of emergence and globalization of vector-borne diseases. *Medical Clinics of North America*, 92, 1323–1343, ix.

Singh, K., et al. (2022) Economic impact of predominant ticks and tick-borne diseases on Indian dairy production systems. *Experimental Parasitology*, 243, 108408.

Stockwell, D. (1999) The GARP modelling system: Problems and solutions to automated spatial prediction. *International Journal of Geographical Information Systems*, 13, 143–158.

Stoddard, S. T., A. C. Morrison, G. M. Vazquez-Prokopec, V. Paz Soldan, T. J. Kochel, U. Kitron, J. P. Elder & T. W. Scott (2009) The role of human movement in the transmission of vector-borne pathogens. *PLoS Neglected Tropical Diseases*, 3, e481.

Stone, B. L., et al. (2017) Brave new worlds: The expanding universe of Lyme disease. *Vector-Borne and Zoonotic Diseases*, 17(9), 619–629.

Sugumaran, R., S. R. Larson & J. P. Degroote (2009) Spatio-temporal cluster analysis of county-based human West Nile virus incidence in the continental United States. *International Journal of Health Geographics*, 8, 43.

Taber, E. D., M. L. Hutchinson, E. A. H. Smithwick & J. I. Blanford (2017) A decade of colonization: The spread of the Asian tiger mosquito in Pennsylvania and implications for disease risk. *Journal of Vector Ecology*, 42, 3–12.

Talbert, A. (2005) Tick-bone relapsing fever in Mvumi Hospital, Tanzania: A retrospective study of laboratory data between 1997 and 2002. *Tanzania Journal of Health Research*, 7, 1.

Tan, L. P., R. H. Hamdan, B. N. H. Hassan, M. F. H. Reduan, I. A.-A. Okene, S. K. Loong, J. J. Khoo, A. S. Samsuddin & S. H. Lee (2021) Rhipicephalus tick: A contextual review for Southeast Asia. *Pathogens*, 10, 821.

Texas_ A&M_AE (n.d.) Lone Star Tick. https://insects.tamu.edu/fieldguide/cimg370.html (last accessed Mar 7, 2014).

Thompson JM, Carpenter A, Kersh GJ, Wachs T, Commins SP & S. JS. (2023) Geographic distribution of suspected alpha-gal syndrome cases - United States, January 2017–December 2022. *Morbidity and Mortality Weekly Report*, 72, 815–820.

Tjaden, N. B., Y. Cheng, C. Beierkuhnlein & S. M. Thomas (2021) Chikungunya beyond the tropics: Where and when do we expect disease transmission in Europe? *Viruses*, 13, 1024.

Todar, K. (2012) Todar's Online Textbook of Bacteriology. bit.ly/1muWuLY (last accessed Feb 5, 2014).

Van der Hoek, W., F. Dijkstra, B. Schimmer, P. M. Schneeberger, P. Vellema, C. Wijkmans, R. Ter Schegget, V. Hackert & Y. Van Duynhoven (2010) Q fever in the Netherlands: An update on the epidemiology and control measures. *Eurosurveillance*, 15, 19520.

van Eer, E. D., G. Bretas & H. Hiwat (2018) Decreased endemic malaria in Suriname: Moving towards elimination. *Malaria Journal*, 17, 56.

Wang, X. & H. Nishiura (2021) The epidemic risk of dengue fever in Japan: climate change and seasonality. *Canadian Journal of Infectious Diseases and Medical Microbiology*, 2021.

Watts, D. M., D. S. Burke, B. A. Harrison, R. E. Whitmire & A. Nisalak (1987) Effect of temperature on the vector efficiency of *Aedes aegypti* for dengue 2 virus. *American Journal of Tropical Medicine and Hygiene*, 36, 143–152.

Weaver, S. C. & A. D. T. Barrett (2004) Transmission cycles, host range, evolution and emergence of arboviral disease. *Nature Reviews Microbiology*, 2, 789–801.

Weaver, S. C. & W. K. Reisen (2010) Present and future arboviral threats. *Antiviral Research*, 85, 328–345.

Weiss, D. J., T. C. D. Lucas, M. Nguyen, A. K. Nandi, D. Bisanzio, K. E. Battle, E. Cameron, K. A. Twohig, D. A. Pfeffer & J. A. Rozier (2019) Mapping the global prevalence, incidence, and mortality of Plasmodium falciparum, 2000–17: A spatial and temporal modelling study. *The Lancet*, 394, 322–331.

Werden, L., I. K. Barker, J. Bowman, E. K. Gonzales, P. A. Leighton, L. R. Lindsay & C. M. Jardine (2014) Geography, deer, and host biodiversity shape the pattern of Lyme disease emergence in the Thousand Islands Archipelago of Ontario, Canada. *PLoS One*, 9, e85640.

WHO (2014) Infographic of Neglected Tropical Vector-Borne Diseases. https://www.who.int/mediacentre/infographic/neglected-tropical-diseases/en/ (last accessed Aug 10 2016).

—— (2016a) Countries, Terrirories and Areas Showing the Distribution of Zika Virus, 2013–2016. https://www.who.int/emergencies/zika-virus/situation-report/Zika-timeline-12-may.jpg (last accessedJan 10 2017).

—— (2016b) Vector-Borne Diseases. https://www.who.int/mediacentre/factsheets/fs387/en/ (last accessed May 2 2024).

—— (2021) Vector Borne Diseases. https://www.who.int/news-room/fact-sheets/detail/vector-borne-diseases#:~:text=Vector-borne%20diseases%20are%20human%20illnesses%20caused%20by%20parasites%2C,Chagas%20disease%2C%20yellow%20fever%2C%20Japanese%20encephalitis%20and%20onchocerciasis. (last accessed Oct 1 2021).

—— (2022) World Malaria Report 2022, 372 p. https://www.who.int/publications/i/item/9789240064898 (last accessed Oct 27 2023).

—— (2023) Malaria. https://www.who.int/news-room/fact-sheets/detail/malaria (last accessed Oct 27 2023).

7

Clustering of Health Risks: Global to Local

Overview

Spatial clustering methods are exploratory tools that are useful for making sense of complex geographic information and patterns. Knowing whether or not a cluster exists and where it is located is useful for making decisions and formulating health policies. The methods discussed in this chapter are useful for exploring spatial data and can be used to confirm or deny the existence of suspected clusters in an efficient and effective manner.

Cluster analysis methods are methods that are used to group a set of objects or information in such a way that objects or information within the same group (called a cluster) are more similar (in some sense or another) to one another than to those in another group (clusters) (Figure 7.1).

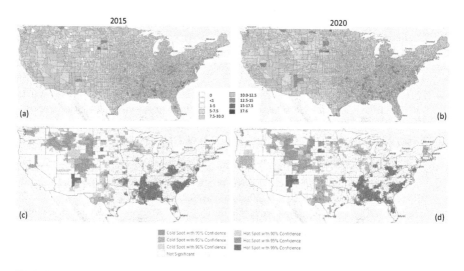

FIGURE 7.1
Diabetes prevalence in the USA in 2015 and 2020. (a and b) Percent adults with diabetes by county and (c and d) cold and hot spots of diabetes. Analysis conducted in ArcGIS Pro using the Hot Spot Getis-Ord G_i^* Analysis. *Data Source*: CDC (2023a). Image created by Blanford (2023).

DOI: 10.1201/9781003435082-7

How Cluster Analysis Has Been Used in Health Studies

Cluster and outlier detection methods are an important set of exploratory analyses in public health and disease surveillance efforts. These methods are useful for (i) the rapid identification of epidemic clusters, (ii) identification of confounders, (iii) generation of research hypotheses (Carpenter, 2001) and (iv) detecting unusual patterns of disease incidence (Carpenter, 2001; Rushton, 2003; Mandl et al., 2004; Cromley and McLafferty, 2012; Fritz et al., 2013; Neethu and Surendran, 2013; Grubesic et al., 2014). Once a cluster has been detected, further epidemiological investigations may be useful to confirm whether the cluster is indeed random or whether it is linked to specific factors that may be environmental, occupational or social. By overlaying cluster maps with other spatial data information (e.g. environment, population, transportation, industry and/or facilities), potential causes or underlying drivers of disease can be explored and variations in health linked to differences in physical and social environments can be examined (Cromley and McLafferty, 2012).

Spatial clustering methods can help provide answers to an array of questions:

- Do any unusual clusters of health events exist in an area?
- What places have an unusually high or low prevalence of a disease?
- Where are the risks of ill health highest or lowest?

Cluster analysis has been extensively used to analyse health-related data for

- **Identifying groups of similarity and variations in prevalence**
  - e.g. spatial clusters (Hamid et al., 2010; Meliker et al., 2009); obesity (Mills et al., 2020)
- **Identifying sub-groups within a group**
  - e.g. (Cassetti et al., 2008)
- **Assessment of quality of the environment**
  - e.g. (Doreena et al., 2012)
- **Identifying social and environmental inequalities affecting health outcomes**
  - e.g. (Padilla et al., 2013)
- **Geographic diffusion and evolution of disease**: Phylogenetic trees: based upon similarities and differences in physical or genetic characteristics.
  - e.g. (Lam et al., 2012; Chen and Holmes, 2010; Si et al., 2009)

- **Examining clustering in space and/or time**
  - e.g. (Jennings et al., 2005; Sugumaran et al., 2009; Barker et al., 2011); geo-social gradients in the UK of COVID-19 (Bowyer et al., 2021); SaTScan (Kulldorff, 1997; Kulldorff, 2005)
- **Monitoring of disease incidence for targeted control and eradication**
  - e.g. (Bousema et al., 2012) and malaria in Suriname (Hiwat et al., 2018; van Eer et al., 2018); anaemia in children in Kenya (Robert et al., 2023), geographical surveillance of diseases (Kulldorff, 1997; Kulldorff, 2005)
- **Profiling disease at a cellular level**: Visualizations, Imaging and spatial analysis methods are disrupting the medical field at the cellular level (e.g. medical imaging, spatial transcriptomics).
  - **Mapping the transcriptome** (Zormpas et al. 2023). Understanding the where, what, why at a cellular level to better understand the relationship between cells, their location within tissues and their role in disease pathology is and will be transformational. Spatial transcriptomics is a molecular profiling method that allows scientists to measure all the gene activity in a tissue sample and map where the activity is occurring.
  - Image analysis using unsupervised and supervised clustering methods for medical imaging and disease detection (e.g. malaria (Maturana et al., 2022,2023)).

Spatial clustering analysis methods offer a way of filtering health information that is useful for identifying unusual occurrences of health events, describing geographical patterns and generating hypotheses (Anselin, 1995).

Clustering Methods

Over the past few decades, there have been major advances in the development of clustering methods. Many different spatial clustering approaches are available; however, four categories stand out: (i) non-hierarchical, (ii) hierarchical, (iii) scan-based and (iv) auto-correlation-based approaches (Grubesic et al., 2014). A brief description of each is provided as follows:

- **Non-hierarchical methods** are classic partitioning based approaches. Non-hierarchical techniques are typically structured as optimization problems intent on identifying k clusters. Specifically, each object is placed based on its most "alike" cluster or group.

- **Hierarchical** approaches can be broken down into two basic types: agglomerative and divisive. Agglomerative approaches start with all objects in their own cluster and then merge objects into clusters. Divisive clustering works in the opposite manner, where all objects are initially assigned to a single cluster and then split into smaller subclusters.

- **Scan-based** approaches for cluster detection are structured using a geographically defined "window" to identify areas with elevated (or deflated) rates of local incidences.

- **Auto-correlation-based** or autocorrelation-based spatial clustering approaches, widely labelled spatial autocorrelation techniques, are generally divided into two families: global and local (Anselin, 1995; Ord and Getis, 1995). Global measures, such as the Moran's I, Geary's C and the general G, summarize the extent to which neighbouring areas are similar. Spatial weight matrices are used to capture proximity, contiguity, or both. Significance testing using Monte Carlo simulation to derive expected values can be included. The Moran's I is structured to measure the correlation of each x_i with all x_j within a specified distance, d. Thus, it captures the degree of covariance within d of all x_i (Getis and Ord, 1992). The major limitation of global approaches is that they cannot identify the specific location of clusters.

A wide variety of clustering methods have been developed to identify spatial and temporal-spatial clustering of disease (Tables 7.1 and 7.2). The clustering methods can be global estimates or identify local clusters. For more information, skim through the following papers (Lawson, 2010; Fritz et al., 2013) to get a sense of how these clustering methods work and how they have been used. Benefits and limitations associated with these methods (Fritz et al., 2013):

- **Using multiple methods to analyse patterns and clusters**: Several papers utilized multiple methods to investigate the spatial phenomena with a closer lens. Adopting one or more exploratory spatial data analysis methods provides the user with different views of the data, enabling them to compare outputs and enhance their understanding of the data and any patterns that are revealed.

- **Visualizations**: A mix of statistical values and visualizations is useful for understanding the patterns that may be within the data.

- **Spatial resolution**: The spatial resolution at which the analysis is conducted can have positive and negative effects on the insights gained, results and subsequent analyses. MAUP is an issue that may arise due to the aggregation and coarseness of the data. Also, first- and second-order trends in the data may change with resolution and aggregation.

TABLE 7.1

Summary of Different Clustering Methods

	Point	Area	Line	Non-Spatial/ Attributes
Space	• Kernel Density Estimate • Nearest Neighbor index • Ripley K function • G function • F function • Cuzick-Edwards test • Density based functions • Scan based functions	• Black-white or join-count test • Ohno method • Moran's I • LISA • Hot Spot Analysis (Getis-Ord G_i^*)	• Randomness of runs & signs; • Nearest neighbour	• Hierarchical clustering • K-means
Space-time	SaTScan DBScan	• Emerging Hotspot Analysis • Hotspot Comparison • SatScan • DBScan		

Source: Carpenter (2001); O'Sullivan and Unwin (2010); Cromley and McLafferty (2012); Grubesic et al. (2014); Baumer et al. (2017); Robertson and Nelson (2010).

TABLE 7.2

List of Different Software Packages and the Variety of Cluster Analysis Methods Available

Software	Description
GeoDa	*Space*
	Spatial cluster analysis software includes methods for Moran's I and LISA Univariate/Bivariate/Differential Moran's I Univariate/Bivariate/Differential Local Moran's I Univariate/Bivariate/Co-location Local Join Count Local G/G* Univariate/Multivariate Local Geary Univariate/Multivariate Quantile LISA Local Neighbor Match Test Spatial Correlogram
	Cluster
	PCA/MDS/t-SNE K Means/Medians/Medoids Spectral/Hierarchical DBScan/HDBScan SC K Means/SCHC/Skater/redcap AZP/max-p
SaTScan	Cluster detection software with several spatial, temporal and space-time scan statistics. Space-time cluster scan with Poisson model

(Continued)

TABLE 7.2 (*Continued*)

List of Different Software Packages and the Variety of Cluster Analysis
Methods Available

Software	Description
R Analysis of Spatial Data	Variety of methods. spatstat – http://spatstat.org/ Cluster – https://cran.r-project.org/web/views/Cluster.html DBScan/HDBScan – https://cran.r-project.org/web/packages/dbscan/dbscan.pdf

ArcGIS	Variety of tools to analyse and map clusters:
	Mapping clusters
	
	• Cluster and outlier analysis (Aneselin Local Moran's I or LISA);
	• Hot Spot Analysis(Getis-Ord G_i^*);
	• Density-based clustering (DBScan and HDBScan)
	• Multivariate clustering
	• Optimized hot spot analysis
	• Optimized outlier analysis
	• Similarity search
	• Spatial outlier detection
	• Spatially constrained multivariate clustering (finds spatially contiguous clusters based on a set of feature attribute values)
	
	Changing clusters
	
	• Space-time pattern
	  • Emerging hotspot analysis
	  • Local outlier analysis
	  • Time series analysis
	• Hotspot comparison
	
	Analysing patterns
	
	• KDE;
	• Average nearest neighbour;
	• Multi-distance spatial cluster analysis (Ripley's k-function);
	• Spatial autocorrelation (Global Moran's I)

Other	ClusterSeer offers a variety of spatial, temporal and space-time clustering methods https://biomedware.com/products/clusterseer/. Space-time cluster scan with Poisson mode
	
	• Spatial clustering
	  • Focused
	  • Local
	  • Global
	• Temporal clustering
	• Space-time clustering
	• Retrospective surveillance

• **Distance-based analyses**: Spatial cluster analysis is useful for gaining insights and understanding of the spatial structure or scale of processes. For processes that are distance-based dependent, it is important to identify the optimal scale at which to perform the analysis and the study area boundary extent, as these can affect the results.

- **Conceptualization of spatial interactions and neighbourhoods**: Varying conceptualizations of space will yield different results for spatial clustering and cluster detection. Likewise, each spatial clustering method will define space with different parameters. For most methods, selecting an appropriate spatial weight conceptualization remains a decision based on researcher discretion and should be heavily considered when synthesizing results.

Point Pattern Analysis (PPA)

Thanks to the availability of cheap GPS technologies, it is now easier than ever to collect precise information about something (e.g., using GPS coordinates through mobile technologies and/or GPS devices). However, it is all very well just mapping the point data (latitude/longitude, X,Y) collected, but being able to gain insights from these data and understand their spatial distribution may provide insights into why they are occurring where they are.

Classical point pattern analysis allows us to explore and determine whether a pattern is evenly-spaced, uniform, clustered or dispersed relative to a null spatial process (usually the independent random process IPR). However, many of these methods do not allow us to say where the pattern is clustered. For most real-world applications, we are interested in knowing where clusters of certain diseases may occur and whether these are always in the same place. In other words, we are interested in knowing if there are disease hotspots. Having statistical confirmation of hotspots or clusters is nice to have so that we know that we are not just imagining that a pattern exists!

Point pattern analysis methods are useful for analysing such point information and have been extensively used for analysing crime information as well as health related events. In fact you already used some of these methods earlier when analyzing the cholera outbreak of 1854. A number of different methods are available for analysing point data (Figure 7.2) that range from exploratory data analysis methods to more quantitative statistical methods, the latter of which take distance into consideration.

Kernel Density Estimation

When plotting points on a map, if points have the same coordinates, they will be plotted on top of one another, making it difficult to see where multiple events are taking place and where the intensity of events is greatest. Kernel density estimation (KDE) helps solve this problem by capturing where events are highest or lowest based on the density of points (where they are distributed) and/or their intensity based on the number of cases orevents.

In summary, a KDE surface is created by fitting a smoothly curved surface over each point, and the number of points that fall within a kernel is summed. The total is then divided by the area of the kernel to provide the grid cell a value. The surface value will be highest at the location of the point

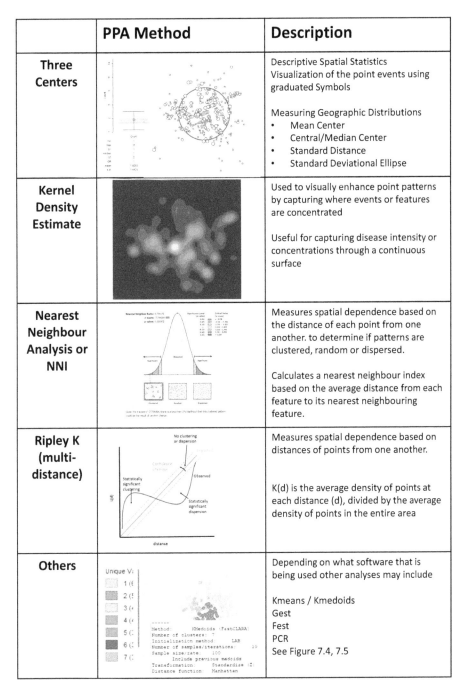

FIGURE 7.2
Summary of different point pattern analysis (PPA) methods available for analysing point data.
Source: Blanford (2023).

and diminish with increasing distance from the point (at the centre), reaching zero at the outer boundary edge of the kernel.

A "neighbourhood" or kernel is defined around each grid cell, consisting of all grid cells with centres within the specified kernel (search) radius.

It is worth pointing out how scale-dependent this method is, as illustrated in the following sequence of maps (Figure 7.3). I have generated a set of maps using varying bandwidths to show how changing the bandwidth affects the spatial pattern. All maps were generated using measles data (1944–1965) from the UK, England (Dalziel et al., 2016).

- **Generalized Impression or over-smoothing**: A larger KDE bandwidth will create a spatial pattern that is a very generalized impression of the density of the events. Furthermore, a large bandwidth will tend to emphasize first-order trend variations (e.g. bandwidth=150 km).

- **Focus on individual events or under-smoothing**: A small KDE bandwidth can be problematic as it focuses too much on individual events and highlights small clusters of events. Depending on how small the bandwidth is, the output may not be much different from the original point pattern and thus will not provide any additional insights beyond the existing point pattern (e.g. bandwidth=20 km).

- **Bandwidth somewhere in between**: An intermediate choice of bandwidth may be of more use as it will allow additional insights into patterns to be gained. A more satisfactory map may be one that enables distinct regions of high and low event density to be visualized and shows how the intensity of events changes across space (e.g. bandwidth=50–100 km).

However, the bandwidths can serve different purposes. Think about how these different bandwidths might be useful or not in making a decision on how to respond. Which of these would you use?

Distance-Based Point Pattern Analyses

The spatstat package in R provides a number of functions that are useful for analysing patterns according to *independent random process (IRP)*, or *complete spatial randomness (CSR)*, where the IPR/CSR events occur with equal probability anywhere; and the place of occurrence of an event is not affected by the occurrence of other events (e.g. see Figure 7.4; Appendix for R code).

- **Nearest neighbour analysis or Nearest neighbour index (NNI)**: Originally developed by plant ecologists (Clark and Evans, 1954) for measuring arrangements in plants. In short, this analysis examines all points and calculates the mean distance to their nearest neighbour.

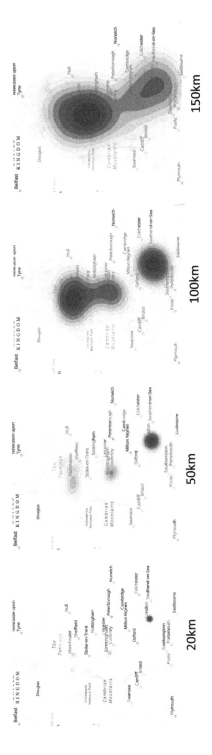

FIGURE 7.3

Kernel Density Estimates. Spatial density analysis of the 1944–1965 measles outbreak in the UK using different bandwidths of 20, 50, 100 and 150km.
Source: Dalziel et al. (2016).

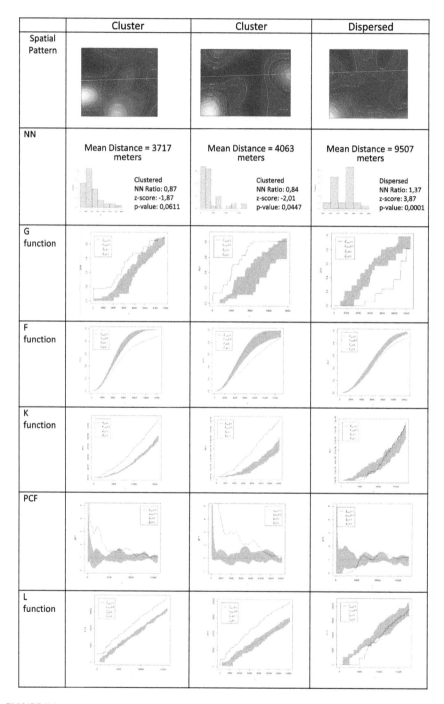

FIGURE 7.4

Results for two different point distributions using a variety of different distance-based analyses in R (G, F, K, PCF, L and Nearest Neighbor). One point distribution represents a clustered distribution and another represents a dispersed distribution. Images created in R, Blanford (2023).

If NNI=0, the distribution is clustered; if NNI=1, the distribution is random; and if NNI>2.14, then the distribution is considered to be uniform.

- **Ripley's K index** quantifies non-random clustering patterns (by estimating the second-order effects of an observed point pattern) (Bailey and Gatrell, 1995). It has been used in plant ecology to gain insights into the distribution of different plant species (e.g. Ferns: Zhang et al., 2010; Acacia: Spooner et al., 2004; and butterflies: Nekola and Kraft, 2002). The K-function considers the intensity of events per unit area for an area R and compares the observed distribution of events with the expected distribution of events for area R. For a region with area *R* and *n* events (e.g. infected cases), Ripley's K index can be used to evaluate clustering. Positive values indicate clustering, and negative values indicate spatial repulsion or non-clustering. Zero values indicate a random spatial distribution.

- **K-function** (Kest in spatstat) analysis (Baddeley, 2023) is based on all inter-event distances, not simply nearest neighbour distances.

- **G function** (Gest in spatstat) is the cumulative frequency distribution of the nearest neighbour distance. This function provides a probability for a specified distance that the nearest neighbour distance to another point will be less than the specified distance (O'Sullivan and Unwin, 2003; Baddeley, 2023).

- **F function** (Fest in spatstat) is the cumulative frequency distribution of the distance to the nearest point from random locations (O'Sullivan and Unwin, 2003; Baddeley, 2023).

- **PCF (PCF in spatstat)** The pair correlation function (PCF) is based on all inter-event distances. It focuses on how many pairs of events are separated by any particular distance. In other words, it describes how likely two events chosen at random will be at some particular distance. Each of these measures can be tested statistically for deviations from the expected values associated with a random point process. The most complex of these tests is the K function, where the additional concept of an L function is introduced to make it easier to detect large deviations from a random pattern.

To illustrate how these different measures can provide insights about a spatial pattern I have used an example from a previously published study on Acacia distributions in Australia by Spooner et al. (2004). The points for each Acacia location was obtained by digitizing points from the maps provided in the paper. Due to the resolution of these some points may have been missed and as such the spatial distribution may vary a little from that in the original paper. Regardless, for our purposes they capture two different spatial patterns; one that is clustered and one that is dispersed. The data was analyzed in R with package spatstat. The NNI, G, F, K, L and PCF analyses were conducted (code is provided at the end of the chapter). Outputs are provided in Figure 7.4.

Interpreting and understanding the graphs and figures?

Dashed red line is the theoretical value of the function for a pattern generated by the IPR/CSR and provides a point of reference.

Grey region (envelope) shows the range of values of the function that occurred across all the simulated realizations of IRP/CSR.

The black line is the function for the observed pattern measured for the dataset.

Interpreting the results: What we are really interested in is whether or not the observed function lies inside or outside the grey envelope." For example, for the PCF, if the black line falls outside the envelope, this tells us that there are more pairs of events at this range of spacings from one another than we would expect to occur by chance. This observation supports the view that the pattern is clustered or aggregated at the stated range of distances. For any distances where the black line falls within the envelope, this means that the PCF falls within the expected bounds at those distances. The exact interpretation of the relationship between the envelope and the observed function is dependent on the function in question.

K-Means and Hierarchical Clustering

Inspired by Baumer et al. (2017) where they clustered cities around the world using k-means in R, I have included them here in case you are also interested in examining some data you may have. Since these methods are already incorporated in GeoDa, running these types of analysis is fairly straight forward making it easy to use and explore. I obtained a data set with the world's largest cities and explored the data using these clustering methods (Figure 7.5).

Spatial Clustering Methods for Areal Information

A lot of health data is available at specified areal units, which can range from municipality, county, district, province, state or country level. From a public health perspective and policy perspective, it is useful to understand if there are particular patterns associated with different disease outcomes and what the drivers of these are so that we can determine if they are of concern and, if so, what needs to be done to further reduce future risks. Similar to PPA introduced earlier, here I have summarized the methods that can be used to explore spatial patterns with areal data (Figure 7.6).

For non-numeric data, the *Joins Count* approach for measuring spatial autocorrelation may be useful. This is infrequently used, and so, are not central to methods used to assess spatial autocorrelation. Again GeoDA does contain these methods so if you need to examine spatial autocorrelation of non-numeric information keep this method in mind.

FIGURE 7.5
Analysis of the world's largest 4,000 cities clustered by (a) the 6-means clustering algorithm and (b) the hierarchical clustering method using ward's-linkage. *Data source*: https://simplemaps.com/data/world-cities. Analysis conducted in GeoDa.

Global Moran's I and Local Moran's I

Although the *Moran's I* (Moran, 1950) looks intimidating, it is very useful for exploring clustering of spatial information and can be used through a variety of tools (GeoDa and ArcGIS Pro (Table 7.2)). It provides a measure of similarity based on surrounding neighbours (e.g. distance or neighbourhood definition – see section on conceptualization of spatial interactions later in the chapter) by including values from the surrounding area into the calculation. The difference measure is summed over all neighbouring pairs (the w_{ij} values from a weights matrix) and adjusted so that the resulting index value is in a standard numerical range.

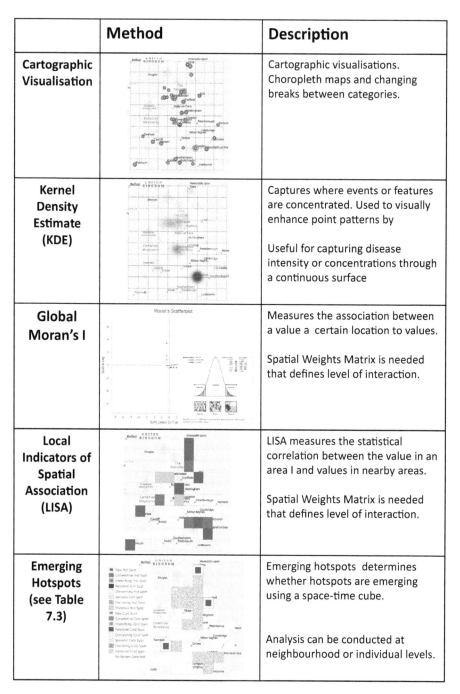

	Method	Description
Cartographic Visualisation		Cartographic visualisations. Choropleth maps and changing breaks between categories.
Kernel Density Estimate (KDE)		Captures where events or features are concentrated. Used to visually enhance point patterns by Useful for capturing disease intensity or concentrations through a continuous surface
Global Moran's I		Measures the association between a value a certain location to values. Spatial Weights Matrix is needed that defines level of interaction.
Local Indicators of Spatial Association (LISA)		LISA measures the statistical correlation between the value in an area I and values in nearby areas. Spatial Weights Matrix is needed that defines level of interaction.
Emerging Hotspots (see Table 7.3)		Emerging hotspots determines whether hotspots are emerging using a space-time cube. Analysis can be conducted at neighbourhood or individual levels.

FIGURE 7.6

Overview of different spatial clustering methods used for analyzing areal data. Illustrations were created using 1944–1965 measles data for the UK. *Source*: Dalziel et al. (2016).

The Local Moran's *I* statistic is calculated as follows (Anselin, 1989, 1995) Eq 7.1:

$$I_i = \frac{x_i - \bar{X}}{S_i^2} \sum_{j=i.j\neq i}^{n} w_{ij} \left(x_j - \bar{X} \right)$$

where x_i is a value for feature *I*, $\bar{X}$ is the mean of the corresponding feature, w_{ij} is the spatial weight between feature *i* and *j*, and Eq 7.2:

$$S_i^2 = \frac{\sum_{j=i.j\neq i}^{n} \left(x_j - \bar{X} \right)^2}{n-1}$$

where *n*=the total number of features.

The z_{Ii} score for the statistics are calculated as Eq 7.3:

$$z_{Ii} = \frac{Ii - E[Ii]}{\sqrt{V[Ii]}}$$

where Eq 7.4:

$$E[Ii] = -\frac{\sum_{j=i.j\neq i}^{n} w_{ij}}{n-1}$$

$$V[Ii] = E\left[I_i^2\right] - E[Ii]^2$$

The product of the differences between each value and the mean is used to determine the similarity measure. A positive result is produced when both the value in the current unit and neighbouring units are higher or lower than the mean. A negative result is produced when the values in the current unit and neighbouring units are on opposite sides of the mean (one higher and one lower). The Moran scatterplot captures the relationship between the value at each location and the average value at neighbouring locations (Figure 7.7).

Cases in the upper-right (HH) and lower-left (LL) confirm positive spatial autocorrelation (Figure 7.7). The upper-right quadrant represents cases where both the value and the local average value of the attribute are higher than the overall average value. Similarly, in the lower-left quadrant, represent cases where both the value and local average value of the attribute are lower than the overall average value. Cases in the lower-right (HL) and upper-left (LH) confirm negative spatial autocorrelation or no spatial autocorrelation. The values for each case are opposite the local average. In summary, these can highlight the following:

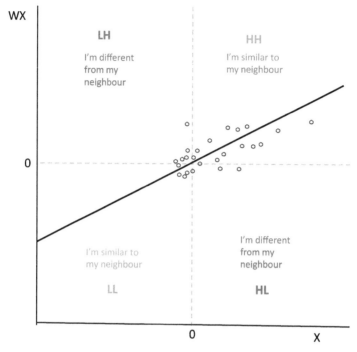

FIGURE 7.7

The four quadrants of the Moran's I scatterplot and how each quadrant relates to clusters of similarity (HH and LL) and differences (HL and LH).

- **Positive autocorrelation** occurs where features are more similar in their location as well as in their attributes than their neighbour;
- **Negative autocorrelation** occurs where features are closer together show greater dissimilarity;
- **Zero autocorrelation** where the location of a feature has no influence.

Global measures are useful for providing us with information about the spatial data and whether the data are spatially autocorrelated or not. However, they do not tell us where clusters are located or which data contributes to the overall spatial autocorrelation trend. In some cases, it may be more important to know which local features are contributing most strongly to the overall pattern and where they are located.

In the context of spatial autocorrelation, the localized phenomena of interest are those areas on the map that contribute particularly strongly to the overall trend (which is usually positive autocorrelation). Methods that enable an analyst to identify localized map regions where data values are strongly positively or negatively associated with one another are collectively known as Local Indicators of Spatial Association (or LISA) (Anselin, 1995). The combination of the cluster map and the significance map allows you to see which locations are contributing most strongly to the global outcome and in which direction.

The LISA statistic address this by capturing how each individual map unit contributes to the overall Moran's I summary measure. A statistical significance level for which each region contributes to the global spatial autocorrelation outcome is determined using a Monte Carlo randomization procedure. In summary, the statistical significance of each value is assessed by comparing the actual value to the value calculated for the same location by randomly reassigning the data among all the areal units and recalculating the values each time. The LISA values are ranked relative to the set of values produced by the randomization process and assessed. If a LISA score is among the top 0.1% (or 1% or 5%) of scores associated with that location under randomization, then it is judged statistically significant at the 0.001 (or 0.01 or 0.05) level.

Hot Spot Analysis (Getis-Ord G_i^*)

The Hot Spot Analysis method calculates the Getis-Ord G_i^* statistic (Getis and Ord, 1992; Ord and Getis, 1995). Each feature is evaluated within the context of neighbouring features. Statistically significant hotspots occur when a feature with a high value is surrounded by other features with high values. The local sum of a feature and its neighbours is compared proportionally to the sum of all features; when the local sum is very different from the expected local sum and when that difference is too large to be the result of random chance, a statistically significant z-score results. For statistically significant positive z-scores, the larger the z-score, the more intense the clustering of high values (hot spot). For statistically significant negative z-scores, the smaller the z-score, the more intense the clustering of low values (cold spots).

The Getis-Ord local statistic is given by Eq 7.5

$$G_i^* = \frac{\sum_{j=1}^{n} w_{i,j} x_j - \bar{X} \sum_{j=1}^{n} w_{i,j}}{S \sqrt{\frac{\left[n \sum_{j=1}^{n} w_{i,j}^2 - \left(\sum_{j=1}^{n} w_{i,j} \right)^2 \right]}{n-1}}}$$

where x_j is the value for feature j, $w_{i,j}$ is the spatial weight between feature i and j and n_i is equal to the total number of features.

$$\bar{X} = \frac{\sum_{j=1}^{n} x_j}{n}$$

$$S = \sqrt{\frac{\sum_{j=1}^{n} x_j^2}{n} - \left(\bar{X}\right)^2}$$

The G_i^* statistic is a z-score (Eqs 7.6, 7.7).

Emerging Hotspot Analysis

The emerging hotspot analysis is useful for identifying trends in the data, such as new trends, intensifying trends, diminishing trends and sporadic hot and cold spot trends, using the Getis-Ord Gi^* statistic (ESRI, 2023). One of several types of patterns can be identified that range from no pattern to one that is changing over time (e.g. intensifying or diminishing; oscillating or is persistent) (Table 7.3).

TABLE 7.3

Hot and Cold Spot Definitions used to Describe Emerging Patterns

Pattern Name	Definition
No pattern detected	Does not fall into any of the hot or cold spot patterns defined below.
New hot spot	A location that is a statistically significant hot spot for the final time step and has never been a statistically significant hot spot before.
Consecutive hot spot	A location with a single uninterrupted run of at least two statistically significant hot spot bins in the final time step intervals. The location has never been a statistically significant hot spot prior to the final hot spot run and less than 90% of all bins are statistically significant hot spots.
Intensifying hot spot	A location that has been a statistically significant hot spot for 90% of the time step intervals, including the final time step. In addition, the intensity of clustering of high counts in each time step is increasing overall and that increase is statistically significant.
Persistent hot spot	A location that has been a statistically significant hot spot for 90% of the time step intervals with no discernible trend in the intensity of clustering over time.
Diminishing hot spot	A location that has been a statistically significant hot spot for 90% of the time step intervals, including the final time step. In addition, the intensity of clustering in each time step is decreasing overall and that decrease is statistically significant.
Sporadic hot spot	A statistically significant hot spot for the final time step interval with a history of also being an on-again and off-again hot spot. Less than 90% of the time step intervals have been statistically significant hot spots and none of the time step intervals have been statistically significant cold spots.
Oscillating hot spot	A statistically significant hot spot for the final time step interval that has a history of also being a statistically significant cold spot during a prior time step. Less than 90% of the time step intervals have been statistically significant hot spots.

(Continued)

TABLE 7.3 (*Continued*)

Hot and Cold Spot Definitions used to Describe Emerging Patterns

Pattern Name	Definition
Historical hot spot	The most recent time period is not hot, but at least 90% of the time step intervals have been statistically significant hot spots.
New cold spot	A location that is a statistically significant cold spot for the final time step and has never been a statistically significant cold spot before.
Consecutive cold spot	A location with a single uninterrupted run of at least two statistically significant cold spot bins in the final time step intervals. The location has never been a statistically significant cold spot prior to the final cold spot run and less than 90% of all bins are statistically significant cold spots.
Intensifying cold spot	A location that has been a statistically significant cold spot for 90% of the time step intervals, including the final time step. In addition, the intensity of clustering of low counts in each time step is increasing overall and that increase is statistically significant.
Persistent cold spot	A location that has been a statistically significant cold spot for 90% of the time step intervals with no discernible trend in the intensity of clustering of counts over time.
Diminishing cold spot	A location that has been a statistically significant cold spot for 90% of the time step intervals, including the final time step. In addition, the intensity of clustering of low counts in each time step is decreasing overall and that decrease is statistically significant.
Sporadic cold spot	A statistically significant cold spot for the final time step interval with a history of also being an on-again and off-again cold spot. Less than 90% of the time step intervals have been statistically significant cold spots and none of the time step intervals have been statistically significant hot spots.
Oscillating cold spot	A statistically significant cold spot for the final time step interval that has a history of also being a statistically significant hot spot during a prior time step. Less than 90% of the time step intervals have been statistically significant cold spots.
Historical cold spot	The most recent time period is not cold, but at least 90% of the time step intervals have been statistically significant cold spots.

Source: ESRI (2023).

Conceptualization of Spatial Interactions – Distance Measures and Concepts

When using spatial clustering methods, it is necessary to consider the relationships between events or objects in physical space. These can be distant-based or more conceptual based on neighbourhoods and adjacency or through social ties.

Distance measures can be calculated in a number of different ways. Here are some of the more common concepts: (i) Euclidean (straight line) distance is often only an approximation, (ii) Spherical, (iii) Network (e.g. using a road network) or Manhattan (Table 7.4).

TABLE 7.4

Different Distance Measures

Distance Measures	Definition	Calculation
Euclidean distance (straight line)	Distance between two points in a straight line	$d_{ij} = \sqrt{(X_i - X_j)^2 + (Y_i - Y_j)^2}$
Spherical distance (latitude and longitude coordinates)	Use for unprojected data, or at world scale	(e.g. Haversine Formula)
Network distance	Distance through a road network	Time/distance/friction (resistance) through a network (also see Chapter 8)
Manhattan metric/city block	The distance between two points in a grid based on a strictly horizontal and/or vertical path (that is, along the grid lines), as opposed to the diagonal or "as the crow flies" distance. The Manhattan distance is the simple sum of the horizontal and vertical components, whereas the diagonal distance might be computed by applying the Pythagorean theorem.	Red, green, yellow = Manhattan distance Blue = Euclidean distance or straight line

Spatial Concepts and Neighbours

Conceptualization of neighbours can be based on adjacency (Figure 7.8) or distance and interactions (Figure 7.9). Spatial methods consider physical distances between objects (Table 7.4, Figure 7.9), such as points or enumeration units, and how close places are in geographical space as defined by a **spatial weights matrix**. A weight is a numerical value that describes the closeness/distance of neighbouring areas or health events and interactions. These can be defined in many ways as captured in Figures 7.8 and 7.9.

Distance-decay and spatial interaction conceptualizations:

- **Inverse distance squared**: impact of one feature on another feature decreases with distance (Figure 7.9a).
- **Fixed distance**: everything within a specified distance of each feature is included in the analysis and everything outside is excluded (Table 7.9b).

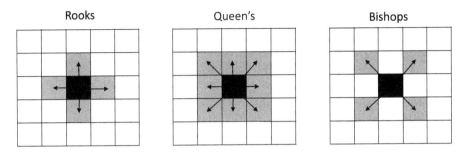

FIGURE 7.8
Rooks, Queen's and Bishop: spatial neighbours based on adjacency or neighbourhood (whether or not areas share a common boundary) and touch the central cell. Image created by Blanford (2023).

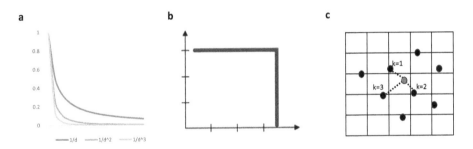

FIGURE 7.9
Spatial neighbours based on (a) inverse distance $1/d$, $1/d^2$, (b) fixed distance and (c) k-nearest neighbours. Image created by Blanford (2023).

- **k-nearest neighbour**: the closest k features are included in the analysis. Each feature is assessed within the spatial context of a specified number of its closest neighbours. If K (the number of neighbours) is 3, then the three closest neighbours to the target features will be included in the computations for that feature (Figure 7.9c).

DBScan – Density-Based Clustering

Density-based clustering methods are designed to discover clusters and noise in data. They are useful for identifying clusters of point features based on their spatial distribution. These methods use unsupervised machine learning clustering algorithms to detect patterns based on spatial location and distance to a specified number of neighbours (Bennett, 2018). DB Scan methods are available in a variety of software (ArcGISPro, GeoDa, R and Python). Three methods are useful for identifying clusters (Bennett, 2018).

- **Defined distance (DBSCAN)**: A specified distance is used to separate dense clusters from noise. All meaningful clusters have similar densities.

- **Self-adjusting (HDBSCAN)**: This method uses a range of distances to separate clusters of varying densities from noise.

- **Multi-scale (OPTICS – ordering points to identify the clustering structure)**: This method uses the distance between neighbouring features to create a reachability plot, which is then used to separate clusters of varying densities from noise. The OPTICS algorithm offers the most flexibility in fine-tuning the clusters that are detected, though it is computationally intensive, particularly with a large Search Distance. Useful for identifying clusters in space and time.

SaTScan

SaTScan™ is a tool developed under the joint auspices of Martin Kulldorff, the National Cancer Institute, and the New York City Department of Health and Mental Hygiene (https://www.satscan.org/). SaTScan™ contains methods useful for analysing data using the spatial, temporal or space-time scan statistics. It was designed to detect spatial or space-time disease clusters and determine if clusters are statistically significant and whether a disease is randomly distributed over space, over time or over space and time. It is used to evaluate the statistical significance of disease clusters and perform repeated time-periodic disease surveillance for early detection of disease outbreaks.
SaTScan uses either:

- a Poisson-based model, where the number of events in a geographical area is Poisson-distributed, according to a known underlying population at risk;

- a Bernoulli model, with 0/1 event data such as cases and controls; a space-time permutation model, using only case data;

- an ordinal model, for ordered categorical data; an exponential model for survival time data with or without censored variables;

- or a normal model for other types of continuous data.

The data may be either aggregated at the census tract, zip code, county or other administrative level (e.g. Chapter 5), or there may be unique coordinates for each observation. SaTScan adjusts for the underlying spatial inhomogeneity of a background population. It can also adjust for any number of categorical covariates provided by the user, as well as for temporal trends, known space-time clusters and missing data. It is possible to scan multiple data sets simultaneously to look for clusters.

Targeting Interventions

Geographical information systems and clustering methods are useful for providing cost-effective ways to monitor and manage disease incidence regionally, nationally and among cross-border migrant populations (e.g. Suriname: Hiwat et al., 2018; Ethiopia: Glendening et al., 2023). Not only are these methods helpful for improving our understanding of disease locations and hotspots (e.g. malaria: Hiwat et al., 2018) or health deficiencies (e.g. anaemia in school children: Robert et al., 2023) but the outputs from these methods are also helpful for guiding interventions.

In particular, SaTScan-based analyses have been useful for identifying malaria clusters from which targeted local control strategies have been implemented (e.g. Coleman et al., 2009; Bostoen and Chalabi, 2006; Hiwat et al., 2018)).

Summary

That is spatial cluster methods in a nutshell! Now it is your turn to explore one of the spatial clustering methods.

Activity – Obesity in the Netherlands. Where Are the Hotspots?

Obesity

Obesity is a problem throughout the world (Malik et al., 2013) and has tripled since 1975 (WHO, 2021). Since 1980, the prevalence of obesity has doubled in more than 70 countries and has continuously increased in most other countries (Collaborators, 2017). More than 1.9 billion adults are overweight, with 650 million considered obese (WHO, 2021). In 2020, 39 million children, age 5 and under, were overweight or obese (WHO, 2021).

In the US, obesity health care costs $147 billion annually (CDC, 2023b). Since 1990, adult obesity rates have increased substantially in the USA. Obesity affects 100.1 million (41.9%) adults and 14.7 million (19.7%) children (CDC, 2023b). Thirty states have obesity rates in excess of 40% (CDC, 2023b). The five states with the highest number of obese counties include Texas, Oklahoma, West Virginia, South Carolina and Virgina (see Figure 7.1; CDC, 2023b).

Obesity has become a problem due to the associated health problems caused by an expanding set of chronic diseases, including cardiovascular disease, diabetes mellitus, chronic kidney disease, many cancers, an array of musculoskeletal disorders (see references in Collaborators, 2017), heart disease, stroke, diabetes, sleep apnoea and complications during pregnancy (CDC, 2010).

A variety of risk factors have been associated with obesity, including behaviour (e.g. physical inactivity and smoking), psychological factors, economic factors, genetics and changes in our environment (e.g. urbanization, cheaper and more accessible food) (Wright and Aronne, 2012; Malik et al., 2013; Safaei et al., 2021). For example, built environments in some geographic locations promote an obesogenic environment, promoting sedentary lifestyles and provide an abundance of processed foods (Charreire et al., 2010; Fraser et al., 2010; Caspi et al., 2012; Cooksey-Stowers et al., 2017; Granheim et al., 2022) and fast-foods (e.g. in the Netherlands: Harbers et al., 2021) which can lead to an increased risk of obesity (Poti et al., 2017; Hall et al., 2019).

Obesity in the Netherlands

Obesity has been steadily rising since 1981 in the Netherlands, increasing from 4.4% to 13.3% (CBS, 2023). Studies examining the obesogenic school environment found that unhealthy food options were more often present than healthier options in schools (Timmermans et al., 2018). For this study, we will examine obesity in the Netherlands and conduct a similar analysis to Qiu et al. (2020) and Frank (2003). Now that you have some background on obesity, it is your turn to assess where obesity could be an issue in the Netherlands.

Overview of Activity

The activity has been broken down into the following steps:

1. Preparing the data for analysis
2. Getting started with GeoDa
3. *Where* are the obesity hotspots in the Netherlands?
   a) Performing the cluster analysis
   b) Analysing the clusters
4. *How* are these results useful?
   a) Assess and interpret the results.
5. Conclusions

Software: ArcGIS Pro and GeoDA
  Relevant Readings:

- Qiu, G., et al. (2020) Geographic clustering and region-specific determinants of obesity in the Netherlands. *Geospatial Health*, 15(1), 839.
- Cooksey-Stowers, K., et al. (2017) Food swamps predict obesity rates better than food deserts in the United States. *International Journal*

of Environmental Research and Public Health, 14, 1366. https://doi. org/10.3390/ijerph14111366.

- Thornton, L. E., A. M. Kavanagh (2012) Association between fast food purchasing and the local food environment. *Nutrition & Diabetes*, 2(12), e53. https://doi.org/10.1038/nutd.2012.27.

Optional Readings:

- Frank, A. I. (2003) *Using Measures of Spatial Autocorrelation to Describe Socio-Economic and Racial Residential Patterns in US Urban Areas*. New York: Taylor & Francis.
- Mills, C. W., et al. (2020) Use of small-area estimates to describe county-level geographic variation in prevalence of extreme obesity among US adults. *JAMA Network Open*, 3(5), e204289.

Data

All the data that will be used for this activity is summarized in Table 7.5.

Obesity Data for the Netherlands: Preparing the Data for Spatial Analysis

The percentage of the population with obesity in the Netherlands is available as a csv file. The information is recorded at a municipality and neighbourhood level. However, to make this data spatial, we will need to join it to a geographic dataset. The obesity dataset contains several codes, one of which can be linked to neighbourhood level information in the CBS dataset wijk-en-buurtkaart.

TABLE 7.5

Summary of Data Sources That Will Be Used to Conduct the Obesity Spatial Clustering Analysis in the Netherlands.

Data set	Description	Date	Source
Obesity in NL	Obesity rates in the Netherlands. Recorded at district and neighbourhood and municipality level	2020	RIVM https://www.vzinfo.nl/ overgewicht/regionaal/obesitas
Neighbourhood boundaries	District and neighbourhood boundaries for the Netherlands.	2022	CBS https://www.cbs.nl/nl-nl/ dossier/nederland-regionaal/ geografische-data/ wijk-en-buurtkaart-2022
Food outlets	Points of interest	2023	OSM https://download.geofabrik.de/ europe/netherlands

Making Obesity Spatial for the Netherlands

In ArcGISPro, join the obesitas_NL2020.csv file with the wijk2022_vch7a shapefile (Figure 7.10).
   Right click on the layer, select Joins and Relates – Add Join

   Input Table: wij2022_vch7a
   Input Join Field: wk_code
   Join Table: Obesitats_NL2022.csv
   Join Table Field: wk

Check that all of the fields were joined correctly (open the attribute table of the shapefile). Were there any problems (e.g. did you notice any <NULL> matches)? Record any issues you found (e.g. number of records that did not match) and why?
   Save as a new feature layer. To create a new feature layer with the obesity information included, export the feature information (right-click on the layer, select Data – Export Features). The new layer will be saved to the project geodatabase.
   Symbolize the map based on the obesity percentages.

What Are the Obesity Rates in the Netherlands?

Explore the obesity map (e.g. Figure 7.10). Where in the Netherlands does obesity occur the highest and lowest? Describe what you found. Explore the data further in GeoDa.

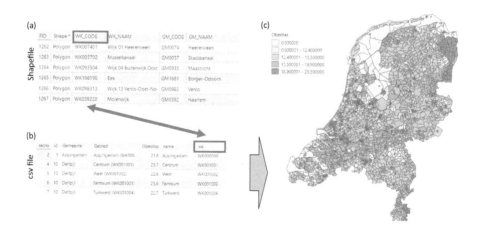

FIGURE 7.10
Illustration of joining (a) geographic information with (b) non-geographic information stored in a table (e.g. csv file) or spreadsheet using a unique geographic identifier to create a (c)map illustrating the distribution of the non-spatial data, in this case obesity in the Netherlands.

GeoDa is free, can be installed from the web (http://geodacenter.github.io/). For an overview see Figure 7.11.

Once installed, open GeoDa. Add the shapefile or the geodatbase (*.gdb) where you saved the feature layer you created.

Explore obesity using the **Map** options available to create different visualizations, such as a quantile (Figure 7.12a), box map or cartogram.

Next **Explore** the values using the different options available. Create a histogram and boxplot (Figure 7.12b,c).

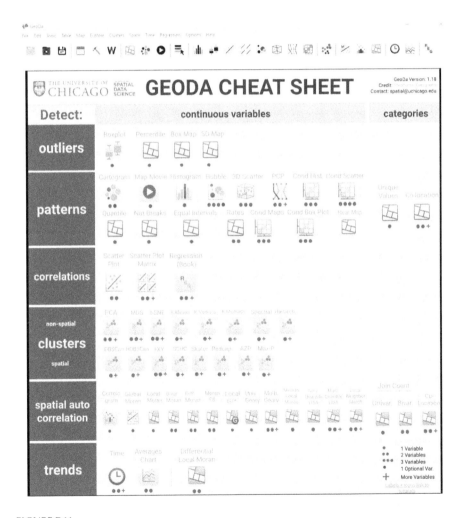

FIGURE 7.11
GeoDa cheatsheet provides a summary of the exploratory data analysis methods that are available in GeoDA and what they can be used to detect. *Source*: Anselin et al., (2022); http://geodacenter.github.io/cheatsheet.html.

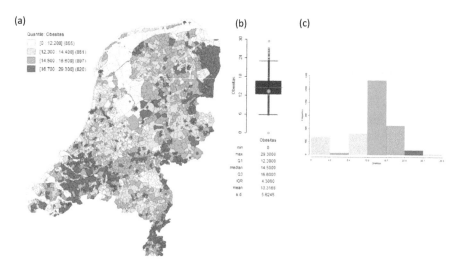

FIGURE 7.12
(a) Spatial distribution of obesity in the Netherlands during 2020 by neighbourhood district.
(b) Boxplot of obesity values and (c) histogram.

In *GeoDa*, each of the windows is linked. Use the linked-brushing options to interact with the data and further develop an understanding of what is taking place where.

Where Are Obesity Hotspots in the Netherlands?

Next, analyse the patterns of obesity in the Netherlands. In Chapter 2, point pattern analyses were used to assess patterns; however, in this case, the data are available in areal units and neighbourhood districts. Instead, the Moran's I and Local Indicators of Spatial Association (LISA) will be used.

Calculate the Global and Local Moran's *I* and Moran Scatter Plot. First, you will need to define a spatial weights matrix (Table 7.6). This defines the representation of the contiguity structure of the map. In other words, this essentially capture the spatial relationship and interactions between the different map units. A number of options are available, as summarized in Figures 7.8 and 7.9. For the Moran's *I* calculation, this provides the w_{ij} values to determine which pairs of values should be included in the correlation calculation.

Go to **Tools-Weights Manager** and create a weights file **(.gal)**:
   Select Id variable (unique identifier): wk_code
   Contiguity Weight (Queen, Rook, etc.): Queen (Order of contiguity = 1)

TABLE 7.6

Spatial Weights Types

Spatial Weights Matrix Type	Options	Sub-Options
Contiguity	Queen	Order of contiguity
	Rook	1...
Distance	Distance metric: Euclidean distance	Specify bandwidth Use inverse distance (set power)
	K-nearest neighbour	Number of neighbours Use inverse distance (set power)
	Kernel	Kernel function (uniform, triangular, Epanechnikov, quartic, Gaussian) Specify bandwidth Use max knn distance as bandwidth

Once the spatial weights matrix has been created, go to **Space**. First, run the **univariate Moran's I** (select Obesitas as the variable). This will provide an overview of whether there is clustering or not.

However, knowing where the clusters are and which features are contributing most strongly to the overall pattern is really what we are interested in so that we can figure out how to respond. To achieve this, next run the **univariate Local Moran's I** (select Obesitas as the variable). Select all of the windows (significance map, cluster map and Moran Scatter Plot). Next explore and interact with the different visualizations and outputs. Use linked-brushing to explore how the scatterplot links to each of the outputs, as illustrated in Figure 7.13. The linked-brushing is useful for revealing the patterns in spatial data more easily and the spatial autocorrelation effects, for exploring cases that may not follow a trend.

Interpreting the results from the map is straightforward. Red regions highlight areas with high values surrounded by neighbours that also have high values (High-High) (Figure 7.13b). Blue regions show areas where values are low, surrounded by neighbours with low values (low-low, Figure 7.13c). Pale blue regions are low-high (Figure 7.13a), and pink areas are high-low (Figure 7.13d), showing where values differ from neighbours and can highlight outliers.

Strongly coloured regions contribute *significantly* to a positive global spatial autocorrelation outcome, while paler colours contribute *significantly* to a negative spatial autocorrelation outcome. The significance map shows how each region contributes to the global spatial autocorrelation outcome.

How to Respond?

Summarize what you found. Did you identify any obesity hotspots in the Netherlands? Where are these located? How can the maps be used to reduce obesity risks?

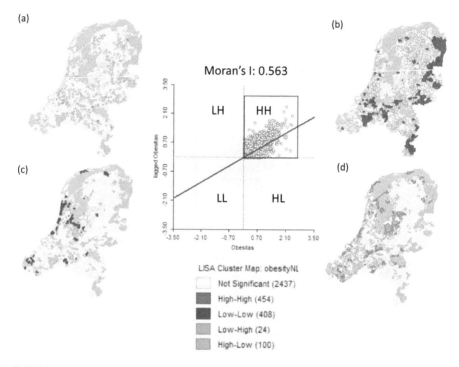

FIGURE 7.13

Obesity clusters identified in the Netherlands. (a) Areas with low-high obesity rates, (b) areas with high obesity rates, (c) areas with low obesity rates and (d) areas with high-low obesity rates. The Moran Scatter Plot shows the relationship between a variable such as obesity and the average value of its neighbours for the same variable.

References

Anselin, L. (1989) What Is Special about Spatial Data? Alternative Perspectives on Spatial Data Analysis. *NCGIA Technical Reports.* https://api.semanticscholar.org/CorpusID:1861693.

—— (1995) Local indicators of spatial association-LISA. *Geographical Analysis*, 27, 93–115.

Anselin, L., X. Li & J. Koschinsky (2022) GeoDa, from the desktop to an ecosystem for exploring spatial data. *Geographical Analysis*, 54, 439–466.

Baddeley, A. (2023) Package 'SPATSTAT': Spatial Point Pattern Analysis, Model-Fitting, Simulation, Tests. *R Topics Documented*, 44 p. https://cran.r-project.org/web/packages/spatstat/spatstat.pdf (last accessed Aug 16 2023).

Bailey, T. C. & A. C. Gatrell (1995) *Interactive Spatial Data Analysis*, 432 p. Essex: Longman Scientific & Technical.

Barker, L. E., K. A. Kirtland, E. W. Gregg, L. S. Geiss & T. J. Thompson (2011) Geographic distribution of diagnosed diabetes in the U.S.: A diabetes belt. *American Journal of Preventive Medicine*, 40, 434–439.

Baumer, B. S., D. T. Kaplan & N. J. Horton (2017) *Modern Data Science with R*. Boca Raton, FL: CRC Press.

Bennett, L. R. (2018) Machine learning in ArcGIS. *ArcUser*, 21, 8–9.

Bostoen, K. & Z. Chalabi (2006) Optimization of household survey sampling without sample frames. *International Journal of Epidemiology*, 35, 751–755.

Bousema, T., J. T. Griffin, R. W. Sauerwein, D. L. Smith, T. S. Churcher, et al. (2012) Hitting hotspots: Spatial targeting of malaria for control and elimination. *PLoS Med*, 9, e1001165.

Bowyer, R. C. E., T. Varsavsky, E. J. Thompson, C. H. Sudre, B. A. K. Murray, et al. (2021) Geo-social gradients in predicted COVID-19 prevalence in Great Britain: Results from 1,960,242 users of the COVID-19 symptoms study app. *Thorax*, 76, 723–725.

Carpenter, T. E. (2001) Methods to investigate spatial and temporal clustering in veterinary epidemiology. *Preventive Veterinary Medicine*, 48, 303–320.

Caspi, C. E., G. Sorensen, S. V. Subramanian & I. Kawachi (2012) The local food environment and diet: A systematic review. *Health & Place*, 18, 1172–1187.

Cassetti, T., F. La Rosa, L. Rossi, D. D'Alo & F. Stracci (2008) Cancer incidence in men: A cluster analysis of spatial patterns. *BMC Cancer*, 8, 344.

CBS (2023) Lengte en gewicht van personen, ondergewicht en overgewicht; vanaf 1981. https://opendata.cbs.nl/#/CBS/nl/dataset/81565NED/table?searchKey words=overgewicht. (last accessed April 27 2024).

CDC (2010) Adult Obesity. Obesity Rising among Adults. *CDC Vitalsign*. https://www.cdc.gov/vitalsigns/adultobesity/ (last accessed Sep 10 2016).

—— (2023a) Diagnosed Diabetes - Total, Adults Aged 20+ Years, Crude Percentage, NaturalBreaks, AllCounties.https://gis.cdc.gov/grasp/diabetes/diabetesatlas-surveillance.html# (last accessed Aug 22 2023).

—— (2023b) HOP 2023: CDC-RFA-DP-23-0013: The High Obesity Program (HOP 2023). https://www.cdc.gov/nccdphp/dnpao/state-local-programs/fundingopp/2023/hop-1809/pdfs/hop-dp-information-call-slides-021423-508.pdf (last accessed April 27 2024).

Charreire, H., R. Casey, P. Salze, C. Simon, B. Chaix, et al. (2010) Measuring the food environment using geographical information systems: A methodological review. *Public Health Nutrition*, 13, 1773–1785.

Chen, R. & E. C. Holmes (2010) Hitchhiking and the population genetic structure of avian influenza virus. *Journal of Molecular Evolution*, 70, 98–105.

Clark, P. J. & F. C. Evans (1954) Distance to nearest neighbor as a measure of spatial relationships in populations. *Ecology*, 35, 73–104.

Coleman, M., M. Coleman, A. M. Mabuza, G. Kok, M. Coetzee, et al. (2009) Using the SaTScan method to detect local malaria clusters for guiding malaria control programmes. *Malaria Journal*, 8, 1–6.

Collaborators, G. B. D. O. (2017) Health effects of overweight and obesity in 195 countries over 25 years. *New England Journal of Medicine*, 377, 13–27.

Cooksey-Stowers, K., M. B. Schwartz & K. D. Brownell (2017) Food swamps predict obesity rates better than food deserts in the United States. *International Journal of Environmental Research and Public Health*, 14, 1366.

Cromley, E. K. & S. L. McLafferty (2012) *GIS and Public Health*. New York: Guilford Press.

Dalziel, B. D., O. N. Bjørnstad, W. G. van Panhuis, D. S. Burke, C. J. E. Metcalf, et al. (2016) Persistent chaos of measles epidemics in the prevaccination United States caused by a small change in seasonal transmission patterns. *PLoS Computational Biology*, 12, e1004655.

Doreena, D., H. Juahir, M. T. Latif, S. M. Zain & A. Z. Aris (2012) Spatial assessment of air quality patterns in Malaysia using multivariate analysis. *Atmospheric Environment*, 60, 172–181.

ESRI (2023) ArcGIS Pro Help. https://pro.arcgis.com/en/pro-app/latest/help/main/welcome-to-the-arcgis-pro-app-help.htm (last accessed Sep 15 2023).

Frank, A. I. (2003) *Using Measures of Spatial Autocorrelation to Describe Socio-Economic and Racial Residential Patterns in US Urban Areas*. New York: Taylor & Francis.

Fraser, L. K., K. L. Edwards, J. Cade & G. P. Clarke (2010) The geography of fast food outlets: A review. *International Journal of Environmental Research and Public Health*, 7, 2290–2308.

Fritz, C. E., N. Schuurman, C. Robertson & S. Lear (2013) A scoping review of spatial cluster analysis techniques for point-event data. *Geospatial Health*, 7, 183–198.

Geo DA. Geo DA Software Version 1.18. The University of Chicago. http://geodacenter.github.io/

Getis, A. & J. K. Ord (1992) The analysis of spatial association by use of distance statistics. *Geographical Analysis*, 24, 189–206.

Glendening, N., Haileselassie, W., Parker, D.M. (2023) *Chapter 5 A conceptual framework for understanding extractive settlements and disease: demography, environment, and epidemiology*. Leiden: Wageningen Academic: 121–160.

Granheim, S. I., A. L. Løvhaug, L. Terragni, L. E. Torheim & M. Thurston (2022) Mapping the digital food environment: A systematic scoping review. *Obesity Reviews*, 23, e13356.

Grubesic, T. H., R. Wei & A. T. Murray (2014) Spatial clustering overview and comparison: Accuracy, sensitivity, and computational expense. *Annals of the Association of American Geographers*, 104, 1134–1156.

Hall, K. D., A. Ayuketah, R. Brychta, H. Cai, T. Cassimatis, et al. (2019) Ultra-processed diets cause excess calorie intake and weight gain: An inpatient randomized controlled trial of ad libitum food intake. *Cell Metabolism*, 30, 67–77.

Hamid, J. S., C. Meaney, N. S. Crowcroft, J. Granerod, J. Beyene, et al. (2010) Cluster analysis for identifying sub-groups and selecting potential discriminatory variables in human encephalitis. *BMC Infectious Diseases*, 10, 364.

Harbers, M. C., J. W. J. Beulens, J. M. A. Boer, D. Karssenberg, J. D. Mackenbach, et al. (2021) Residential exposure to fast-food restaurants and its association with diet quality, overweight and obesity in the Netherlands: A cross-sectional analysis in the EPIC-NL cohort. *Nutrition Journal*, 20, 56.

Hiwat, H., B. Martínez-López, H. Cairo, L. Hardjopawiro, A. Boerleider, et al. (2018) Malaria epidemiology in Suriname from 2000 to 2016: Trends, opportunities and challenges for elimination. *Malaria Journal*, 17, 1–13.

Jennings, J. M., F. C. Curriero, D. Celentano & J. M. Ellen (2005) Geographic identification of high gonorrhea transmission areas in Baltimore, Maryland. *American Journal of Epidemiology*, 161, 73–80.

Kulldorff, M. (1997) A spatial scan statistic. *Communications in Statistics - Theory and Methods*, 26, 1481–1496.

Lam, T. T., H. S. Ip, E. Ghedin, D. E. Wentworth, R. A. Halpin, et al. (2012) Migratory flyway and geographical distance are barriers to the gene flow of influenza virus among North American birds. *Ecology Letters*, 15, 24–33.

Lawson, A. B. (2010) Hotspot detection and clustering: Ways and means. *Environmental and Ecological Statistics*, 17, 231–245.

Malik, V. S., W. C. Willett & F. B. Hu (2013) Global obesity: Trends, risk factors and policy implications. *Nature Reviews Endocrinology*, 9, 13–27.

Mandl, K. D., M. Overhage, M. M. Wagner, W. B. Lober, P. Sebastiani, et al. (2004) Implementing syndromic surveillance: A practical guide informed by the early experience. *Journal of the American Medical Informatics Association*, 11, 141–150.

Maturana, C. R., A. D. de Oliveira, S. Nadal, B. Bilalli, F. Z. Serrat, et al. (2022) Advances and challenges in automated malaria diagnosis using digital microscopy imaging with artificial intelligence tools: A review. *Frontiers in Microbiology*, 13.

--- (2023) iMAGING: a novel automated system for malaria diagnosis by using artificial intelligence tools and a universal low-cost robotized microscope. *Frontiers in Microbiology*, 14

Meliker, J. R., G. M. Jacquez, P. Goovaerts, G. Copeland & M. Yassine (2009) Spatial cluster analysis of early stage breast cancer: A method for public health practice using cancer registry data. *Cancer Causes Control*, 20, 1061–1069.

Mills, C. W., G. Johnson, T. T. K. Huang, D. Balk & K. Wyka (2020) Use of small-area estimates to describe county-level geographic variation in prevalence of extreme obesity among US adults. *JAMA Network Open*, 3, e204289.

Moran, P. A. P. (1950) Notes on continuous stochastic phenomena. *Biometrika*, 37, 17–23.

Neethu, C. V. & S. Surendran (2013) Review of spatial clustering methods. *International Journal of Information Technology Infrastructure*, 2, 15–24.

Nekola, J. C. & C. E. Kraft (2002) Spatial constraint of peatland butterfly occurrences within a heterogeneous landscape. *Oecologia*, 130, 53–61.

O'Sullivan, D. & D. Unwin (2003) *Geographic Information Analysis*. New York: John Wiley & Sons.

Ohno, Y. & K. Aoki (1981) Cancer deaths by city and county in Japan (1969±1971): A test of significance for geographic clusters of disease. *Statistics in Medicine*, 15D, 251–258.

Ohno, Y., K. Aoki & N. Aoki (1979) A test of significance for geographic clusters of disease. *International Journal of Epidemiology*, 8, 273–281.

Ord, J. K. & A. Getis (1995) Local spatial autocorrelation statistics: Distributional issues and an application. *Geographical Analysis*, 27, 286–306.

Padilla, C. M., S. Deguen, B. Lalloue, O. Blanchard, C. Beaugard, et al. (2013) Cluster analysis of social and environment inequalities of infant mortality. A spatial study in small areas revealed by local disease mapping in France. *Science of the Total Environment*, 454–455, 433–441.

Poti, J. M., B. Braga & B. Qin (2017) Ultra-processed food intake and obesity: what really matters for health-processing or nutrient content? *Current Obesity Reports*, 6, 420–431.

Qiu, G., X. Liu, A. Y. Amiranti, M. Yasini, T. Wu, et al. (2020) Geographic clustering and region-specific determinants of obesity in the Netherlands. *Geospatial Health*, 15, 839.

Robert, B. N., A. Cherono, E. Mumo, C. Mwandawiro, O. Collins, et al. (2023) Spatial variation and clustering of anaemia prevalence in school-aged children in Western Kenya. *PLoS One*, 18, e0282382.

Robertson, C. & T. A. Nelson (2010) Review of software for space-time disease surveillance. *International Journal of Health Geographics*, 9, 16.

Rushton, G. (2003) Public health, GIS, and spatial analytic tools. *Annual Review of Public Health*, 24, 43–56.

Safaei, M., E. A. Sundararajan, M. Driss, W. Boulila & A. Shapi'i (2021) A systematic literature review on obesity: Understanding the causes & consequences of obesity and reviewing various machine learning approaches used to predict obesity. *Computers in Biology and Medicine*, 136, 104754.

SaTScan™. Software for the Spatial Space Time Statistics. SaTScan v6.0. https://www.satscan.org/.

Si, Y., A. K. Skidmore, T. Wang, W. F. De Boer, P. Debba, et al. (2009) Spatio-temporal dynamics of global H5N1 outbreaks match bird migration patterns. *Geospatial Health*, 4, 65–78.

Spooner, P. G., I. D. Lunt, A. Okabe & S. Shiode (2004) Spatial analysis of roadside Acacia populations on a road network using the network K-function. *Landscape Ecology*, 19, 491–499.

Sugumaran, R., S. R. Larson & J. P. Degroote (2009) Spatio-temporal cluster analysis of county-based human West Nile virus incidence in the continental United States. *International Journal of Health Geographics*, 8, 43.

Timmermans, J., C. Dijkstra, C. Kamphuis, M. Huitink, E. Van der Zee, et al. (2018) 'Obesogenic' school food environments? An urban case study in the Netherlands. *International Journal of Environmental Research and Public Health*, 15, 619.

van Eer, E. D., G. Bretas & H. Hiwat (2018) Decreased endemic malaria in Suriname: Moving towards elimination. *Malaria Journal*, 17, 56.

Walter, S. D. (1993) Assessing spatial patterns in disease rates. *Statistics in Medicine*, 12, 1885–1894.

WHO (2021) Obesity. https://www.who.int/news-room/fact-sheets/detail/obesity-and-overweight (last accessed 29 Oct 2023).

Wright, S. M. & L. J. Aronne (2012) Causes of obesity. *Abdominal Radiology*, 37, 730–732.

Zhang, L., S. Nurvianto & R. Harrison (2010) Factors affecting the distribution and abundance of Asplenium nidus L. in a tropical lowland rain forest in Peninsular Malaysia. *Biotropica*, 42, 464–469.

Appendix

R Code for Point Pattern Analysis

Libraries Needed in R

```
library(maptools)
library(spatstat)
library(sf)
library(terra)
library(rgdal)
```

Setting Up the Files for Analysis in R

```
#Setting up the data
filedirgis <-"C:/directory /"
geofilebnd <-paste(filedirgis,"acaciabnd.shp", sep="")
geofile <-paste(filedirgis,"acacia1.shp", sep="")

#check that R can find the file and that there are no path
name problems.
file.exists(geofile)
setwd(system.file("shapes", package="maptools"))

#read in the shapefile
acaciabnd <- readShapeSpatial(geofilebnd)
s.sfbnd <- st_read(geofilebnd)
acaciabndogr <- readOGR(filedirgis, "acaciabnd")
acbnd <- as.owin(acaciabndogr)

#convert data to spatstat format
#Set up the analysis environment by creating an analysis
window
acacia1shp <- readShapeSpatial(geofile)
s.sf <- st_read(geofile)
acaciaogr <- readOGR(filedirgis, "acacia1")
ac1<-as.ppp(acaciaogr)

#check the shapefile will plot correctly
plot(acbnd)
plot(ac1$x, ac1$y)
```

Kernel Density Analysis in R

Kernel density visualization is performed in spatstat using the density() function. The second (optional) parameter in the density function is the bandwidth (sigma). R's definition of bandwidth is the standard deviation of a Gaussian (i.e., normal) kernel function and is actually only around 1/2 of the radius across which events will be 'spread' by the kernel function.

R provides a function that will suggest an optimal bandwidth to use. However, still explore different bandwidths.

```
#Density Default
KDE1 <-density(ac1) # Using the default bandwidth
plot(KDE1, main=NULL, las=1)
contour(KDE1, add=TRUE, col = "white")
#add a label to the plot
mtext("default bandwidth")
#Density-optimal bandwidth
KDE_opt <-bw.diggle(ac1)
# Convert the bandwidth vector
diggle_bandwidth <-toString(KDE_opt, width =5)

#Density using optimal bandwidth
KDE3 <-density(ac1, sigma=KDE_opt) # Using the diggle
bandwidth
plot(KDE3, main=NULL)
contour(KDE3, add=TRUE, col = "white")
mtext(diggle_bandwidth)

#Density-changing bandwidth (to change the bandwidth change
the value of sigma)
KDE2 <-density(ac1, sigma=500) # set different bandwidths. This
data is in projected meters.
plot(KDE2, main=NULL)
contour(KDE2, add=TRUE, col = "white")
mtext("500 meters")
```

Mean Nearest Neighbor Distance Analysis in R

The spatstat nearest neighbor function nndist.ppp() returns a list of all the nearest neighbor distances in the pattern. For a quick statistical assessment, you can also compare the mean value to the expected for an IRP/CSR pattern of the same intensity

```
nnd <-nndist.ppp(ac1)
hist(nnd)
summary(nnd)
#mean distance (change k=2,3 etc.)
obs_ann <- mean(nndist(ac1,k=1))
obs_ann
```

Quadrat Analysis in R

In spatstat, quadratcount() and quadrat.test() can be used to perform the quadrat analysis. The test will report a *p*-value, which can be used to determine whether the pattern is statistically different from one generated by IRP/CSR.

```
# the grid (4 rows × 8 columns) can be adjusted.
q <-quadratcount(ac1, 4, 8)
plot(q)
#Add intensity of each quadrat
plot(intensity(q, image=TRUE), main=NULL, las=1)
#perform the significance test
quadrat.test(ac1, 4, 8)
```

Distance-Based Analysis with Monte Carlo Assessment

The distance-based functions: G, F, K (and L) and the pair correlation function are all supported in spatstat using the built-in functions, Gest(), Fest(), Kest(), Lest() and pcf().

```
g_env <-Gest(ac1)
plot(g_env)

#Add an envelope
#For the first run set the nsim to a low number. For the final
analysis remember to change nsim=99

#initializing and plotting the G estimation
g_env <-envelope(ac1, Gest, nsim=5, nrank=1)
plot(g_env)

#initializing and plotting the F estimation
f_env <- envelope(ac1, Fest, nsim=5, nrank=1)
plot(f_env)

#initializing and plotting the K estimation
k_env <-envelope(ac1, Kest, nsim=5, nrank=1)
plot(k_env)

#initializing and plotting the L estimation
l_env <- envelope(ac1, Lest, nsim=5, nrank=1)
plot(l_env)

#initializing and plotting the pcf estimation
# To control the range of values displayed in a plot's axes
use xlim= and ylim= parameters
pcf_env <-envelope(ac1, pcf, nsim=5, nrank=1)
plot(pcf_env, ylim=c(0, 5))
```

Determining the Three Centers in R

In ArcGIS these descriptive spatial statistics are easy to calculate. In R however, you need to do the calculations by hand. I have included the necessary code below.

```
#------MEAN CENTRE
#calculate mean centre of the crime locations
xmean <- mean(ac1$x)
ymean <-mean(ac1$y)


#------MEDIAN CENTRE
#calculate the median centre of the crime locations
xmed <- median(ac1$x)
ymed <- median(ac1$y
#to access the variables in the shapefile, the data needs to
be set to a data.frame
newhom_df<-data.frame(ac1)
#check the definition of the variables.
str(newhom_df)


#If the variables you are using are defined as a factor then
convert the variables to an integer
newhom_df$FREQUENCY <- as.integer(newhom_df$FREQUENCY)
newhom_df$OBJECTID <- as.integer(newhom_df$OBJECTID)


#create a list of the x coordinates. This will be used to
define the number of rows
a=list(ac1$x)


#------WEIGHTED MEAN CENTRE
#Calculate the Weighted mean
d=0
sumcount = 0
sumxbar = 0
sumybar = 0
for(i in 1:length(a[[1]])){
  xbar <- (ac1$x[i]  * newhom_df$FREQUENCY[i])
  ybar <- (ac1$y[i]  * newhom_df$FREQUENCY[i])
  sumxbar = xbar + sumxbar
  sumybar = ybar + sumybar
  sumcount <- newhom_df$FREQUENCY[i] + sumcount
}
xbarw <- sumxbar/sumcount
ybarw <- sumybar/sumcount


#------STANDARD DISTANCE OF EVENT
# Compute the standard distance of an event
```

```
#Std_Dist <- sqrt(sum((ac1$x - xmean)^2 + (ac1$y - ymean)^2) /
nrow(ac1$n))

#Calculate the Std_Dist
d=0
for(i in 1:length(a[[1]])){
  c<-((ac1$x[i] - xmean)^2 + (ac1$y[i] - ymean)^2)
  d <-(d+c)
}
Std_Dist <- sqrt(d /length(a[[1]]))

# make a circle of one standard distance about the mean center
bearing <- 1:360 * pi/180
cx <- xmean +Std_Dist * cos(bearing)
cy <- ymean +Std_Dist * sin(bearing)
circle <- cbind(cx, cy)

#------CENTRAL POINT
#Here is some code I put together to identify the most central
point:
#Calculate the point with the shortest distance to all points
#sqrt((x2-x1)^2 + (y2-y1)^2

sumdist2 = 1000000000
for(i in 1:length(a[[1]])){
  x1 = ac1$x[i]
  y1= ac1$y[i]
  recno = newhom_df$OBJECTID[i]
  #print(recno)
  #check against all other points
    sumdist1 = 0
    for(j in 1:length(a[[1]])){
      recno2 = newhom_df$OBJECTID[j]
      x2 = ac1$x[j]
      y2= ac1$y[j]
      if(recno==recno2){
      }else {
      dist1 <-(sqrt((x2-x1)^2 + (y2-y1)^2))
          sumdist1 = sumdist1 +dist1
          #print(sumdist1)
          }
    }
    #print(«test»)
    if (sumdist1 < sumdist2){
          dist3<-list(recno, sumdist1, x1,y1)
          sumdist2 = sumdist1
          xdistmin <- x1
```

```
            ydistmin <- y1
        }
}

#------MAP THE RESULTS
#Plot the different centers with the crime data
plot(acbnd)
points(ac1$x, ac1$y)
points(xmean,ymean,col="red", cex = 1.5, pch = 19) #draw the
mean center
points(xmed,ymed,col="green", cex = 1.5, pch = 19) #draw the
median center
points(xbarw,ybarw,col="blue", cex = 1.5, pch = 19) #draw the
weighted mean center
points(dist3[[3]][1],dist3[[4]][1],col="orange", cex = 1.5, pch
= 19) #draw the central point
lines(circle, col='red', lwd=2)
```

8

Accessibility Methods: Spatial Accessibility to Health Services and Essential Healthcare

Overview

Despite advances in healthcare, accessibility to healthcare is still a major problem. An important part of diagnosis, treatment and recovery includes accessibility to healthcare services in a timely manner. We often take accessibility to different health services and facilities for granted, yet 400 million people have no basic healthcare and more than 15 million people are waiting for HIV treatment (UN, 2023a). Two billion people around the world do not have access to clean and safe drinking water, and approximately 3.6 billion people (46% of the world's population) lack adequate sanitation services (UNESCO, 2023), which results in 829,000 people dying each year from diarrhoea as a result of unsafe drinking water, sanitation and hand hygiene (UN, 2021). In 2022, 419 million people practiced open defecation, 2.2 billion people lacked safely managed drinking water, with 703 million without basic water service and 2 billion lacking basic handwashing facilities with soap and water at home (UN, 2023b). In low-income countries, where services are scarcer, 1 in 41 women dies from a maternal cause (WHO, 2019). Having access to key infrastructure such as health services can be life-saving.

Introduction

Every healthcare system works in different ways. However, if we are going to achieve universal healthcare – *where all people have access to the full range of quality health services they need, when and where they need them, without financial hardship* (WHO, 2023b), then we need to understand what services are available, where service gaps are and what inequalities exist so that something can be done. For example, availability of vaccinations and how this changed over time (Figure 8.1).

Before moving forward with accessibility, take a moment to think about the healthcare services you access. Think about your willingness to travel to

DOI: 10.1201/9781003435082-8

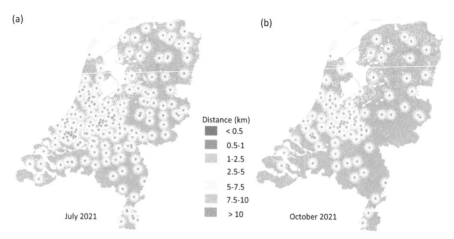

Vaccination Center Type	Description	
Fixed	Centers are located in permanent buildings (e.g. local community centers, sports centers; religious centers).	During the vaccination campaign fixed centers were used by the municipal health services (GGD) to administer 300,000 doses a week, scaling up to 1.5 million doses per week within 6 weeks.
Mobile	Centers are located in units that are mobile and can be moved to different locations (e.g. mobile vaccination unit or jab bus)	
Pop-ups	Centers that are located in buildings that are operational for a set number of weeks, on certain days and for a set number of hours.	Mobile and Pop-ups were used to improve vaccination rates in areas with a lower-than-average vaccination up-take.

FIGURE 8.1

Change in accessibility to vaccination centres during the mass COVID-19 vaccination campaign in the Netherlands. Open vaccination centers in (a) July 2021 (N=197) and (b) October 2021 (N=99). For additional details, see Al-Huraibi et al. (2023).

visit a doctor or hospital. What are some of the factors you consider? How do you travel to see your doctor? How far are you willing to travel?

What factors do you consider when selecting a health provider? Are your decisions based on costs, health insurance, or convenience (e.g. close proximity based on distance or time) or some other factor such as quality of service? Are you able to select a health provider or are you assigned a health provider? What if any challenges do you face when accessing healthcare services?

Five Dimensions of Accessibility

Accessibility is central to the performance of healthcare systems around the world (Levesque et al., 2013). However, accessibility to healthcare remains a complex notion that is multidimensional. It is governed by five dimensions, known as the 5As, that include (Accessibility, Availability, Affordability, Acceptability and Accommodation) (Penchansky and Thomas, 1981; Peters et al., 2008). In summary, these are defined as and include (Penchansky and Thomas, 1981; McLaughlin and Wyszewianski, 2002; Peters et al., 2008):

- **Accessibility** refers to geographic or spatial accessibility. It is determined by how easily one can reach the location where the health service or provider is located. This can include the physical distance or travel time from the service delivery point (the location of the health service provider; e.g. the doctor's office or hospital location) to the home location of the client.

- **Availability** captures the resources, services and technologies available to the healthcare provider to meet the needs of the client. In other words, it is the supply of services and having the right type of care available for those who need it, such as hours of operation and waiting times that meet the demands of those who would use the care, as well as having the appropriate type of services, medical treatment and medication.

- **Affordability** or financial accessibility refers to the relationship between the price of services and the willingness and ability to pay for those services, as well as being protected from the economic consequences of health costs. In other words, this is the price of services and the ability to pay for services influenced by income and insurance coverage.

- **Acceptability** captures the extent to which the client is comfortable with the provider, and vice versa. These can include how responsive health service providers are to the social and cultural expectations of individual users and communities. Characteristics may include the age, sex, social class and ethnicity of the provider (and of the client), as well as the diagnosis and type of coverage of the client. Barriers may be linked to gender, culture, ethnicity, sexual orientation and stigmas associated with disease type.

- **Accommodation** reflects the way services are organized and whether they meet the preferences of the client. These can include hours of operation, communication and how this is accomplished (e.g. telephone communications and how these are handled vs text messages, email), an online booking system and the ability to receive care without prior appointments (e.g. walk-in vs phone appointments).

Over time, some variations have been suggested by others, including Levesque et al. (2013) with dimensions of approachability, acceptability, availability/accommodation, affordability and appropriateness. The framework takes into account the health system and the patient's perspective and behaviour regarding accessibility. In an ideal world, having these perspectives are useful for making improvements (Cu et al., 2021). However, supplementary information is not always readily available or easily obtainable. Regardless, the Levesque et al.'s (2013) framework provides a comprehensive overview of the different considerations and captures the different dimensions from multiple perspectives. Thus, in summary, accessibility is based on the interaction between supply and demand (Figure 8.2).

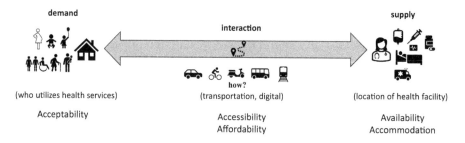

FIGURE 8.2

Measures of accessibility are governed by the interaction (mobility) between the supply of health services and demand for these services. *Source*: Image created by Blanford (2023).

The Role of GIS/SDS/Remote Sensing for Mapping Accessibility of Healthcare Services

Geographic information and the use of GIS and spatial data science methods have been extensively used to map accessibility to healthcare facilities and better understand services and how they relate to the population or different needs within a population. The literature is very rich, and the topics covered are diverse. To put this chapter together, I have compiled examples from a variety of sources (e.g. Park and Goldberg, 2021; Cromley and McLafferty, 2012; Guagliardo, 2004; Neutens, 2015; Higgs, 2004; among others), including the work I have conducted (Blanford et al., 2012; Al-Huraibi et al., 2023). For the remainder of this section, I will highlight how geographic information and spatial analysis methods have been used to better understand different aspects of the accessibility of healthcare services. To do so, I have organized the chapter into the following sections:

- **Supply**: mapping where services are located (what is available)
- **Demand**: mapping healthcare needs and services (planning – what is needed where; identifying inequalities and gaps)
- Optimization of location of healthcare (maximizing services with needs; targeted interventions).

Supply: Mapping Where Services Are Located (What Is Available)

Knowing what services are available and where these are located is important not just to the public but also to healthcare providers. As a healthcare provider, knowing what services to provide and where they should be

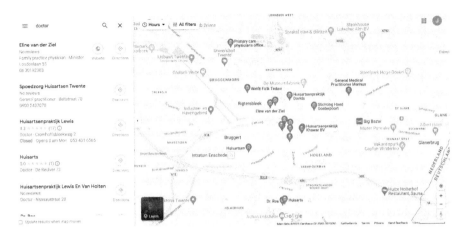

FIGURE 8.3
Google map showing the location of different health facilities and services available in the Enschede area. *Source*: https://www.google.com/maps.

located to satisfy the needs of the population is important. But before these types of evaluations can occur, it is first necessary to create a dataset and database that contain the necessary information about **where** health facilities are located so that the services can be located by different mapping services as shown by the example in (Figure 8.3) where GoogleMaps is used to locate different health facilities and health services in the town of Enschede, Netherlands.

Mapping Health Facilities

Creating a dataset of where different types of health facilities are located can be conducted for specific projects (e.g. Al-Huraibi et al., 2023) or collected and maintained by public health organizations. A number of ongoing health facilities mapping projects are developing global health facilities datasets (e.g. Global Health Facilities Dataset (WHO, 2023a), the health facilities spatial database for sub-Saharan Africa (SSA) (Maina et al., 2019) and the Global Healthsites Mapping project (https://healthsites.io/)).

The Geolocated Health Facilities Data Initiative by the World Health Organization (WHO) GIS Centre for Health is aiming to create the world's first central and accessible public database of health facility locations that is interoperable. The goals of this initiative are to strengthen healthcare planning and decision-making with accurate and accessible health facility master lists (WHO, 2023a). The information includes three key attributes:

- Locations (latitude, longitude coordinates, address)
- Names (name of facility)
- Types (as defined by each country – e.g. hospital, clinic, doctor, dentist, maternity, public/government, private, etc.)

By 2027, the WHO aims to have all 194 WHO Member States regularly contributing information to the global database. The GHFD initiative is a key activity of WHO's GIS Centre for Health.

In addition, the Humanitarian OpenStreetMap (OSM) Team is also mapping the location of health facilities. The core datasets in OSM will include different public and private healthcare infrastructures such as clinics, dentists, doctors, hospitals, pharmacies and healthcare boundaries. The Global Healthsites Mapping Project's aim is to create an online map of every health facility in the world and make the details of each location easily accessible (https://healthsites.io/) via OSM (https://www.hotosm.org/impact-areas/public-health/) and the Humanitarian Data Exchange (HDX) Portal (Figure 8.4). Related OSM datasets also include administrative information (e.g. buildings, community centres, regional boundaries), transportation (e.g. roads, public transport, railways, airports, waterways/seaways), water and sanitation (e.g. solid waste management, water management, drinking water) and environment (waterways and land use).

Note: since this is collected by citizens, the data will change and may be incomplete (refer to the chapter on data and refer to the section on citizen science and volunteered geographic information).

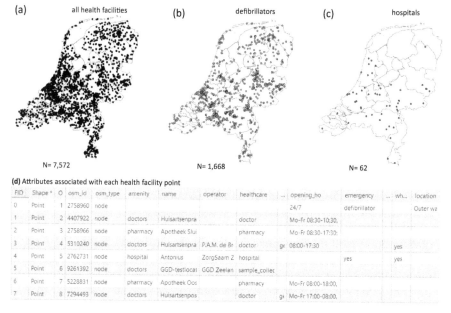

FIGURE 8.4

Example of health facilities available through the OSM healthsites Mapping Project. Data are for the Netherlands and show the distribution of (a) all facilities, (b) defibrillators and (c) hospitals. The attributes collected are captured in the (d) attributes associated with each health facility point. Data *Source*: OSM healthsites Mapping Project.

Mapping Healthcare Needs and Services (Planning – What Is Needed Where?)

Mapping the availability, demand and potential service areas of health facilities (Figure 8.5) is useful for a number of different reasons as follows:

- **Service availability**: Knowing what healthcare services are available and where they are located. Showing the location of services (service hubs) and different providers (e.g. Tanser et al., 2001; McLafferty and Grady, 2004) (Figure 8.5a).

- **Service areas**: Showing the reach or service area of different healthcare facilities or services can help with infrastructure planning and understanding who and how services are utilized (e.g. Figure 8.5d–f). For example, in South Africa, Tanser et al. (2001) was able to delineate service areas (e.g. Figure 8.5f) for different health services by mapping the clinic usage of each household (e.g. Figure 8.5e and f). Thus, service area can be delineated based on a set of criteria such as travel time, distance or location-allocation evaluations (e.g. Minimize Weighted Impedance (P-Median) (Oppong and Hodgson, 1994); maximal coverage (Church and ReVelle, 1974; Indriasari et al., 2010)).

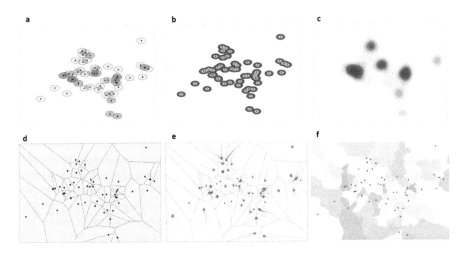

FIGURE 8.5
Mapping health care availability, demand and potential service areas. (a) Buffers at a specified distance from a health facility, (b) multiple buffers each representing different distances from a health facility, (c) density of health facilities, (d) potential service areas for each health facility where all areas within the boundaries are closest to that health facility, (e) hypothetical patients and how far they travel to each health facility and (f) adjusted potential service areas for each health facility using patient distribution from (e).

- **Addressing healthcare needs**: Evaluating healthcare needs by visualizing spatial matches between service availability and where the
  demand for these services are located (Figure 8.5e) is useful for identifying service gaps and enables for mismatches to be addressed. (e.g.
  patient-doctor ratio; patient needs and service availability; cancer
  hotspot and treatment availability; age-related services; identifying
  unequal access to services; Cattaneo et al., 2021; Blanford et al., 2012).
  Examining the concentration of services (density-based analysis)
  providers (McLafferty and Grady, 2004) such as the number of services per unit area, it becomes clear where services are concentrated
  (e.g. Figure 8.5c). When these are evaluated against where demands
  are concentrated mismatches between supply and demand may
  become more apparent (e.g. McLafferty and Grady, 2004; Al-Huraibi
  et al. 2023). Population-to-provider ratios vary by geographic location (Rosenthal et al., 2005); however, in the US, federal regulations
  stipulate that for primary medical care, the population-to-provider
  ratio must be at least 3,500 to 1 (3,000 to 1 if there are unusually high
  needs in the community) (HRSA, 2023; Albert and Butar, 2005).

A variety of methods can be used to capture geographic availability and
delineate healthcare service areas, as illustrated by the examples provided in
Figure 8.5. Some of these are fairly straightforward to perform and, at a minimum, require the location of health facilities (e.g. buffers, Thiessen Polygons
or Kernel Density Estimates) while others might need a road network (e.g.
networked service area) or population information (e.g. provider-to-population ratios). To get you thinking about how to capture service availability
and the areas served by these facilities, I have provided an overview of some
methods that have been used in a variety of studies (Table 8.1). Most of these
methods are readily available in many GIS packages.

Mapping Spatial/Physical Accessibility of Health Services

The four major accessibility measurement approaches are opportunity-based,
gravity-type, utility-based and space-time measures (Kwan, 1998; Geurs and
Van Wee, 2004; Liu and Zhu, 2004; Benenson et al., 2017; Lee and Miller, 2018).
Accessibility can be determined using a variety of methodologies and tools
(see Neutens, 2015). Potential accessibility can be characterized using *distance, time* and *density* measures. One of the simplest ways to determine
access to health services is to calculate distance from the population in need
to the nearest service provider, as captured in the previous section. Although
distance is a fundamental indicator of geographical access, travel time, cost,
transportation access and perceived distance are often much more relevant
to healthcare utilization. Travel time can be estimated along road networks,
taking into account average speeds and speed limits along different classes
of roads for different modes of transportation (e.g. Blanford et al., 2012).

TABLE 8.1

Overview of Measurements and Methods Used for Assessing Availability of Services and Determining Service Areas of Facilities

Measurement Type	Method	Description	Source and/or Example
Service availability	Provider-to-population ratio	Useful in highlighting differences between administrative boundaries and identification of gaps in service provision Ratio of physicians to population needs. Also referred to as supply ratios, are computed within bordered areas.	Neutens, (2015); Cromley and McLafferty, (2012); McLafferty and Grady (2004); Guagliardo et al. (2004)
	Kernel density estimate (KDE)	Useful for understanding the density of services available and where these are concentrated.	McLafferty and Grady (2004)
Service areas	Buffer	Creates a service area by buffering a point using a specified distance. The circle that is created shows the area within the specified distance.	ESRI (2023)
	Euclidean distance	Straight line of travel from a point of residence/address location to the location of a health service provider.	Guagliardo (2004); Noor et al. (2009); Noor et al. (2006)
	Thiessen polygons	Thiessen polygon – represents the potential area of influence around a point. That is all locations within the Thiessen Polygon are closest to the point than to any other point.	Schuurman et al. (2006); Tanser et al. (2001); ESRI (2023)
	KDE	KDE is used to determine a 'sphere of influence' whose radius is the bandwidth of the kernel density.	Schuurman et al. (2010)
	Networked service areas	The service areas created around a health facility represent the maximum distance that can be travelled along a network within a specified distance and includes all areas that can be reached by roads. This can be used on distance or travel time.	Tanser (2006); Noor et al. (2009)
	Patient preference service areas	Utilization of health facilities are used to delineate service area around a health facility.	Tanser et al. (2001)
	Maximal covering location problem (MCLP)	Aims to locate a number of facilities so that the population covered is maximized. Other forms may include minimize weighted impedance (P-Median); maximize attendance and capacitated coverage	Church and ReVelle (1974); Indriasari et al. (2010); Oppong (1996)

Factors Associated with the Utilization of Health Facilities

Accessibility measures based on travel time, distance and density offer a partial view of access to services. In reality, people trade off geographical and non-geographical factors in deciding what services to use (e.g. Figure 8.2).

Location and distance have a significant effect on the willingness and ability of patients to use health services. For example, in rural areas, in Kogi state, Nigeria (Awoyemi et al., 2011), household size, distance and the total cost of seeking healthcare affected the utilization of government and private hospitals. In addition, the rate at which health facilities are utilized has been found to decrease with distance (Al-Taiar et al., 2010; Noor et al., 2003), quality of transportation (Blanford et al., 2012) and road conditions. During the wet season, roads may become impassable, particularly motor transport (Poku-Boansi et al., 2010; Oppong, 1996; Blanford et al., 2012), thus reducing accessibility to health services. Seasonal variation in accessibility to a range of healthcare facilities has been examined in Niger (Blanford et al., 2012), Ghana (Oppong, 1996) and maternal health services in southern Mozambique (Makanga et al., 2017) and Zambia (Mroz et al., 2023).

Physical accessibility using different modes of transport (e.g. walking, biking, public transport and private cars) (Blanford et al., 2012; Al-Huraibi et al., 2023) and a mix of transportation modes (Mao and Nekorchuk, 2013) is the reality and likely the norm. How far availability and 'attractiveness' influence accessibility to healthcare can vary by any *impedance function* where the impedance captures the ease of access across a surface and, when coupled with cost-distance-based calculations, is useful for capturing the rate at which access decreases with increasing distance, travel time and cost from one geographic location to another, as well as by time of day.

In reality, accessibility is not statistic but changes dynamically at different temporal scales – seasonally, daily and hourly. How accessibility is affected will vary by geographic location. Several studies have included temporal variations in accessibility (e.g. Kwan, 2013; Park and Goldberg, 2021), including the availability of quick-response time-critical health services such as defibrillators (e.g. Sun Christopher et al., 2016) and also how accessibility may change in conflict zones (e.g. West Bank (Eklund and Martensson, 2012); Ukraine (Dzhus and Golovach, 2023)). Access in these areas can be severely impacted due to security concerns, restricted mobility, broken supply chains, the removal of health facilities and services, mass displacement and changing demands (Dzhus and Golovach, 2023).

In summary, geographic accessibility can be influenced by a variety of factors as follows:

- Distance, time and mobility (e.g. public transport and traffic)
- Seasonal climate effects (e.g. flooding)
- Hours of operation and availability of services (opening, closing times and mobility of services)

- Conflict
- Acceptability (e.g. cultural and other influences)

I have already highlighted a few straightforward methods that are useful for capturing the availability of health services and obtaining an overview of what is available and where. These essentially capture one or two dimensions that can affect accessibility. In reality, the demand for care and services is also of interest, as is how to realistically capture the factors that may influence geographic accessibility.

Gravity models provide the most valid measures of spatial accessibility by representing the potential interaction between any population point and all service points within a reasonable distance, discounting the potential with increasing distance or travel impedance. Because gravity measures take into account all alternative service points, they are sometimes referred to as cumulative opportunity measures. However, a main limitation of the gravity model is that it only models supply with no adjustment for demand. In addition, data on the location of all health services is not always available at the resolution needed for gravity-based estimations.

Two-step floating catchment area (Luo and Wang, 2003) addressed this problem in the two-step floating catchment area method as follows:

- **Step 1**: In the first step, a supply-to-population ratio is estimated for each supply location (ZIP code centroid). The number of providers assigned to the ZIP centroid is divided by the population living within that centroid's 30-minute drive time catchment. The supply-to-population ratio is then assigned to the entire catchment area. The ZIP-centred catchment ratio computation is repeated for all ZIP centroids (in essence, the focus of the calculation is "floated" over all ZIP centroids, hence the method's name).

- **Step 2**: In the second step, for each population point, a value is obtained by summing the provider-to-population ratios of all the first-step provider catchments that overlie the point. Depending on what resolution is available, population points might include household, tract centroids or ZIP centroids. The summed supply ratios are assigned to the entire area represented by the population point. Thus, all population areas are assigned a value.

(Luo and Wang, 2003) recognize that the method has limitations. While geopolitical borders are well handled, the drive-time catchment borders are themselves artificially sharp. Spatial accessibility near the periphery of the catchment is as high as at the centre and drops to zero just over the line. Limitations of the 2SFCA is that it assumes that all locations have equal access to supply locations within a catchment, no access to supply locations outside of the catchment, each doctor has the same attraction to care seekers and that each care seeker's need/demand for service is the same. To address these shortcomings, an enhanced 2SFCA (E2SFCA) method (Luo and Qi, 2009) introduces weights to differentiate travel time zones to account for the stepwise decay of accessibility within each catchment. A variable 2SFCA (V2SFCA) method (Luo and Whippo, 2012) introduced an adaptive size of the catchment area determined according to the service capacity of the supply location. To consider the overestimation of demand on supply locations in the E2SFCA, competition among supply locations was introduced in a three-step floating catchment area (3SFCA) method (Wan et al., 2012). The M2SFCA (Delamater, 2013) model also improves on the E2SFCA as it is more accurate in dealing with systems with sub-optimal configurations as it takes into account the absolute distance separation between demand (the population) and supply (health facilities) locations. As you can see, this is a model that continues to be used and improved. I have only highlighted a few, as there are many other adaptations.

Cost-friction, cost-distance or impedance from each health facility. This measures accessibility from a health facility's location (point). Travel cost can be measured in Euclidean distance, travel distance along the surface or transportation networks or be based on estimated travel times (Time=Distance / Speed). This has been extensively used to explore geographic accessibility to facilities (e.g. Ray and Ebener, 2008; Blanford et al., 2012; Macharia et al., 2017; Weiss et al., 2018; Weiss et al., 2020; Macharia et al., 2021; Ouma et al., 2021; Ray et al., 2022; Al-Huraibi et al., 2023). At a minimum, this method only requires two data inputs: the location of a health facility and a transportation network.

The method uses a *friction surface* that represents the characteristics of the landscape. Since this surface represents the landscape, it can be adapted to capture the real world at different spatial and temporal scales at a local or global level. For example, this can be adapted to capture different spatial mobility patterns locally using different modes of transportation (e.g. walking, biking, public transportation and private vehicles) and also different temporal mobility patterns due to variations in behaviour and the local environment (e.g. cultural differences between and within a country) or changes in the availability of services (e.g. operating schedule of public transportation services). In the study by Blanford et al. (2012), four *friction surfaces* were created, capturing (i) the walking travel speed during the dry season, (ii) the walking travel speed (with impediments due to flooding) during the wet season, (iii) vehicle travel speed along roads and walking speed for all other surface types during the dry season and (iv) vehicle travel speed along roads and walking speed for all other surface types (with impediments due to flooding) during the wet season. These friction surfaces then serve as inputs from which to calculate the path with the least cost. In other words, this will determine the fastest route across the surface by calculating the linear and

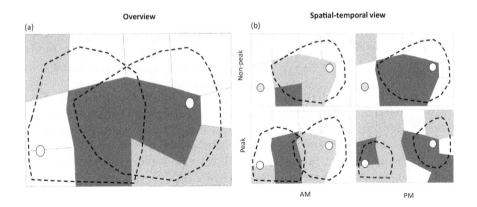

FIGURE 8.6
(a) Overview of supply of health services, service area of each health facility and demand; (b) Hypothetical dynamically changing supply and demand for health services in a 24-hour period. Supply of health services is illustrated by the location (dots) and service area (---). The status of availability is captured by the colour of the dot (yellow dot=open; blue dot=closed). The coverage of the service area (---) captures how this changes based on travel time from the health facility location. Demand for health services is captured in the underlying polygons (white=no demand, dark grey=highest demand).

diagonal accumulated least cost of getting from the nearest health facility to each grid cell for the entire study area. This allows for any route between locations to be considered, not just those along established roads (provided this is captured in the friction surfaces), and allows for the examination of spatial-temporal changes in accessibility that may result (e.g. distance, time and mobility; changes in the availability of public transportation services or changes in travel speed due to traffic; seasonal climate effects (e.g. flooding); hours of availability (opening, closing times and mobility of services); conflict) (see Figure 8.6).

Spatial-temporal spatial accessibility methods. In reality, accessibility is not statistic and changes dynamically for any number of reasons (temporary, seasonally, daily and hourly), as shown in Figure 8.6. One way of capturing changes over space and time is to use methods that are flexible enough to capture variations in the system (e.g. cost-friction method) or use space-time cubes that enable the partitioning of space and time into segments representative of the system being examined (e.g. Lee and Miller, 2018). In essence, being able to capture changes in a system boils down to three things – availability of information, the time required to construct and maintain the information (if it is not readily available) needed to perform the analysis and the computational power for performing the analysis.

For example, Lee and Miller (2018) examined the impact of a new transit system would have on accessibility to health services using a high-resolution space-time accessibility measure. To do so, they (i) constructed a multimodal transport network of travel times based on transit networks and walking; (ii) delineated accessible areas using space-time prisms by developing a multi-modal network prism concept that captures the *potential multimodal network area* (PMNA) based on 30-minute (local accessibility) and 60- minute (regional accessibility) time budgets. They then determined (iii) space-time accessibility for four different departure times (8 a.m., 1 p.m., 6 p.m. and 9 p.m.) and two representative days (Thursday and Sunday) to capture different mobility effects from which they constructed (iv) space-time accessibility outputs that determined the space-time constrained cumulative opportunity measure by counting the number of healthcare services that can be reached based on travel from designated origins given a time budget via walking and public transit. For healthcare, the 8 a.m. and 1 p.m. departure times on weekdays were used to reflect the availability of healthcare services. The results are useful for gaining insights into how accessibility can change across space and time.

Other studies have used the density of Out-Of-Hospital-Cardiac Arrests (OHCA) and the availability of defibrillators to evaluate gaps in the distribution of these life-saving devices and the potential impact they may have (Hansen et al., 2013; Sun et al., 2016; Aeby et al., 2021).

In summary, knowing the coverage of services available (Figure 8.6a) and the availability of these services and how these may change accessibility and potential service areas (Figure 8.6b) are useful for addressing inequalities in healthcare. To do so, it is important to obtain different perspectives on healthcare accessibility, as illustrated in Figure 8.6, where the optimized services are shown in (a) and the impact of different factors and how they can change the availability of services and the area serviced by each health facility (b). When combined with demand, further evaluations can be used to explore gaps and mismatches between demand and supply.

To close this section, see the summary provided in Table 8.2 of the different methods that can be used for determining spatial accessibility to healthcare facilities.

Optimization of Locations of Healthcare: Evaluating Varying Spatial and Temporal Needs for Designing Intervention Strategies (Maximizing Services, Spatial Targeting and Vaccination Campaigns) and Evaluating Impact

At the start of this chapter, I asked several questions about visiting a doctor or hospital. What are some of the factors you consider? How do you travel to see your doctor? How far are you willing to travel for medical help or advice? If you could not travel to a healthcare facility, how would this affect your health?

TABLE 8.2

Overview of Measurements and Methods Used for Determining Spatial Accessibility to Healthcare Facilities

Measurement Type	Method	Description	References
Spatial/Geographic accessibility	Gravity models	Gravity model is a combination of supply and demand across an area. – Availability and accessibility across defined spatial units It controls for "capacity" of a facility, competition between facilities and ability to estimate gravity values using numerous methods	Neutens (2015); Luo (2004)
Spatial/Geographic accessibility (Supply × Demand)	Two-step floating catchment area methods (2SFCA)	This method requires **data on locations and capacities** of service providers (supply) and **data on location of population in need** of services (demand). **Accessibility score:** The resulting value is the accessibility score for the population at each place. Higher values represent better access.	Luo and Wang (2003)
	Modified and enhanced 2SFCA	• Enhanced two-step floating catchment area (E2SFCA) method • Modified two-step floating catchment area (M2SFCA) method • 3SFCA • And other modifications and enhancements	Delamater (2013); Ni et al. (2015); Luo (2004); Wan et al. (2012); McGrail and Humphreys (2015); Luo and Qi (2009); McGrail (2012); Bryant Jr and Delamater (2019)
Service areas	Network analysis Network distance/time	Network analysis entails the use the actual transport/travel routes to compute either travel time or distance to the nearest service provider	Al-Huraibi et al. (2023); Noor et al. (2006); Noor et al. (2009)

(Continued)

TABLE 8.2 (*Continued*)

Overview of Measurements and Methods Used for Determining Spatial Accessibility to Healthcare Facilities

Measurement Type	Method	Description	References
Spatial/Geographic accessibility	Travel cost or impedance from nearest provider. Cost-distance analysis	Friction surface and adaptation to create realistic travel scenarios across the earth's surface. Influence of transport services on accessibility and barriers of travel such as impedance by land use, road network and elevation or due to seasonal changes or due to different modes of transportation.	Ray and Ebener (2008); Blanford et al. (2012); Ray et al. (2022); Weiss et al. (2020)
Dynamically changing spatial/geographic accessibility	Spatial-temporal spatial accessibility models	Friction surfaces; high-resolution space-time accessibility measure. A variety of data and methods used to capture changes in accessibility. These can take into account demand by patients, availability of healthcare services (opening/closing times), availability of transportation services (hours of operation, transit time, variations in services (weekday/weekend)); different start times.	Park and Goldberg (2021); Widener et al. (2015); Sun Christopher et al. (2016); Lee and Miller (2018); Kwan (2013)

In the previous section, I talked about methods that can be used to assess and evaluate different aspects of accessibility. The methods that you select to perform your analysis will be governed by the availability of data and what questions you are trying to answer or what you are examining. At a minimum, the data that is needed to explore accessibility are the location of health facilities and a road network, all of which are available today thanks to technological improvements, citizen science, open science and data repositories and hubs.

Now that you have a better understanding of how to map service availability and explore accessibility to health facilities, let's think about how to use this information to address inequalities and service gaps to achieve universal healthcare. The absence of effective access to health services is a key contributor to poor health outcomes in many places. What is needed to be able to evaluate the impact and address changing needs? At the end of the previous section, I discussed dynamically changing needs and how this can be addressed using spatial-temporal analyses. I provided some examples of different types of healthcare needs that range from routine demands (routine) to emergency-related demands (emergency), some of which require time-critical responses (emergency time-critical response requirements) (e.g. Table 8.3). If you were to evaluate accessibility for these different types of needs, what criteria would you use? Accessibility based on time, response time, lives saved versus not saved, population with and without

TABLE 8.3

Examples of Varying Spatial and Temporal Healthcare Accessibility Needs

Routine	Emergency	Emergency Time-Critical Response Requirements
Routine – regular and ad hoc visit to a doctor. Non-vital (e.g. annual check-up, vaccination)	Hospital emergency 24/7 services	AED's/Defibrillators
Routine – maternity care	Pandemic and outbreaks (e.g. intensive care needs, isolation)	Ambulance
Mobile medical units	Mass vaccinations	Snakebite, accident (e.g. ski, car)
Disease eradication (the last-mile approach)	Rapid testing/pop-ups and/or mobile units for short-term emergency services	
Conflict zones with known disruptions		

Note: Demand can vary spatially and temporally for a variety of reasons and may require (a) regular/routine services, (b) emergency responses some of which are (c) time-critical response.

accessibility, etc. In Table 8.4, I have provided a variety of criteria extracted from the literature (this is not comprehensive) that have been used to evaluate accessibility from which recommendations and decisions have been made.

Here are some examples of different evaluations and how they have been used:

- By examining intensive care use in the EU (Bauer et al., 2020), hospitals are able to share resources during times of critical need.
- Mass vaccinations in NL (Markhorst et al., 2021)
- Mapping settlements, estimating populations and identifying vaccination gaps from remotely sensed imagery helped in the planning of various vaccination campaigns (Chabot-Couture, 2017; Takane et al., 2016; Corbane et al., 2021; Weber et al., 2018), as well as detecting communities and households to ensure broad coverage (Bruzelius et al., 2019).

TABLE 8.4

Healthcare Accessibility Criteria for Different Services

Healthcare Accessibility Need	Criteria	Source
Routine access to health facilities	< 1 hour travel time	WHO; Noor et al. (2006)
Palliative care units	< 1 hour travel time	Cinnamon et al. (2008)
Minimum set of health services available to the general population	5 km < 1-hour walk to a health centre The Ministry of Health in Niger aims to provide a "Minimum Activity Package," for the population to access a health centre	WHO (2011); Blanford et al. (2012)
Emergency – hospital	<30 minute travel time standard	Bosanac et al. (1976)
Emergency – obstetric complications	<2 hour travel time limit to reach healthcare	WHO; Maine (2009)
Emergency – snake bite	30 minute to >6 hour depending on the snake species and the associated clinical syndrome; <1 hour for the neurotoxic syndrome	Alirol et al. (2010); Ochoa et al. (2023)
Emergency – AED's/ defibrillators – out of hospital cardiac arrest	<6 minute zone; every future OHCA victim is reached within 6 minute; Reached within 3–5 minutes	The Dutch Heart foundation (HartslagNu, www.hartslagnu.nl); Thannhauser et al. (2022); Perkins et al. (2015)

(Continued)

TABLE 8.4 (*Continued*)

Healthcare Accessibility Criteria for Different Services

Healthcare Accessibility Need	Criteria	Source
	To be placed ≤1.5 minute of brisk walking distance to Out of Hospital Cardiac Arrest locations which translates to a straight-line distance of 100 m	American Heart Association Aufderheide et al. (2006)
Primary care for malaria treatment	<2 hour walk (to minimize hospitalization) <4 hours walk (to reduce mortality)	O'Meara et al. (2009); Schoeps et al. (2011)
Normal accessibility to the nearest urban centre	1.0-hours for the vehicle only; 1.5-hours for the walking-motorcycle-vehicle to the nearest urban centre	Macharia et al. (2021)
Hard-to-reach populations; missed populations	Mobility of healthcare services (supply); changing supply strategies (e.g. last-mile transport options).	McCoy (2013)

Note: Examples of varying maximum response times for different health needs – normal/routine to time-critical response.

- Achieving universal electrification of rural healthcare facilities in SSA with decentralized renewable energy technologies (Moner-Girona et al., 2021). This study showed that by improving the electricity supply to over 50,000 healthcare facilities in rural SSA, 281 million people could reduce their travel time to healthcare facilities by an average of 50 minutes.

- In rural clinics where health workers previously walked or used bicycles to visit hard-to reach villages, through the use of motorcycles, they were able to increase the number of health interventions by reaching the last-mile populations. Due to these efforts, in the Gambia, there has been an increase in infant vaccination coverage from 62% to 73%, and in a district of Zimbabwe, there has been a 20% reduction in mortality from malaria (McCoy, 2013).

A variety of tools are available for evaluating accessibility, as captured in Table 8.5.

TABLE 8.5

Summary of Different Accessibility Tools

Tools	Description	Source
ArcGISPro	Network analysis to create service areas Or raster analysis based on Speed = Distance/Time	ESRI (2023)
AccessMod 5.0	Stand-a-lone tool that can be used to evaluate accessibility.	https://www.accessmod.org/
Accessibility Tool in R		Accessibility – Transport Accessibility Measures. https://cran.r-project.org/web/packages/accessibility /vignettes/accessibility.html https://cran.r-project.org/web/packages/accessibility/accessibility.pdf
Healthcare Accessibility		Data – Nelson et al. (2019); Weiss et al. (2020)

Beyond Accessibility for Healthcare

In addition to healthcare needs, accessibility is also vital for gaining insights into a range of essential infrastructure needs that contribute to the health and well-being of a population. These can range from access to clean water and bathrooms to healthy food environments. Below are some examples that capture how these methods can be further extended to improve health beyond just accessibility to health care services:

- **Water**: Around 2 billion people (26% of the worlds population) around the world does not have access to clean and safe drinking water, and approximately 3.6 billion people (46% of the world's population) lack adequate sanitation services (UNWWDR, 2023). Geospatial technologies coupled with accessibility measures are useful for finding the closest water tap (e.g. Find a water tap (https://dopper.com/water-tap-maps#find-a-water-tap).
- **Bathrooms**: With over 400 million people practising open defecation (UN, 2023b), having access to bathrooms and other services is vital for reducing open defecation (e.g. in San Francisco (Amato et al., 2022); The National Public Toilet Map in Australia (https://toiletmap.gov.au/)).
- **Food**: Mapping the evolution of 'food deserts' in a Canadian city: Supermarket accessibility in London, Ontario, 1961–2005 (Larsen and Gilliland, 2008). Food swamps predict obesity rates better than food deserts in the United States (Cooksey-Stowers et al., 2017). Evaluating the changing landscape of food deserts (Karpyn AE et al., 2019) and rating food landscapes (Thornton and Kavanagh, 2012).

Summary

Now that you have a good overview of the different aspects of accessibility, it is your turn to explore how accessible health services are in Kenya.

Activity: Evaluate Accessibility of Health Facilities in Kenya

If we are to achieve universal healthcare coverage, then we need to start by addressing accessibility to healthcare services for all.

In SSA, one-sixth of the population lives more than 2 hours away from a public hospital, and one in eight people is no less than 1 hour away from the nearest health centre (Falchetta et al., 2020). For reproductive-age women in SSA, access to healthcare was found to be 42.56% (Tessema et al., 2022). In rural SSA, over 50,000 healthcare facilities lack electricity supply (Moner-Girona et al., 2021). By electrifying health facilities, 281 million people could reduce their journey time by 50 minutes on average (Moner-Girona et al., 2021) and improve healthcare access for women who reside in the countryside (Tessema et al., 2022). If commonly accepted universal healthcare accessibility targets are to be met in sub-Saharan African countries, approximately 6,200 new facilities will need to be built by 2030 (Falchetta et al., 2020). However, to do so, it is necessary to examine quality of healthcare and different ways of accessing healthcare (e.g. access using the walking speed of children; Watmough et al., 2022; by bicycle Al-Huraibi et al., 2023) so that inequalities can be addressed (e.g. monitoring subnational healthcare quality; Allorant et al., 2023). By measuring rates of coverage and spatial access to healthcare services, it will be possible to inform policies for development (Florio et al., 2023).

Having access to good health facilities in a timely manner matters, as captured by the different criteria in Table 8.4. For example, in malaria-risk areas, having access to healthcare facilities within a 2–4 hour walk can be crucial for minimizing mortality and enhancing recovery, as captured in the studies of O'Meara et al. (2009) and Schoeps et al. (2011). During this activity, you will be able to put into practice some of the things you read about in this chapter. The exercise has been broken down into the following components:

Overview of activity:

1. Distribution and service area of healthcare facilities
   a. *Availability of services* – density of health facilities
   b. *Service Area/Area of Influence* – straight-line vs road network distance.

2. Mapping access to healthcare facilities: distance versus time
   a. *Geographic/Spatial Accessibility* based on straight-line distance
   b. *Geographic/Spatial Accessibility* – distance vs time (different modes of transport)

3. Evaluating the availability and accessibility of healthcare
   a. Defining the criteria

4. Planning for response – evaluating populations at risk
5. Conclusions and recommendations?

Data

All of the data that will be used is available in the public domain and was obtained from various sources, as captured in Table 8.6.
Relevant Readings:

- Blanford, J. I., S. Kumar, W. Luo & A. M. MacEachren (2012) It's a long, long walk: accessibility to hospitals, maternity and integrated health centers in Niger. *International Journal of Health Geographics*, 11, 24. https://ij-healthgeographics.biomedcentral.com/articles/10.1186/1476-072X-11-24.
- Al-Huraibi, A., S. Amer, J. I. Blanford (2023) Cycling to get my vaccination: How accessible are COVID-19 vaccination centers in the Netherlands? *AGILE: GIScience Series*, 4, 16, 2023. https://doi.org/10.5194/agile-giss-4-16-2023.
- Macharia, P. M., E. Mumo, E. A. Okiro (2021) Modelling geographical accessibility to urban centres in Kenya in 2019. *PLoS One* 16(5), e0251624. https://doi.org/10.1371/journal.pone.0251624.

Distribution and Service Area of Healthcare Facilities in Kenya

In Kenya, there are different types of healthcare facilities. These include hospitals, clinics, health centres and dispensaries. Select one type of facility to analyse (Figure 8.7).

TABLE 8.6

Summary of Data and Data Sources Used to Examine Accessibility in Kenya

Data set	Description	Date	Source
Hospital and clinic locations in Kenya	Point location of different healthcare facilities in Kenya. These include hospitals, clinics, health centres, dispensaries in Kenya. healthfacilitiesSSA	2020	https://data.kimetrica.com/dataset/ kenya-health-facilities-shapefile/resource/ f1642061-4cb0-4dd8-a9d8-7718530e3488; Noor et al. (2009) or https://data.humdata. org/dataset/ health-facilities-in-sub-saharan-africa; Maina et al. (2019)
Malaria	MAP of Malaria risk (MAP)	2020	https://malariaatlas.org/
Population	Population density	2020	https://www.worldpop.org/
Admin Level 0	Country Boundary KEN_adm0	2020	https://data.humdata.org/dataset/ cod-ab-ken
Admin Level 1	County (formerly districts) KEN_adm1_prj	2020	Runfola et al. (2020); https://data.humdata. org/dataset/ geoboundaries-admin-boundaries-for-kenya
Admin Level 2	Sub-County KEN_adm2_prj	2020	Runfola et al. (2020)
Admin Level 3	Sub-County KEN_adm3_prj	2020	Runfola et al. (2020)
OSM data	Roads Roadosm_prj	2020	OSM – OpenStreetMap http://download.geofabrik.de/osm/africa/ kenya-latest.shp.zip
OSM data	POIs	2020	OSM – OpenStreetMap
World Topographic Map	Background map service available in ArcGISPro	2023	ESRI

Assessing the Distribution of Health Facilities in Kenya by Administrative Level

Map the distribution of healthcare services throughout Kenya and assess if there are any potential health care service gaps. Are there any areas without health facilities? Where are the areas with the least number of health facilities and where are the areas with the most?

Determine the number of health facilities located in each administrative boundary level 2 using the summary statistic tool. Map the distribution of health facilities by administrative boundary level 2. Hint: To do so, join the summary table to the KEN_adm2 and create a choropleth map and insert the map into your write-up.

(a)   (b)   (c)

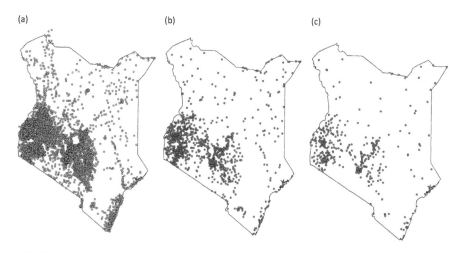

FIGURE 8.7
Distribution of (a) dispensaries, (b) health centres or clinics and (c) hospitals in Kenya. Map created in ArcGISPro. Image created by Blanford (2023).

Analysis Toolbox – Overlay – Spatial join

Target feature: healthfacilitiesSSA_prj

Join feature: KEN_adm2_prj

Join operation: leave as default one_to_one

Keep all target features

Match option: Completely_within

Service Area/Area of Influence – Straight-Line vs Road Network Distance

Set the raster processing environment: The processing environment can be set in each tool (ENVIRONMENT settings). In processing extent, set the extent to KEN_adm1 and, in raster analysis, set the mask to KEN_adm1. Set the cell size to 1,000 so that each grid cell represents 1km.

OR

From the dropdown menu on the main toolbar, **Geoprocessing – Environment**. In processing extent, set the extent to KEN_adm1 and, in raster analysis, set the mask to KEN_adm1. Set the cell size to 1,000 so that each grid cell represents 1km.

Health facility: Select a type of health facility (e.g. hospital, maternity, health centre/clinic) and examine the area serviced by that health facility. Select a type of health facility to analyse and create a new layer to use for the remainder of the analysis.

To save a map as an image on the main menu, go to **Share – Output**

Service Area – Euclidean Distance

Accessibility Based on Straight-Line Distance

Buffer a distance around each health facility to represent how far you would travel. Select a distance to buffer. If you had to visit your local doctor, how far would you travel? Create two maps: one based on how far you would be willing to walk in a day and the other on how far you would be willing to travel by car.

Buffers (Figure 8.8a)

Analysis Tools – Proximity – Multiple Ring Buffer

Input features: Health Facility layer (in this case maternity health facilities)

Distances: 1,000, 10,000, 20,000, 50,000

Distance unit: meters

Buffer distance field name: distance

Distance allocation (Figure 8.8b)

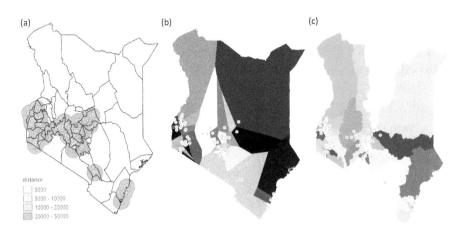

FIGURE 8.8
Service areas based on (a) Buffers at 1,000, 10,000, 20,000 and 50,000 meters for maternity facilities in Kenya. (b) Thiessen polygons or distance allocation. Each unique colour represents the catchment or service area for each maternity health facility. (c) Distance allocation based on a road network. Map created in ArcGIS Pro. Image created by Blanford (2023).

To create a map showing the area of influence of each health facility based on the Euclidean distance: **Spatial Analyst Tools – Distance – Distance Allocation.**

Input raster or feature source data: healthfacilitiesKE_prj (select the health facility type you are going to analyse (hospital, maternity or clinic). Only pick one and keep using this throughout the rest of the analysis).

Source field: recno

Recno represents a unique value, so when you create the service area, each service area will be assigned a unique value for each maternity facility.

Set the processing extent for all of Kenya, click on the ENVIRONMENTS tab in the tool and, for the **Processing Extent,** select KEN_adm1_prj dataset. Also set the cell size to 1,000.

Service Area – Road Network (Figure 8.8c)

To create a more realistic service area for each health facility, first add in roads since most people will travel along a road or path to get to a health facility. Next, create a cost surface, also known as *a friction surface,* that essentially creates a digital twin of mobility and captures the impedance (or ease) of travel across a surface (see Blanford et al., 2012), Table 8.7 for details. Each segment of the road is assigned an impedance value. For example, a highway has a low impedance value since it is easier to travel on than a track, which would have a high impedance value. In the roadosm_prj layer, several impedance values have been added for this activity (Table 8.7).

Where each attribute in the table contains the following information:

Type 2 = road type

Speed = average speed of travel with a vehicle on the road type

Impedance = the impedance value associated with each road type.
  1 = less impedance (e.g. motorway/hiway) and 100 = higher imped-ance (e.g. track or path)

TABLE 8.7

Summary of Impedance Values Associated with Different Road Types

Type2	Speed	Type4	Impedance	Times_w	Times_car
Primary	80	1	1	720	45
Secondary	50	1	2	720	60
Tertiary	40	1	3	720	90
Other Road	35	1	5	720	103
Road	30	1	4	720	120
Street	30	1	4	720	120
Track	25	1	9	720	144
Path	0	1	100	720	0
Other	0	1	100	720	0

Source: Travel Times are from (Blanford et al., 2012).

Type 4 = a value of 1 to use when calculating distance so that each cell is assigned the same value.

Times_w = walking speed in seconds to represent the time it takes to walk 1 km (representing 5 km/hour)

Times_car = speed it takes a vehicle to travel across a cell (representing average speed)

Service Area – Road Network

The impedance values have already been added to the roadsosm_prj data, as described in Table 8.7. Adjust these if needed. The data is in vector format and will need to be converted to a raster format to conduct the remainder of the analysis. For the purpose of this activity, set the cell size to 1,000 so that each cell represents 1,000 × 1,000 m (=1 × 1 km). For other cell sizes, see Figure 8.9. The smaller the cell size, the more computational power is needed to perform the analysis later, as there will be more cells to work through. However, this does come at a cost so you need to decide what level of accuracy you need.

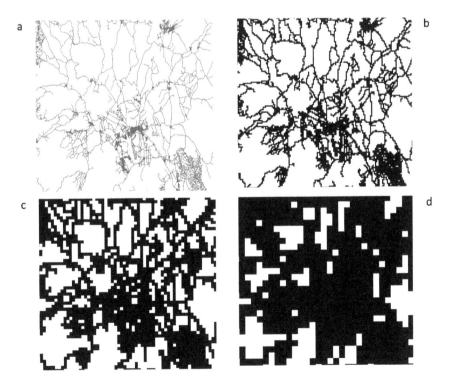

FIGURE 8.9
Illustration of the (a) vector road network and rasterised using different cell sizes of (b) 1,000 m, (c) 2,500 m and (d) 5,000 m. Maps created in ArcGIS Pro. Image created by Blanford (2023).

Convert roads to raster: **Conversion Tools – To Raster – Polyline to Raster**

Input features: roadsosm_prj

Value field: type4

Output raster dataset: roads_g

Cell size: 1,000 (so that each cell represents 1 km)

To change the NODATA values, reclassify NODATA to a high value of 5,000. **Spatial Analyst Tools – Reclass –Reclassify.** Before you run the cost allocation, reclassify the NODATA to a value of 5,000 so that you can get a continuous surface that delineates boundaries rather than just the roads. Change value=1–1,000 so that each cell represents 1,000 meters.

Input raster=roads_g

Reclass field=value

Reclassification (to populate values click on unique below the table). Manually change the values in the New column

Output raster=roads_greclass

Create a new service area map based on the road network: **Spatial Analyst Tools – Distance – Distance Allocation**

Input raster or feature source data=healthfacilities

Source field=recno

Output distance allocation raster=servicearea_roads

Input cost raster=roads_greclass

Set the processing extent for all of Kenya, click on the ENVIRONMENTS tab in the tool and, for the **Processing Extent,** select KEN_adm1_prj dataset. Also set the cell size to 1,000.

Mapping Access to Healthcare Facilities: Distance vs Time

Access to healthcare facilities is important for any number of reasons. One of the easiest ways to understand the accessibility is to map physical access. This can be as simple as determining the distance from a healthcare facility. However, accessibility may vary depending on how you travel. To account for this variation travel times by different modes of transportation, such as walking, biking, vehicle transport or even public transport can be incorporated. To do this see how travel times may vary (e.g. times_w vs times_car) Table 8.7.

Accessibility – Distance vs Time (Different Modes of Transport)

Create two different accessibility maps. One based on distance and one based on travel time (select either walking speed or vehicle speed, Table 8.7).

- **Distance**: Type 4 has a value of 1, so all roads are assigned the same weight. Earlier, you already created a distance surface of the roads (roads_greclass) (Figure 8.10a). The final map will look something like Figure 8.11a. To perform the distance calculations follow the steps in Table 8.8.
- **Travel by foot**: Times_w = time it takes to walk 1 km (720 seconds). The final map will look something like Figure 8.11b. To perform the travel times by walking calculations follow the steps in Table 8.9.
- **Travel by car**: Times_car = time it takes to travel by car 1 km (time in seconds) varies by road type (Figure 8.10b). The final map will look something like Figure 8.11c. To perform the travel times by walking calculations follow the steps in Table 8.10.

Or perhaps you are interested in examining different modes of transport. For travel by bike, see Al-Huraibi et al. (2023)).

(a) (b)

FIGURE 8.10
Rasterized road network: (a) distance and (b) different travel speeds by car.

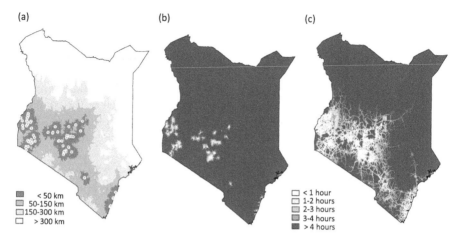

FIGURE 8.11

Accessibility surfaces based on (a) Distance, (b) walking speed and (c) vehicle travel.

TABLE 8.8

Summary of Steps Required to Perform the Distance Calculations (Figure 8.11a)

1. Convert Roads to Raster	2. Perform the Cost-Distance Calculation
Create a new cost surface where each grid cell is equal to 1 km.	Create a new output that captures the distance from each health facility
Convert roads to raster (already did this earlier)	**Spatial Analysis Tools – Distance – Distance Accumulation**
Convert roads to raster: **Conversion Tools – To Raster – Polyline to Raster**	**Input raster or feature**: Health Facility layer you created
Input Features: roadsosm_prj	**Input cost raster**: new cost surface (roads_greclass)
Value field: type4	**Distance method**: geodesic
Output raster dataset: roads_g	
Cell size: 1,000 (so that each cell represents 1 km)	
Create a continuous surface. Assign NODATA values a high impedance value	
Input raster=roads_g	
Reclass field=value	
Reclassification (to populate values click on unique below the table). Manually change the values in the New column Reclass value 1= 1,000; NODATA=5,000	
Output raster=roads_greclass	

TABLE 8.9

Summary of Steps Required to Perform Travel Times by Walking (Figure 8.11b)

1. Convert Roads to Raster for Walking Times	2. Perform the Cost-Distance Walking Calculation
Create a new cost surface where each grid cell is equal to 1 km.	Create a new output that captures the distance from each health facility
Convert roads to raster	**Spatial Analyst Tools – Distance – Distance Accumulation**
Convert roads to raster: **Conversion Tools – To Raster – Polyline to Raster**	**Input raster or feature:** Health Facility layer you created
Input Features: roads_osm_prj	
Value field: times_w	**Input cost raster:** roads_walk_rcl
Output raster dataset: roads_walk	**Output raster:** CDroads_walk
Cell size: 1,000 (so that each cell represents 1 km	**Distance method:** geodesic
Create a continuous surface. Assign NODATA values a high impedance value	Convert this into units so that you can set the legend for the map to capture, 1 hour, 2 hour, 3 hour, 4 hours, 4 hour+ of travel
Input raster=roads_walk	**Spatial Analyst Tools – Map Algebra – Raster Calculator**
Reclass field=value	
Reclassification	= (CDroads_walk/1,000)/60
Reclass value 720= 720; NODATA=5,000	Symbolize the map
Output raster=roads_walk_rcl	

TABLE 8.10

Summary of Steps Required to Perform Travel Times by Vehicle (Figure 8.11c)

1. Convert Roads to Raster for Vehicle Travel Times	2. Perform the Cost-Distance Vehicle Calculation
Create a new cost surface where each grid cell is equal to 1 km.	Create a new output that captures the distance from each health facility
Convert roads to raster	**Spatial Analyst Tools – Distance – Cost Accumulation**
Convert roads to raster: **Conversion Tools – To Raster – Polyline to Raster**	**Input raster or feature:** Health Facility layer you created
Input Features: roads_osm_prj	
Value field: times_car	**Input cost raster:** roads_car_rcl
Output raster dataset: roads_car	**Output raster:** CD_roads_car
Cell size: 1,000 (so that each cell represents 1 km	**Distance method:** geodesic
Create a continuous surface. Assign NODATA values a high impedance value	Convert this into units so that you can set the legend for the map to capture, 1 hour, 2 hour, 3 hour, 4 hours, 4 hour+ of travel
Input raster=roads_walk	**Spatial Analyst Tools – Map Algebra – Raster Calculator**
Reclass field=value	
Reclassification	= (CD_roads_car/1,000)/60
Click unique values	Symbolize the map
Reclass value copy values across; Set NODATA=5,000 and 0=5,000	
Output raster=roads_car_rcl	

Evaluating Availability of Healthcare

In the previous section, you created several outputs that show accessibility based on distance and time using different modes of transportation. Now assess what percentage of the population that doesn't have available healthcare using walking or driving time.

Before beginning the evaluation, define the criteria to use to determine what is considered to be acceptable access. Should everyone be within an hour of a health facility? In the paper by Blanford et al. (2012), access *"greater than 4-hours walk from a health facility was considered to be inadequate."* For the purpose of this exercise, examine accessibility within 1 hour, 2 hours, 4 hours and >4 hours to determine the percentage of the population that falls within each of these criteria.

Reclassify the walking time map and the driving time map you created so that the values fall into five categories (Table 8.11): **Spatial Analyst Tools – Reclass – Reclassify**

The old values shown here represent the legend used. The values are in minutes.

To find the total number of people that fall within each category, use the zonal statistic. Once you have the summaries, you will be able to calculate the percentage of the population that falls within each of the zones defined in Table 8.11.

Spatial Analyst Tools – Zonal – Zonal Statistics as Table

Input raster or feature zone data: **reclass_timecar**

Zone field: **value**

Input value raster: **afr2010KE_prj**

Ignore NoData in calculations: **tick the box**

Statistics type: **ALL**

TABLE 8.11

Classification Values

Old Values	New Values
0	1
0–15	1
15–30	1
30–45	1
45–60	1
60–120	2
120–180	3
180–240	4
240–7,000	9
No Data	NoData

Planning for Healthcare Services – Evaluating Populations at Risk to Malaria

Now that you know what percentage of the population doesn't have access to healthcare based on different modes of transportation, you can evaluate the populations without adequate accessibility to health care and possibly the population most at risk to health issues.

For the purpose of this exercise, evaluate populations at risk from malaria. Consider malaria risk should any risk value greater than ≥0.005 as highlighted in Table 8.12 (or revisit the malaria activity from Chapter 6 and use the same criteria). Based on Schoeps et al. (2011), the risk of dying from malaria increases if it takes longer than 4 hours to walk to a health facility, so use this as one of the criteria.

Combine the two reclassified maps: **Spatial Analyst Tools – Local – Combine**

Open the attribute table of the combined layer and add a field: descript (text, 20) and add in the following descriptions with the associated locations (Table 8.13).

Recalculate the population values within each of the 4 zones using zonal statistics. Use the descriptive field as the zone field.

To examine the association between access to healthcare and disease incidence, in this case, exposure to malaria fill in the table below using the output from the zonal statistics calculation in the previous step (Table 8.14).

What did you find out about the distribution of health facilities in relation to malaria risk and the distribution of the population?

TABLE 8.12

Reclassify Walking Travel Time

Walking	Reclass Value	Malaria	Reclass Value
<4 hours	10	>=0.005	1
>4 hours	20	No Data	2
		<0.005	

TABLE 8.13

Description of Combined Classes

Descript
<4 hour and malaria
<4 hour and no malaria
>4 hour and malaria
>4 hour and no malaria

TABLE 8.14

Odds Ratio of Accessibility and Malaria Risk

	Disease		
	Malaria Yes	Malaria No	Total
Yes within 4 hours walk of health facility	a	b	a+b
Not within 4 hours walk of health facility	c	d	c+d

References

Aeby, D., P. Staeger & F. Dami (2021) How to improve automated external defibrillator placement for out-of-hospital cardiac arrests: A case study. *PLoS One*, 16, e0250591.

Al-Huraibi, A., S. Amer & J. I. Blanford (2023) Cycling to get my vaccination: How accessible are COVID-19 vaccination centers in the Netherlands?. *AGILE GIScience Series*, 4, 16.

Al-Taiar, A., A. Clark, J. C. Longenecker & C. J. M. Whitty (2010) Physical accessibility and utilization of health services in Yemen. *International Journal of Health Geographics*, 9, 1–8.

Albert, D. P. & F. B. Butar (2005) Estimating the de-designation of single-county HPSAs in the United States by counting naturopathic physicians as medical doctors. *Applied Geography*, 25, 271–285.

Alirol, E., S. K. Sharma, H. S. Bawaskar, U. Kuch & F. Chappuis (2010) Snake bite in South Asia: A review. *PLoS Neglected Tropical Diseases*, 4, e603.

Allorant, A., N. Fullman, H. H. Leslie, M. Sarr, D. Gueye, et al. (2023) A small area model to assess temporal trends and sub-national disparities in healthcare quality. *Nature Communications*, 14, 4555.

Amato, H. K., D. Martin, C. M. Hoover & J. P. Graham (2022) Somewhere to go: Assessing the impact of public restroom interventions on reports of open defecation in San Francisco, California from 2014 to 2020. *BMC Public Health*, 22, 1673.

Aufderheide, T., M. F. Hazinski, G. Nichol, S. S. Steffens, A. Buroker, et al. (2006) Community lay rescuer automated external defibrillation programs: Key state legislative components and implementation strategies: A summary of a decade of experience for healthcare providers, policymakers, legislators, employers, and community leaders from the American heart association emergency cardiovascular care committee, council on clinical cardiology, and office of state advocacy. *Circulation*, 113, 1260–1270.

Awoyemi, T. T., A. A. Obayelu & H. I. Opaluwa (2011) Effect of distance on utilization of health care services in rural Kogi State, Nigeria. *Journal of Human Ecology*, 35, 1–9.

Bauer, J., D. Brüggmann, D. Klingelhöfer, W. Maier, L. Schwettmann, et al. (2020) Access to intensive care in 14 European countries: A spatial analysis of intensive care need and capacity in the light of COVID-19. *Intensive Care Medicine*, 46, 2026–2034.

Benenson, I., E. Ben-Elia, Y. Rofé & D. Geyzersky (2017) The benefits of a high-resolution analysis of transit accessibility. *International Journal of Geographical Information Science*, 31, 213–236.

Blanford, J. I., S. Kumar, W. Luo & A. M. MacEachren (2012) It's a long, long walk: Accessibility to hospitals, maternity and integrated health centers in Niger. *International Journal of Health Geographics*, 11, 24.

Bosanac, E. M., R. C. Parkinson & D. S. Hall (1976) Geographic access to hospital care: A 30-minute travel time standard. *Medical Care*, 14, 616–623.

Bruzelius, E., M. Le, A. Kenny, J. Downey, M. Danieletto, et al. (2019) Satellite images and machine learning can identify remote communities to facilitate access to health services. *Journal of the American Medical Informatics Association*, 26, 806–812.

Bryant Jr, J. & P. L. Delamater (2019) Examination of spatial accessibility at micro-and macro-levels using the enhanced two-step floating catchment area (E2SFCA) method. *Annals of GIS*, 25, 219–229.

Cattaneo, A., A. Nelson & T. McMenomy (2021) Global mapping of urban-rural catchment areas reveals unequal access to services. *Proceedings of the National Academy of Sciences of the United States of America*, 118, e2011990118.

Chabot-Couture, G. (2017) Scale in Disease Transmission, Surveillance, and Modeling. In *Integrating Scale in Remote Sensing and GIS*, pp. 375–408. Boca Raton, FL: CRC Press.

Church, R. & C. ReVelle (1974) The maximal covering location problem. *Papers of the Regional Science Association*, 32, 101–118.

Cinnamon, J., N. Schuurman & V. A. Crooks (2008) A method to determine spatial access to specialized palliative care services using GIS. *BMC Health Services Research*, 8, 1–11.

Cooksey-Stowers, K., M. B. Schwartz & K. D. Brownell (2017) Food swamps predict obesity rates better than food deserts in the United States. *International Journal of Environmental Research and Public Health*, 14, 1366.

Corbane, C., V. Syrris, F. Sabo, P. Politis, M. Melchiorri, et al. (2021) Convolutional neural networks for global human settlements mapping from Sentinel-2 satellite imagery. *Neural Computing and Applications*, 33, 6697–6720.

Cromley, E. K. & S. L. McLafferty (2012) *GIS and Public Health*. New York: Guilford Press.

Cu, A., S. Meister, B. Lefebvre & V. Ridde (2021) Assessing healthcare access using the Levesque's conceptual framework-a scoping review. *International Journal for Equity in Health*, 20, 116.

Delamater, P. L. (2013) Spatial accessibility in suboptimally configured health care systems: A modified two-step floating catchment area (M2SFCA) metric. *Health & Place*, 24, 30–43.

Dzhus, M. & I. Golovach (2023) Impact of Ukrainian- Russian war on health care and humanitarian crisis. *Disaster Medicine and Public Health Preparedness*, 17, e340.

Eklund, L. & U. Martensson (2012) Using geographical information systems to analyse accessibility to health services in the West Bank, occupied Palestinian territory. *EMHJ-Eastern Mediterranean Health Journal*, 18, 796–802.

ESRI (2023) ArcGIS Pro Help. https://pro.arcgis.com/en/pro-app/latest/help/main/welcome-to-the-arcgis-pro-app-help.htm (last accessed Sep 15 2023).

Falchetta, G., A. T. Hammad & S. Shayegh (2020) Planning universal accessibility to public health care in sub-Saharan Africa. *Proceedings of the National Academy of Sciences*, 117, 31760–31769.

Florio, P., S. Freire & M. Melchiorri (2023) Estimating geographic access to healthcare facilities in Sub-Saharan Africa by degree of urbanisation. *Applied Geography,* 160, 103118.

Geurs, K. T. & B. Van Wee (2004) Accessibility evaluation of land-use and transport strategies: Review and research directions. *Journal of Transport Geography,* 12, 127–140.

Guagliardo, M. F. (2004) Spatial accessibility of primary care: Concepts, methods and challenges. *International Journal of Health Geographics,* 3, 1–13.

Guagliardo, M. F., C. R. Ronzio, I. Cheung, E. Chacko & J. G. Joseph (2004) Physician accessibility: An urban case study of pediatric providers. *Health & Place,* 10, 273–283.

Hansen, C. M., M. Wissenberg, P. Weeke, M. H. Ruwald, M. Lamberts, et al. (2013) Automated external defibrillators inaccessible to more than half of nearby cardiac arrests in public locations during evening, nighttime, and weekends. *Circulation,* 128, 2224–2231.

Higgs, G. (2004) A literature review of the use of GIS-based measures of access to health care services. *Health Services and Outcomes Research Methodology,* 5, 119–139.

HRSA (2023) Designated Health Professional Shortage Areas Statistics. Fourth Quarter of Fiscal Year 2023, Designated HPSA Quarterly Summary, 15 p. https://data. hrsa.gov/default/generatehpsaquarterlyreport (last accessed Aug 10 2023).

Indriasari, V., A. R. Mahmud, N. Ahmad & A. R. M. Shariff (2010) Maximal service area problem for optimal siting of emergency facilities. *International Journal of Geographical Information Science,* 24, 213–230.

Karpyn AE, Riser D, Tracy T, Wang R & S. YE. (2019) The changing landscape of food deserts. *UNSCN Nutrition,* 44, 46–53.

Kwan, M. P. (1998) Space-time and integral measures of individual accessibility: A comparative analysis using a point-based framework. *Geographical Analysis,* 30, 191–216.

Kwan, M.-P. (2013) Beyond space (as we knew it): Toward temporally integrated geographies of segregation, health, and accessibility: Space-time integration in geography and GIScience. *Annals of the Association of American Geographers,* 103, 1078–1086.

Larsen, K. & J. Gilliland (2008) Mapping the evolution of 'food deserts' in a Canadian city: Supermarket accessibility in London, Ontario, 1961–2005. *International Journal of Health Geographics,* 7, 16.

Lee, J. & H. J. Miller (2018) Measuring the impacts of new public transit services on space-time accessibility: An analysis of transit system redesign and new bus rapid transit in Columbus, Ohio, USA. *Applied Geography,* 93, 47–63.

Levesque, J.-F., M. F. Harris & G. Russell (2013) Patient-centred access to health care: Conceptualising access at the interface of health systems and populations. *International Journal for Equity in Health,* 12, 18.

Liu, S. & X. Zhu (2004) Accessibility analyst: An integrated GIS tool for accessibility analysis in urban transportation planning. *Environment and Planning B: Planning and Design,* 31, 105–124.

Luo, W. & Y. Qi (2009) An enhanced two-step floating catchment area (E2SFCA) method for measuring spatial accessibility to primary care physicians. *Health & Place,* 15, 1100–1107.

Luo, W. & F. Wang (2003) Measures of spatial accessibility to health care in a GIS environment: Synthesis and a case study in the Chicago region. *Environment and Planning B: Planning and Design,* 30, 865–884.

Luo, W. & T. Whippo (2012) Variable catchment sizes for the two-step floating catchment area (2SFCA) method. *Health & Place*, 18, 789–795.

Macharia, P. M., P. O. Ouma, E. G. Gogo, R. W. Snow & A. M. Noor (2017) Spatial accessibility to basic public health services in South Sudan. *Geospatial Health*, 12, 510.

Macharia, P. M., E. Mumo & E. A. Okiro (2021) Modelling geographical accessibility to urban centres in Kenya in 2019. *PLoS One*, 16, e0251624.

Maina, J., P. O. Ouma, P. M. Macharia, V. A. Alegana, B. Mitto, et al. (2019) A spatial database of health facilities managed by the public health sector in sub Saharan Africa. *Scientific Data*, 6, 134.

Maine, D. (2009) *Monitoring Emergency Obstetric Care: A Handbook*. Geneva, Switzerland: World Health Organization.

Makanga, P. T., N. Schuurman, C. Sacoor, H. E. Boene, F. Vilanculo, et al. (2017) Seasonal variation in geographical access to maternal health services in regions of southern Mozambique. *International Journal of Health Geographics*, 16, 1.

Mao, L. & D. Nekorchuk (2013) Measuring spatial accessibility to healthcare for populations with multiple transportation modes. *Health & Place*, 24, 115–122.

Markhorst, B., T. Zver, N. Malbasic, R. Dijkstra, D. Otto, et al. (2021) A data-driven digital application to enhance the capacity planning of the COVID-19 vaccination process. *Vaccines*, 9, 1181.

McCoy, J. H. (2013) Overcoming the Challenges of the Last Mile: A Model of Riders for Health. In *Handbook of Healthcare Operations Management: Methods and Applications*, pp. 483–509. Cham: Springer.

McGrail, M. R. (2012) Spatial accessibility of primary health care utilising the two step floating catchment area method: An assessment of recent improvements. *International Journal of Health Geographics*, 11, 1–12.

McGrail, M. R. & J. S. Humphreys (2015) Spatial access disparities to primary health care in rural and remote Australia. *Geospatial Health*, 10, 358.

McLafferty, S. & S. Grady (2004) Prenatal care need and access: A GIS analysis. *Journal of Medical Systems*, 28, 321–333.

McLaughlin, C. G. & L. Wyszewianski (2002) Access to care: Remembering old lessons. *Health Services Research*, 37, 1441–1443.

Moner-Girona, M., G. Kakoulaki, G. Falchetta, D. J. Weiss & N. Taylor (2021) Achieving universal electrification of rural healthcare facilities in sub-Saharan Africa with decentralized renewable energy technologies. *Joule*, 5, 2687–2714.

Mroz, E. J., T. Willis, C. Thomas, C. Janes, D. Singini, et al. (2023) Impacts of seasonal flooding on geographical access to maternal healthcare in the Barotse Floodplain, Zambia. *International Journal of Health Geographics*, 22, 17.

Nelson, A., D. J. Weiss, J. van Etten, A. Cattaneo, T. S. McMenomy, et al. (2019) A suite of global accessibility indicators. *Scientific Data*, 6, 266.

Neutens, T. (2015) Accessibility, equity and health care: Review and research directions for transport geographers. *Journal of Transport Geography*, 43, 14–27.

Noor, A. M., D. Zurovac, S. I. Hay, S. A. Ochola & R. W. Snow (2003) Defining equity in physical access to clinical services using geographical information systems as part of malaria planning and monitoring in Kenya. *Tropical Medicine & International Health*, 8, 917–926.

Noor, A. M., A. A. Amin, P. W. Gething, P. M. Atkinson, S. I. Hay, et al. (2006) Modelling distances travelled to government health services in Kenya. *Tropical Medicine & International Health*, 11, 188–196.

Noor, A. M., V. A. Alegana, P. W. Gething & R. W. Snow (2009) A spatial national health facility database for public health sector planning in Kenya in 2008. *International Journal of Health Geographics*, 8, 1–7.

O'Meara, W. P., A. Noor, H. Gatakaa, B. Tsofa, F. E. McKenzie, et al. (2009) The impact of primary health care on malaria morbidity--defining access by disease burden. *Tropical Medicine & International Health*, 14, 29–35.

Ochoa, C., M. Rai, S. B. Martins, G. Alcoba, I. Bolon, et al. (2023) Vulnerability to snakebite envenoming and access to healthcare in the Terai region of Nepal: A geospatial analysis. *The Lancet Regional Health-Southeast Asia*, 9, 100103.

Oppong, J. R. (1996) Accommodating the rainy season in third world location-allocation applications. *Socio-Economic Planning Sciences*, 30, 121–137.

Oppong, J. R. & M. J. Hodgson (1994) Spatial accessibility to health care facilities in Suhum District, Ghana. *The Professional Geographer*, 46, 199–209.

Ouma, P., P. M. Macharia, E. Okiro & V. Alegana (2021) Methods of Measuring Spatial Accessibility to Health Care in Uganda. In *Practicing Health Geography: The African Context*, ed. P. T. Makanga, pp. 77–90. Cham: Springer International Publishing.

Park, J. & D. W. Goldberg (2021) A review of recent spatial accessibility studies that benefitted from advanced geospatial information: Multimodal transportation and spatiotemporal disaggregation. *ISPRS International Journal of Geo-Information*, 10, 532.

Penchansky, R. & J. W. Thomas (1981) The concept of access: Definition and relationship to consumer satisfaction. *Medical Care*, 19, 127–140.

Perkins, G. D., A. J. Handley, R. W. Koster, M. Castrén, M. A. Smyth, et al. (2015) European resuscitation council guidelines for resuscitation 2015: Section 2. Adult basic life support and automated external defibrillation. *Resuscitation*, 95, 81–99.

Peters, D. H., A. Garg, G. Bloom, D. G. Walker, W. R. Brieger, et al. (2008) Poverty and access to health care in developing countries. *Annals of the New York Academy of Sciences*, 1136, 161–171.

Poku-Boansi, M., E. Ekekpe & A. A. Bonney (2010) Combating maternal mortality in the Gushegu district of Ghana: The role of rural transportation. *Journal of Sustainable Development in Africa*, 12, 274–283.

Ray, N. & S. Ebener (2008) AccessMod 3.0: Computing geographic coverage and accessibility to health care services using anisotropic movement of patients. *International Journal of Health Geographics*, 7, 63.

Ray, N., S. Ebener & F. Moser (2022) AccessMod 5: Supporting Universal Health Coverage by Modelling Physical Accessibility to Health Care. User Manual & Tutorial, Version 5.8, 144 p. https://www.accessmod.org/ (last accessed Aug 17 2023).

Rosenthal, M. B., A. Zaslavsky & J. P. Newhouse (2005) The geographic distribution of physicians revisited. *Health Services Research*, 40, 1931–1952.

Runfola D, Anderson A, Baier H, Crittenden M, Dowker E, et al. (2020) geoBoundaries: A global database of political administrative boundaries. *PLoS One*, 15, e0231866.

Schoeps, A., S. Gabrysch, L. Niamba, A. Sie & H. Becher (2011) The effect of distance to health-care facilities on childhood mortality in rural Burkina Faso. *American Journal of Epidemiology*, 173, 492–498.

Schuurman, N., R. S. Fiedler, S. C. W. Grzybowski & D. Grund (2006) Defining rational hospital catchments for non-urban areas based on travel-time. *International Journal of Health Geographics*, 5, 1–11.

Schuurman, N., M. Berube & V. A. Crooks (2010) Measuring potential spatial access to primary health care physicians using a modified gravity model. *The Canadian Geographer/Le Geographe Canadien*, 54, 29–45.

Sun, C. L. F., D. Demirtas, S. C. Brooks, L. J. Morrison & T. C. Y. Chan (2016) Overcoming spatial and temporal barriers to public access defibrillators via optimization. *Journal of the American College of Cardiology*, 68, 836–845.

Sun Christopher, L. F., D. Demirtas, C. Brooks Steven, J. Morrison Laurie & C. Y. Chan Timothy (2016) Overcoming spatial and temporal barriers to public access defibrillators via optimization. *Journal of the American College of Cardiology*, 68, 836–845.

Takane, M., S. Yabe, Y. Tateshita, Y. Kobayashi, A. Hino, et al. (2016) Satellite imagery technology in public health: Analysis of site catchment areas for assessment of poliovirus circulation in Nigeria and Niger. *Geospatial Health*, 11, 462.

Tanser, F. (2006) Methodology for optimising location of new primary health care facilities in rural communities: A case study in KwaZulu-Natal, South Africa. *Journal of Epidemiology & Community Health*, 60, 846–850.

Tanser, F., V. Hosegood, J. Benzler & G. Solarsh (2001) New approaches to spatially analyse primary health care usage patterns in rural South Africa. *Tropical Medicine & International Health*, 6, 826–838.

Tessema, Z. T., M. G. Worku, G. A. Tesema, T. S. Alamneh, A. B. Teshale, et al. (2022) Determinants of accessing healthcare in Sub-Saharan Africa: A mixed-effect analysis of recent demographic and health surveys from 36 countries. *BMJ Open*, 12, e054397.

Thannhauser, J., J. Nas, R. A. Waalewijn, N. van Royen, J. L. Bonnes, et al. (2022) Towards individualised treatment of out-of-hospital cardiac arrest patients: An update on technical innovations in the prehospital chain of survival. *Netherlands Heart Journal*, 30, 345–349.

Thornton, L. E. & A. M. Kavanagh (2012) Association between fast food purchasing and the local food environment. *Nutrition & Diabetes*, 2, e53–e53.

UN (2021) The United Nations World Water Development Report 2021: Valuing Water, 206 p. https://unesdoc.unesco.org/ark:/48223/pf0000375724 (last accessed Aug 5 2023).

—— (2023a) The SDGs in Action. https://www.undp.org/sustainable-development-goals#good-health (last accessed Aug 5 2023).

—— (2023b) The Sustainable Development Goals Report 2023: Special edition. Towards a Rescue Plan for People and Planet, 80 p. https://unstats.un.org/sdgs/report/2023/The-Sustainable-Development-Goals-Report-2023.pdf (last accessed Aug 5 2023).

UNWWDR (2023) United Nations World Water Development Report 2023. https://www.unesco.org/reports/wwdr/2023/en/way-forward. (last accessed May 4 2024)

Wan, N., B. Zou & T. Sternberg (2012) A three-step floating catchment area method for analyzing spatial access to health services. *International Journal of Geographical Information Science*, 26, 1073–1089.

Watmough, G. R., M. Hagdorn, J. Brumhead, S. Seth, E. Delamónica, et al. (2022) Using open-source data to construct 20 metre resolution maps of children's travel time to the nearest health facility. *Scientific Data*, 9, 217.

Weber, E. M., V. Y. Seaman, R. N. Stewart, T. J. Bird, A. J. Tatem, et al. (2018) Census-independent population mapping in northern Nigeria. *Remote Sensing of Environment*, 204, 786–798.

Weiss, D. J., A. Nelson, H. S. Gibson, W. Temperley, S. Peedell, et al. (2018) A global map of travel time to cities to assess inequalities in accessibility in 2015. *Nature*, 553, 333–336.

Weiss, D. J., A. Nelson, C. A. Vargas-Ruiz, K. Gligoric, S. Bavadekar, et al. (2020) Global maps of travel time to healthcare facilities. *Nature Medicine*, 26, 1835–1838.

WHO (2011) Financial Sustainability Plan of the EPI-Nige, 35 p.

—— (2019) Uneven Access to Health Services Drives Life Expectancy Gaps: WHO. https://www.who.int/news/item/04-04-2019-uneven-access-to-health-service s-drives-life-expectancy-gaps-who (last accessed Aug 17 2023).

—— (2023a) Geolocated Health Facilities Data Initative: Strengthening Planning and Decision-Making with Accurate and Accessible Health Facility Master Lists. https://www.who.int/data/GIS/GHFD (last accessed Aug 10 2023).

—— (2023b) Universal Health Coverage. https://www.who.int/health-topics/univers al-health-coverage#tab=tab_1 (last accessed Aug 20 2023).

Widener, M. J., S. Farber, T. Neutens & M. Horner (2015) Spatiotemporal accessibility to supermarkets using public transit: An interaction potential approach in Cincinnati, Ohio. *Journal of Transport Geography*, 42, 72–83.

9

Geographic Information for Planetary Health

> Climate change is 'the single biggest health threat facing humanity' in the 21st century
>
> *(WHO, 2023)*

Overview

The environment in which we live can influence our health in many ways (WHO, 2016; Brusseau et al., 2019). In this chapter, I provide a range of examples of the health effects of exposures in our environment (Figure 9.1).

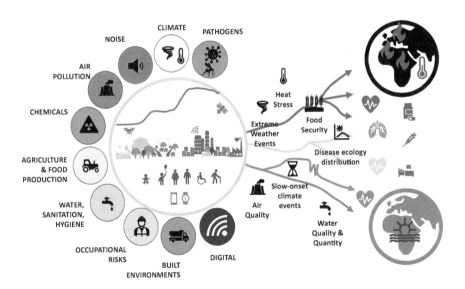

FIGURE 9.1
The environment in which we live can influence our health in many ways. This will be further influenced by the climate change trajectory that lies ahead. (Image created Blanford (2023).)

DOI: 10.1201/9781003435082-9

State of Our Environment

Climate change affects the social and environmental determinants of health – clean air, safe drinking water, sufficient food and secure shelter. Between 2030 and 2050, climate change is expected to cause approximately 250,000 additional deaths per year, from malnutrition, malaria, diarrhoea and heat stress (WHO, 2023).

The climate crisis (see IPCC reports; Lee et al., 2023) is causing frequent forest and wild fires (see World Fire Atlas for latest updates (ESA)), heatwaves (Thompson et al., 2022; Ullah et al., 2022; Kumar et al., 2022; Puley et al., 2022), floods (theGFD) and extreme drought (NDMC, 2023b; Toreti et al., 2023; International, 2023; NASA, 2022). Biodiversity is rapidly declining (Isbell et al., 2023); environmental pollution has reached every corner of the Earth; deforestation (Ritchie and Roser, 2021) is continuing at an alarming rate; and freshwater sources are being depleted and are drying up (Rojanasakul et al., 2023). Billions of people are at health risk around the world if these global environmental changes continue. Many millions will die each year due to many of the effects the climate crisis is having. These include, but are not limited to, heat stress, infectious diseases, malnutrition, flooding and mental disorders as a result of having to cope with these rapid and impactful effects. Some human communities will find their very existence threatened as their environment is flooded or scorched by heat or drought. During the summer of 2023, we are already witnessing the devastation these effects are having on many communities. To name a few:

- **Wildfires** are on the rise. In Hawaii, fires decimated a town; wildfires in Greece have devasted vast areas and destroyed livelihoods; smoke from fires in Canada (e.g. fires in Nova Scotia have crossed borders, causing poor air quality in many locations in the US, including New York, New Jersey and Pennsylvania; NASA, 2023).

- **Extreme heat** and hotter than usual temperatures have been experienced globally, with temperatures reaching 40°C in many parts of the world for several days (West Africa (Carrington and Borràs, 2024); South East Asia (Ratcliffe 2024a,b,c); India and Pakistan (Zachariah et al., 2023); Europe (Niranjan, 2023); Spain (Kassam, 2022).

- **Excess rainfall** The intensity of storms has changed causing record rainfall levels to fall, leading to flooding (e.g. Mozambique, Zimbabwe and Malawi (e.g. Cyclone Idai Mar 2019 (Phiri et al., 2021); Netherlands, Germany, Belgium and Luxemburg (Copernicus, 2021)) and Cyclone Kenneth (Cambaza et al., 2019); several hurricanes caused extensive flooding in California (e.g. Hurricane Hilary (Jones, 2023)) and Florida (Hurricane Idalia, 2023 (Bello and Cardona, 2023)).

These are but a few examples to highlight the many changes that are occurring in different parts of the world. Global environmental changes are already affecting human health in many ways. More people are dying of heat stress; severe drought is causing widespread hunger and affecting nutritional uptake; and infectious diseases are spreading to new geographic locations or re-emerging in locations where they have been absent for 50 or more years (e.g. refer back to the chapter on vector-borne diseases).

Transformative changes will be necessary in all sectors (e.g. energy supply, transportation and food production), including changes in our behaviours (consumption patterns and expectations), but most importantly in the surveillance of our planet so that we can better understand how the environment and changes in the environment are affecting our health and well-being. As I was putting this chapter together and searching for data, it struck me how difficult it is to find the data in the right format. With all of the different sensors collecting a ridiculous amount of information about our environment, some at micro-seconds, we need to get better at making these data available in a variety of formats and through central hubs. We have to make it easier to do the research and conduct the analyses needed so that we can make effective decisions.

Environment Exposure and Health Outcomes: The Role of the Environment in Health Outcomes

Where we live plays a fundamental role in shaping our health, be it through direct contact with our environment (e.g. altitude, climate, greenness, hazards and pollutants), our behaviour (e.g., social networks, risk behaviours and lifestyles) or economic condition (e.g., access – to health care; healthy environments (e.g. quality of food; physical activity spaces; quality housing)). Knowing how different factors affect our health is important for determining how to respond through education/policies and identifying the necessary services needed to enable recovery (e.g. timely treatment for infectious and non-infectious diseases and health issues). To do so, we need to understand how these factors affect health outcomes.

Understanding why a disease is occurring (e.g. epidemiological triad, Chapter 2) in a particular location or in a population is important for determining how to respond and what mitigations or adaptations are needed for recovery (new medicines, technologies, changing the environment, funding, policy changes, etc.). To get you thinking, I have provided a variety of examples that capture how exposure to elements in the environment or changes in the environment and climate can influence different health and disease outcomes.

The Role of Geographic Location and Behaviour in the Transmission of Diseases

Tuberculosis

Tuberculosis (TB) is an airborne bacterial infection caused by the organism *Mycobacterium tuberculosis* (*Mtb*) and affects 2–3 billion people worldwide (Ma et al., 2018). The reproductive number for TB estimates range from 0.24 in the Netherlands (during 1933–2007) to 4.3 in China in 2012 (Ma et al., 2018). For some diseases that are transmitted through close contact (e.g. COVID-19, TB and sexually transmitted infections), transmission can take place in spaces that are contaminated with poor ventilation. To provide an example, in Houston, Texas, USA, during initial investigation using conventional outbreak detection strategies (e.g. the steps used for investigating and managing an outbreak, Figure 2.5, Chapter 2), it was revealed that there were few contacts among 37 patients with an identical strain of TB. Further investigations revealed 40 places (including many bars) through which the patients who otherwise had nothing in common could be linked. Eventually, the transmission location was revealed to be a bar with poor ventilation (Klovdahl et al., 2001). A similar study, by Godoy et al. (2017) also found bars were an important transmission location for TB. Similar work by Jolly and Wylie (2002) and Jolly and Wylie (2013) also identified the source location (a hotel) for the transmission of a gonorrhoea outbreak in Canada (Figure 1.10b). Logan et al. (2016) incorporated geographic location to further understand the role of "hangout" spaces in the transmission of sexually transmitted diseases (Figure 1.10a). For TB, this can also be extended to living conditions, as illustrated in the analysis by Zürcher et al. (2016) who examined TB in Bern, Switzerland, during 1856–1950.

Measles

Measles is a highly contagious, acute viral illness that can lead to complications such as pneumonia, encephalitis and death. The basic reproduction number (R_0) of measles ranges between 12 and 18 (Guerra et al., 2017). Measles is highly transmittable AND is vaccine-preventable. In many countries, measles cases are at low levels including the US where it was eliminated (defined as interruption of year-round endemic transmission) in 2000 (Katz et al., 2004). However, importations from countries where measles continue to persist can lead to localized outbreaks and clusters of measles cases when imported to other geographic regions such as in the United States in 2015 (e.g. Clemmons et al., 2015). Outbreaks are influenced by seasonal human movement in unvaccinated populations (e.g. Niger, Ferrari et al., 2009). In the chapter on vector-borne diseases, you were introduced to the framework for human movements and their relevance to vector-borne pathogen transmission (Stoddard et al., 2009). Have a quick look at this framework (Figure 6.4, Chapter 6), and as you read through the following examples, evaluate how measles can circulate locally and globally.

MEASLES OUTBREAK IN THE US 2008, 2015

On January 13, 2008, an unvaccinated 7-year-old boy returned home to San Diego from a family vacation in Switzerland. He and his family were unaware at the time that he had been infected with measles during their trip. He became sick within a week of arriving home, and only received a diagnosis of measles the following week. Public health officials scrambled to assess the situation and ultimately determined that, by unintentionally importing the virus causing measles, he had exposed 839 people in the San Diego area to it, of whom 11 also developed the disease, including a hospitalized infant who was too young to be vaccinated. (Peeples 2019)

On January 5, 2015, the California Department of Public Health (CDPH) was notified about a suspected measles case. The patient was a hospitalized, unvaccinated child, aged 11 years with rash onset on December 28. The only notable travel history during the exposure period was a visit to one of two adjacent Disney theme parks located in Orange County, California. On the same day, CDPH received reports of four additional suspected measles cases in California residents and two in Utah residents, all of whom reported visiting one or both Disney theme parks during December 17–20. By January 7, seven California measles cases had been confirmed. In addition, 15 cases linked to the two Disney theme parks have been reported in seven other states: Arizona (seven), Colorado (one), Nebraska (one), Oregon (one), Texas (one), Utah (three), and Washington (two), as well as linked cases reported in two neighboring countries, Mexico (one) and Canada (10). (Zipprich et al., 2015)

MEASLES OUTBREAK IN SUB-SAHARAN AFRICA

In the Sahel, measles outbreaks oscillate from year to year (Figure 9.2) and are also associated with seasonal human movement but for quite different reasons – survival. In the Sahel during the dry season (Ferrari

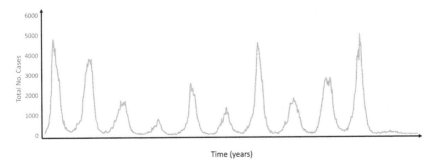

FIGURE 9.2
Measles outbreaks between 1995 and 2005, Niger. (Data from Blake (2020).)

et al., 2008), seasonal migration occurs where families will move into urban areas temporarily to find work and access food and nutrition centres (Tomaszewski et al., 2011), returning to their land at the start of the rainy season (Ferrari et al., 2008).

The Role of Natural Elements in the Environment on Health

There are many naturally occurring elements in the environment that we are exposed to on a day-to-day basis, some through direct contact with the soil, air or water. Over the next few pages, I will highlight different examples where prolonged exposure from the environment and elements in the environment can affect health. By mapping where these diseases are occurring and using many of the methods covered throughout this book (see Figures 1.5 and 3.20 for overview), will make it easier to determine if diseases are clustered or not so that further investigations can be conducted to better understand the underlying drivers.

Podoconiosis

Podoconiosis is a neglected tropical disease that can be eliminated by wearing shoes (Harter, 2022; Deribe et al., 2018). An estimated 4 million people in Africa, Latin America and Asia are affected by podoconiosis (Deribe et al., 2018), a non-infectious form of lymphoedema, also known as elephantiasis, that causes extreme swelling in the feet and legs (Harter, 2022). Podoconiosis is a painful, disfiguring condition that is caused by extended exposure to certain soils rich in minerals. In Ethiopia, the condition is linked to barefoot farming on red clay soil in the central highlands, where about 11 million farmers go without shoes (Deribe et al., 2013). Since 2012, there has been an increase in education about this condition and an increase in the purchase of shoes (Harter, 2022). Deribe et al. (2018) describe podoconiosis as *"one of the few diseases that could be eliminated within one generation."*

Radon

Radon is a naturally occurring radioactive noble gas that can be found in rocks and soil. For a view of where radon is distributed, see the radonmap (https://radonmap.com/). Radon is an established human lung carcinogen, and in miners, it is the second leading cause of lung cancer death after tobacco smoke (Tracy et al., 2006; Al-Zoughool and Krewski, 2009). In North America, a positive association between residential radon and lung cancer has been found, and it was estimated that some 6%–16% of all lung cancer deaths in Canada could be attributable to residential radon (Tracy et al., 2006). Long-term exposure can lead to lung cancer, but exposure can be reduced through ventilation and installation of radon mitigation systems.

Arsenic

Excess arsenic exposure may occur from groundwater, as is the case in some regions of India and Bangladesh (Nordstrom, 2002). Several health effects from long-term exposure to arsenic have been identified (Bhowmick et al., 2018). These range from an increased likelihood of lung and bladder cancers and skin lesions (Nrc, 2001) as well as associations with diabetes mellitus and hypertension (Nrc, 2001) (Tseng et al., 2002). In children, arsenic can affect memory and attention and can be negatively associated with IQ scores (Wasserman et al., 2004; Rosado et al., 2007). Furthermore, malnourished individuals, especially women of reproductive age and young children, may be more vulnerable to adverse health effects (Kordas et al., 2007). Thus, knowing how arsenic affects the health of different populations and where populations are most at risk is useful for determining mitigation strategies (Bhowmick et al., 2018).

Air Quality, Particles and Pollution

> Ten million deaths a year is a hundred million a decade
> *(Wallace-Wells 2021)*

Environmental pollution at local and regional scales has been a long-standing issue of concern in public health (Manisalidis et al., 2020). Air pollution has become a worldwide problem that has major impacts on human health (Dockery et al., 1993; Kampa and Castanas, 2008; Bauer et al., 2019; Manisalidis et al., 2020) (Table 9.1), leading to 10 million deaths a year (Wallace-Wells, 2021). During 2023, the WHO provided new guidelines to achieve air quality that protects public health (Table 9.1).

TABLE 9.1

Air Pollutants, Health Effects and Recommended Guideline Value for Each Pollutant

Pollutant	Guideline Value	Averaging Time	Health Effect
PM$_{2.5}$	5 µg/m³ 15 µg/m³	Annual 24-hour	Is capable of penetrating deep into the lungs and enter the bloodstream causing cardiovascular (ischaemic heart disease), cerebrovascular (stroke) and respiratory impacts. Both long-term and short-term exposure to particulate matter is associated with morbidity and mortality from cardiovascular and respiratory diseases. Long-term exposure has been further linked to adverse perinatal outcomes and lung cancer.
PM$_{10}$	15 µg/m³ 45 µg/m³	Annual 24-hour	As PM2.5

(Continued)

TABLE 9.1 (*Continued*)

Air Pollutants, Health Effects and Recommended Guideline Value for Each Pollutant

Pollutant	Guideline Value	Averaging Time	Health Effect
Carbon monoxide	$4\,mg/m^3$	24-hour	Exposure to carbon monoxide can cause difficulties breathing, exhaustion, dizziness and other flu-like symptoms. Exposure to high levels of carbon monoxide can be deadly.
Nitrogen dioxide	$10\,\mu g/m^3$ $25\,\mu g/m^3$	Annual 24-hour	irritate airways and aggravate respiratory diseases. NO_2 is an important ozone precursor, a pollutant closely linked to asthma and other respiratory conditions
Sulphur dioxide	$40\,mg/m^3$	24-hour	Associated with asthma hospital admissions and emergency room visits.
Formaldehyde	$0.1\,mg/m^3$	30-minute	Short-term exposure to formaldehyde can lead to eye, nose and throat irritation and increased allergic sensitization. Long-term exposure has been associated with nasopharyngeal cancer.
Polycyclic aromatic hydrocarbons	8.7×10^{-5} per ng/m^3		Short-term exposure can irritate eyes and breathing passages. Long-term exposure to PAH has been linked to lung cancer.
Radon	$100\,Bq/m^3$		lung cancer
Lead	$0.5\,\mu g/m^3$	Annual	For children exposed to lead include behaviour and learning problems, lower IQ and hyperactivity, slowed growth, hearing problems and anaemia. In rare cases, ingestion of lead can cause seizures, coma and even death. For pregnant women, health risks include reduced growth of the foetus and premature birth. Adults exposed to lead also have a higher risk of cardiovascular effects increased blood pressure, the incidence of hypertension, decreased kidney function and risk of reproductive problems in both men and women.
Black carbon (soot)			Major source of PM2.5. Is emitted from anthropogenic (e.g. diesel vehicles, biomass cookstoves) and natural (e.g. wildfires) sources. Short- and long-term exposure to black carbon has been associated with cardiovascular health effects and premature mortality
Mould			Allergens and irritants can cause asthma attacks and irritate the eyes, skin, nose, throat and lungs.

Source: Compiled from WHO (2021); WHO (2010); WHO (2000); (WHO).

Smoke, Burning, Wildfires and Fossil Fuel Burning

Although tremendous progress has been made in the development and availability of renewable energy sources, many households still do not have access to electricity, with only 17% of the population having access to clean cooking and 2.5 million premature deaths annually from cooking smoke (IEA, 2020). In households without reliable electricity, poor indoor quality of air from fossil fuels/wood smoke can affect health through respiratory illnesses, asthma and acute respiratory infection; heat stress can affect cardiorespiratory disease and death; and climate change-induced weather can influence the adverse health impacts of aeroallergens and air pollution (Graham et al., 2005; Conway et al., 2019; Brown-Luthango et al., 2017; West et al., 2020; Flores Quiroz et al., 2021).

Fine Particles – Sand/Dust Storms

Anthropogenic climate change is an important driver of the generation of dust (Shepherd, 2017). Many regions that are currently dusty will likely become drier and contribute more atmospheric dust. There are positive and negative associations of dust.

- **Positive effects**: Mineral dust deposition provides nutrients such as iron and other trace elements to terrestrial and marine ecosystems in different parts of the world (Yu et al., 2015; Wang et al., 2017; Garrison et al., 2003; Shepherd, 2017).

- **Negative effects**: Dust can harm animals and humans. For humans, inhaling fine particles can generate and aggravate asthma, bronchitis, emphysema and silicosis (Derbyshire, 2007; Shepherd, 2017). Finer dust can also deliver a range of pollutants, spores, bacteria, fungi and allergens. These may contribute to eye infections, skin irritations, Valley Fever (Shepherd, 2017) and Q Fever (Tozer et al., 2014). In the Sahel, dust arriving from the Sahara has been found to strongly coincide with meningitis outbreaks (García-Pando et al., 2014) across several countries (e.g. Molesworth et al., 2002). Chronic exposure to fine dust has also been found to contribute to premature death from respiratory and cardiovascular diseases, lung cancer and acute lower respiratory infections (Goldstone, 2015).

During droughts and dry seasons, sand and dust storms may increase. These may occur alongside other effects related to water shortages and food security (e.g. production and price), which may result in increased migration (e.g. Tomaszewski et al., 2011). These can result in a multitude of health effects ranging from mental health and anxiety, nutrition, exposure to a variety of infectious diseases (e.g. measles; Tomaszewski et al., 2011) and other diseases associated with migration (e.g. HIV Sznajder et al., 2024), including access to health care (Bellizzi et al., 2020; Tomaszewski et al., 2011), some of which may result in increased mortality.

Other Fine Particles – Pollen

Globally, pollen allergy is a major public health problem (Lake Iain et al., 2017) that has increased rapidly in recent decades, and it is now recognized as a major global epidemic (Pawankar, 2014; Platts-Mills, 2015), with 400 million people suffering from allergic rhinitis and 300 million from asthma (WHO, 2007). The economic burden of allergic disease is considerable, costing 19.7 billion USD in the US during 2007 and between 55 and 151 billion EUR in Europe (Zuberbier et al., 2014). In the study by Lake et al. (2017), they examined the potential impacts of climate change on ragweed plant distribution, plant productivity, and pollen production and dispersal, as well as the resulting impacts on pollen concentrations and allergic sensitization. With climate change, they estimated that individuals may experience more severe symptoms as a consequence of higher ragweed pollen levels and an extended pollen season that will last into September and October across much of Europe (Lake Iain et al., 2017).

Chemicals

Health effects from chemical spraying are well documented (Bassil et al., 2007; Conway and Pretty, 2013) and can range from birth defects, a variety of cancers (Bassil et al., 2007) to chronic kidney disease, as noted from drinking contaminated well water (Jayasumana et al., 2015) from spraying of glyphosate and other pesticides in paddy fields (Jayasumana et al., 2015) and agricultural practices in different geographic locations (Hansson et al., 2021).

Water

By 2050, more than 50% of the population will reside in water stressed regions (UN, 2021) due to a variety of reasons, ranging from higher temperatures and a reduction in precipitation to changes in intensity and frequency of weather-related events.

Water quantities and levels are diminishing. Water is essential for all life, yet resources are being depleted at alarming rates in different parts of the world. Groundwater levels are at low levels in some parts of the US (Rojanasakul et al., 2023) and have run dry in others (e.g. South Africa (Mahr, 2018); US (Cagle, 2020)). In addition to rising temperatures, glaciers and snow cover are disappearing quickly. These are important sources of water for buffering against drought events (e.g. crop loss, wildfires, lack of drinking water and power shortage) (Pritchard, 2019; Pritchard, 2017; Van Loon et al.,

2014; Talukder et al., 2021). These changes have profound effects on local populations (Mark et al., 2010; Kaenzig and Piguet, 2013; Brugger et al., 2013).

Access to water varies by household; however, there are many households still lacking indoor plumbing in many parts of the world, including the US. For example, in the US, almost half a million people do not have indoor plumbing (Lakhani et al., 2021). In areas where running water is not available within households or has run dry, there has been a rise in water vending machines (e.g. in Africa (Banning-Lover, 2014) and in California, US (Koran, 2019)). For vulnerable populations such as the homeless access to water can also be problematic (Anthonj et al., 2024a,b).

Water quality varies and may be contaminated naturally (e.g. arsenic in Bangladesh and India, as previously mentioned) or by anthropogenic influences (e.g. pollutants of varying concentrations, drugs and medication) or by pathogens (e.g. cholera (Escobar et al., 2015), leptospirosis (Morgan et al., 2002; Radl et al., 2011; Wynwood et al., 2014; Ifejube et al., 2024)). For example, wastewater (e.g. wastewater-based epidemiology) is increasingly used (Sims and Kasprzyk-Hordern, 2020; Safford et al., 2022) to monitor the prevalence of diseases (e.g. COVID-19 (see Chapter 1)), polio, antibiotic resistant, bacteria, medicine residues and Mpox (RIVM, 2023)). Rising temperatures are also changing water temperatures and creating environments suitable for the growth of bacteria such as vibrio bacteria (cholera) (Escobar et al., 2015).

Water-borne diseases were rampant in the 1950's in Switzerland. In 1963, Switzerland suffered a typhoid outbreak (Bernard, 1965), but thanks to the policy changes starting in 1967, followed by improvements in waste water infrastructure that led to a reduction in raw sewage and other pollutants being dumped into lakes and waterways. Fifty years later (2023) Switzerland has some of the cleanest water in the world that is safe to swim in (Ammann, 2017).

Climate Change Impacts on Human Health

Climates are changing. Temperatures are rising locally (e.g. Figure 9.3) and globally around the world (Campbell, 2021). Many geographic locations are experiencing abnormally high temperatures on land (Niranjan, 2023; Tripathy and Mishra, 2023; Oliver and Sainato, 2023) and in the oceans (Cheng et al., 2022). The year 2023 was reported as the hottest year on record (Copernicus, 2023). For example, during the summer of 2023, many places in Europe experienced high temperatures that hit and exceeded 40°C (e.g. Rome, Seville and Athens reached 40°C, 41°C and 42°C, respectively (Niranjan, 2023)). In Phoenix, Arizona, temperatures were forecast to reach 47.7°C with temperatures expected to reach 17 consecutive days of 40°C temperatures (Oliver and Sainato, 2023). The intensity, frequency and duration of heatwaves are likely to increase (Meehl and Tebaldi, 2004).

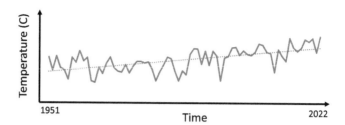

FIGURE 9.3

Mean annual temperature for the Twente region, Netherlands 1951–2022 (Data from KNMI). (Image created by Blanford (2023).)

There are a number of heatwave characteristics that can be used to gain insights into heatwaves and evaluate their effects (e.g. in Mozambique (Marghidan et al., 2023; Sharma et al., 2022)). Heatwave characteristics as defined by Perkins and Alexander (2013) include the following:

- **Number of heatwaves (HWN)**: the number of individual heatwaves per year
- **Heatwave frequency (HWF)**: the total number of days that contribute to heatwaves per year
- **Heatwave duration (HWD)**: the length of the longest yearly heatwave
- **Heatwave magnitude (HWM)**: average temperature across all heatwaves identified by HWN
- **Heatwave amplitude (HWA)**: Hottest day of the hottest yearly heatwave

In addition, the excess heat factor (EHF), an intensity measure that categorizes heatwaves by their severity (Nairn and Fawcett, 2015) has also been used. This index is based on a 3-day averaged daily T_{mean}, in relation to the 95th percentile of long-term average temperatures and the recent (prior 30-day) temperatures for a particular geographic location. It can be used to determine the average excess heat felt across all heatwave events and the excess heat felt on the hottest day of the hottest yearly heatwave (Loridan et al., 2016). The outputs can be used to create a heatwave severity risk map (e.g. Loridan et al., 2016). Heatwave indices are based on the intensity of heat, frequency of occurrence, duration and accumulated intensity. For an overview of the different heatwave indices that are used, see Barriopedro et al. (2023) and how these are used for heat-health action plans see (Casanueva et al., 2019).

Each of these measures provides insights into the different characteristics of a heatwave and can be used to evaluate mitigation and adaptation strategies. In brief, a heatwave is an extended period of hot weather relative to the expected conditions of the area at that time of year (UKMetOffice, 2023). In Australia, the Bureau of Meteorology defines a heatwave as "the maximum and the minimum temperatures are unusually hot over a 3-day period" in relation to the local climate and past weather (BOM, 2023).

Heatwaves can be classified into three types based on their intensity and health impacts (BOM, 2023). These include the following:

- **Low-intensity heatwaves** are more frequent, particularly during the summer months. Most people can cope during these heatwaves.
- **Severe heatwaves** are less frequent and are likely to be more challenging for vulnerable people, such as the elderly, particularly those with medical conditions.
- **Extreme heatwaves** are rare. They are a problem for people who don't take precautions to keep cool – even for people who are healthy. People who work or exercise outdoors are also at greater risk of being affected.

Climate change has negatively impacted ecosystems, food and water security, and urban infrastructure and has already adversely affected humans through physical and mental health, heat-related mortality and morbidity, and various infectious diseases (Pörtner et al., 2022; IPCC, 2023). The relationship between human health and the environment are complex, with many direct and indirect pathways, some of which are already well documented (e.g. examples provided earlier in this chapter; Table 2.4), while others are still unknown. With regards to climate change, the effects of climate change on human health have been well documented (e.g., McMichael et al., 2006; Mora et al., 2022; Haines and Ebi, 2019; Ebi and Hess, 2020; Watts et al., 2021; Pörtner et al., 2022; KNAW, 2023) and are summarized in Table 9.2.

TABLE 9.2

Climate Change Impacts on Health

Pathways	Examples	Health Effects
Extreme Weather Events	• Landslides • Wildfires • Floods • Storms	• Injuries • Fatalities • Mental health effects
Heat Stress	• Rise in average temperatures, • Extreme hot days, • Heatwaves • Heat island effect	• Heat-related illness • Death • Respiratory and cardiovascular conditions, allergies • Decrease productivity • Populations with pre-existing conditions and elderly are more vulnerable

(Continued)

TABLE 9.2 (*Continued*)

Climate Change Impacts on Health

Pathways	Examples	Health Effects
Air Quality	• Rise in air pollutants	• Asthma • Respiratory diseases • Respiratory allergies
Water Quality & Quantity	• Contamination of water sources • Flooding • Change in rainfall patterns	• Undernutrition and malnutrition • Food poisoning • Foodborne illnesses • Mycotoxin effects • Decrease in crop yield and nutritional value
Food Security	• Changes in seasonality and rainfall have negative agricultural effects • Rise in temperature	• Undernutrition and malnutrition • Food poisoning • Foodborne illnesses • Mycotoxin effects • Decrease in crop yield and nutritional value
Vector ecology distribution	• Shifts in habitat range of pathogen • Emergence of pathogens in new geographic locations	• Malaria • Dengue, chikungunya, Zika, WNV, RVF • Lyme disease • Encephalitis
Slow-onset climate events	• Drought • Salinization • Glacial retreat • Desertification • Sea level rise	• Physical and mental health • Food and water insecurity • Poverty • Forced migration • Conflict

Source: Adapted from Tye and Waslander (2021); Haines and Ebi (2019).

Although some of the human health effects are known, there is considerable uncertainty about the magnitude of these effects (Pörtner et al., 2022).

- For the medium to long term (2041–2100) and depending on the effectiveness of mitigation and adaptation actions, climate change is expected to cause escalating damage to human and natural systems and result in heat-related mortality and morbidity, mental health problems and an increase or decrease in climate-sensitive infectious diseases depending on the pathogen (e.g. Table 9.3).

- Under a high-emissions scenario, over 9 million climate-related deaths are expected annually worldwide by 2,100. A further 3.5 billion people are estimated to be highly vulnerable to the effects of climate change and be exposed to food and water insecurity, flood risk, displacement and poverty.

Pathogens Affected by Climate

TABLE 9.3

Climate-Sensitive Pathogens

Transmission Mechanism	Pathogen	Disease	Epidemic Potential	Climatic or Environmental Transmission Drivers	Transmission Mechanism
Air-borne	Bacterial	Meningococcal meningitis	Yes	Aridity, dust, low relative humidity, temperature	Aerosol
Air-borne	Viral	Influenza	Yes	Humidity, temperature	Aerosol
Food-borne	Bacterial	Gastroenteritis	No	temperature	Inappropriate food handling
Vector-borne	Bacterial	Lyme disease	No	Rainfall, temperature, NDVI	Ticks Ixodes
Vector-borne	Filarial	Onchocerciasis	No	Rainfall, temperature, wind, NDVI, wind, river discharge	Blackflies
Vector-borne	Filarial	African eye worm	No	forest soils, NDVI	Chrysops
Vector-borne	Parasitic	Malaria	Yes	Rainfall, humidity, temperature, surface water puddles, river margins, irrigation, altitude, NDVI	Mosquitoes
Vector-borne	Parasitic	Schistosomiasis	No	Surface water, NDVI, temperature, rainfall, elevation	Snails
Vector-borne	Parasitic	African Trypanosomiasis		Gallery forests, savannah, wood-land, temperature, NDVI	Tsetse
Vector-borne	Viral	Yellow fever	Yes	Rainfall, temperature	Mosquitoes, Aedes
Vector-borne	Viral	Rift valley fever	Yes	Rainfall, humidity, surface water, temperature, NDVI	Mosquitoes, Aedes
Vector-borne	Viral	Dengue and dengue Haemorrhagic fever; Zika; chikungunya	Yes	Temperature, rainfall, humidity	Mosquitoes, Aedes

(Continued)

TABLE 9.3 (*Continued*)

Climate-Sensitive Pathogens

Transmission Mechanism	Pathogen	Disease	Epidemic Potential	Climatic or Environmental Transmission Drivers	Transmission Mechanism
Water-borne	Bacterial	Cholera	Yes	Water and air temperature, water depth, rainfall and conductivity, algal blooms, flooding, sunlight, Sea Surface Temperature	Faecal/oral route and filth flies via mechanical transmission
Water-borne	Viral	Gastroenteritis	Yes	Humidity, cool/winter, dry months, low rainfall, water shortages, flood	Faecal/oral route and filth flies via mechanical transmission
Water-borne	Bacterial	Trachoma	No	Aridity, dust, low relative humidity, temperature	Flies

Source: Adapted from Thomson et al. (2018; Thomson and Mason, (2018).

Heat-Islands Effects

With these changes, heatwaves will negatively affect the health and wellbeing of different populations (Sharma et al., 2022). Impacts can range from mental health, adverse health effects (Palinkas et al., 2022) to increased human mortality (Mora et al., 2017; Sheehan, 2022; Gosling et al., 2009), resulting in excess deaths. For example, the European heatwave of 2003 resulted in an estimated 35,000–70,000 *excess deaths* across Europe (e.g. Conti et al., 2005; Robine et al., 2008), with over 14,800 excess deaths in France alone (e.g., Fouillet et al., 2006). The impact of heatwaves is often felt in built up areas where the climate is often modified, creating urban heat island effects that result in areas that are significantly warmer than surrounding rural areas, as illustrated in Figure 9.4.

By combining heatwave severity with heat island risks, population characteristics and urban typologies, we can start to determine heat health risks and identify heat vulnerabilities in different geographic locations (e.g. mapping heat island effects and vulnerability in the Dhaka Metropolitan Area (Abrar et al., 2022); heat health risk assessment strategies in Birmingham (Tomlinson et al., 2011); determining vulnerabilities in cities under different climate change scenarios (Ortiz et al., 2021); mapping vulnerabilities to weather extremes such as heat and flooding (Hamstead and Sauer, 2021))

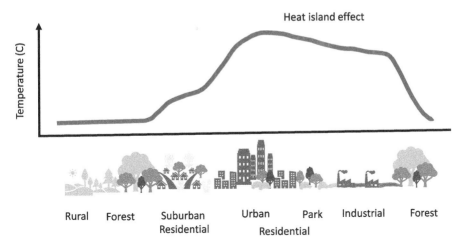

FIGURE 9.4

Illustration of heat island effects. The red line shows the potential difference in temperature between urban (built up area) and rural (more green) areas.

Changing Environments: Urbanization, Agriculture and Biodiversity Loss

Population growth, biodiversity losses and deforestation are occurring at alarming rates. Since 1990, an estimated 420 million hectares of forests have been converted to other land uses (UNEP, 2020). Agricultural expansion is one of the main drivers of deforestation and forest fragmentation. For example, large-scale commercial farming (e.g. cattle ranching and the cultivation of soya bean and oil palm) accounted for 40% of tropical deforestation between 2000 and 2010, and local subsistence agriculture for another 33% (UNEP, 2020). Health effects associated with changes in food production and the environment may also lead to increases in non-communicable diseases and nutrient deficiencies (WHO, 2015).

Along with this, cities are growing. More than 4 billion people (half of the world) live in urban areas (UNEP, 2020). This number will only continue to grow as the global population continues to rise (Habitat, 2022). Many of the people living in and close to urban areas will reside in settlements without access to basic needs and infrastructure such as water and sanitation, further contributing to the transmission of diseases. Changes in the environment can affect health in many ways, not only by increasing exposure to new and different pathogens but also due to climate-related changes through heat island effects (e.g. Figure 9.4), greenness and sound levels.

Soundscapes and Green Spaces

Anthropogenic activities (e.g. transportation and construction) are elevating sound pressure levels in many parts of the world (e.g. BTS, 2023). For example, in the US, more than 80% of the US has elevated sound pressure levels as seen on the National Parks Sound Map (https://www.nps.gov/subjects/sound/soundmap.htm).

Prolonged exposure to noise can affect human health negatively (Barbaro, 2022) (e.g. raised blood pressure levels resulting in sleep disturbance and annoyance, hearing loss, stress and heart disease (Hammer et al., 2014)), as well as diabetes, depression and anxiety (SA, 2023). In Europe, 22 million people suffer from chronic noise annoyance, resulting in premature deaths and ischemic heart disease (Barbaro, 2022). Such negative effects can be costly and affect Disability Adjusted Life Years (DALYs) (Jephcote et al., 2023).

As our society continues to urbanize, the risk of prolonged exposure to loud anthropogenic sounds will rise and will require policies to be developed to minimize health impacts. Preserving natural sounds is important for improving human cognition (Abbott et al., 2015), enhancing positive moods (Benfield et al., 2018) and general health and well-being.

Geographic information is increasingly being used to document noise levels (Jephcote et al., 2023) and map the distribution of natural sound levels on landscapes (Mennitt et al., 2014) and how transportation (Cai et al., 2015; Ko et al., 2011; Aguilar et al., 2023) and construction noise (Hong and Jeon, 2014; Lee et al., 2008) dissipate across different landscapes. By knowing where noise levels are high, necessary planning (Aguilar et al., 2023; Bild et al., 2016), mitigation strategies (Aletta and Kang, 2015) can be implemented to regulate noise levels (Murphy and King, 2010) and minimize negative impacts (e.g. Hong and Jeon, 2014; Cai et al., 2015; Liu et al., 2013) by planning sound barriers and preserving natural areas (e.g. Ferguson et al., 2024).

Droughts and Floods

As climates change, precipitation patterns will change (e.g. Figure 9.5) and may oscillate between droughts and flooding with varying frequency and intensity. The impact varies but may result in disruptions, economic damages, migration and loss of life (Asmall et al., 2021; Sheehan, 2022; Butsch et al., 2023). Depending on the geographic composition of places, flood risk may be due to one or multiple types of flooding events that may be due to fluvial, pluvial, groundwater or coastal flooding (Ortiz et al., 2021):

- **Fluvial/lakeshore flooding**: Flooding that occurs when the stage of rivers, streams or other freshwater bodies rises above the lake banks and/or the height of levees or flood protection infrastructure.

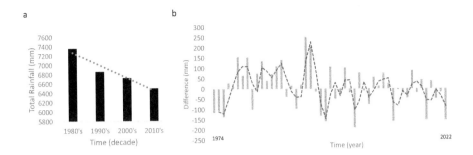

FIGURE 9.5

Variable annual rainfall (mm) for Twente, Netherlands between 1974 and 2022. Data show (a) total annual rainfall per decade (1980–2020) and (b) differences in the total annual rainfall from the mean total rainfall for all years (1974–2022) with drier (–ve) and wetter (+ve) than average years (Data from KNMI). (Image created by Blanford (2023).)

- **Coastal flooding**: Flooding that occurs when tide levels exceed an elevation threshold and results in the inundation of infrastructure or disruption of socioeconomic activities.
- **Pluvial flooding**: Flooding that occurs when precipitation rates exceed the rate of natural or engineered drainage, resulting in overland inundation and/or flow.
- **Groundwater flooding**: Flooding that occurs when groundwater tables rise above a threshold level that results in the inundation of infrastructure or disruption of socioeconomic activities.

Drought

Not only should floods be considered but also droughts. In Figure 9.5, we can see that in the Twente region, not only can there be times of excess rainfall but also reduced rainfall, which can result in droughts. Droughts are defined based on the approaches used to measure them. These include *meteorological, hydrological, agricultural* and *socioeconomic* (Wilhite and Glantz, 1985), where the first three approaches measure drought as a physical phenomenon and the last deals with drought in terms of supply and demand and the effects of water shortages through socioeconomic systems. In summary, droughts vary from geographic location to geographic location and can be defined as:

- **Meteorological drought**: a period of abnormally dry weather sufficiently prolonged for the lack of water to cause serious hydrologic imbalance in the affected area (Huschke, 1959). Definitions of meteorological drought must be considered as region-specific since the atmospheric conditions that result in deficiencies of precipitation are highly variable from region to region (NDMC, 2023a).

- **Hydrologic drought**: a period of below-average water content in streams, reservoirs, groundwater, aquifers, lakes and soils (Yevjevich et al., 1978). Hydrological drought is associated with the effects of periods of precipitation (including snowfall) and shortfalls on surface or subsurface water supply (i.e., streamflow, reservoir and lake levels and groundwater). The frequency and severity of hydrological droughts are often defined on a watershed or river basin scale. Although all droughts originate with a deficiency of precipitation, hydrologists are more concerned with how this deficiency plays out through the hydrologic system (NDMC, 2023a).
- **Agricultural drought:** Agricultural drought links various characteristics of meteorological (or hydrological) drought to agricultural impacts, focusing on precipitation shortages, differences between actual and potential evapotranspiration, soil water deficits, reduced groundwater or reservoir levels, and so forth (NDMC, 2023a).
- **Socioeconomic drought:** Socioeconomic definitions of drought associate the supply and demand of some economic goods with elements of meteorological, hydrological and agricultural drought. Examples may include impact hydroelectric power production (NDMC, 2023a).
- **Ecological drought:** a prolonged and widespread deficit in naturally available water supplies – including changes in natural and managed hydrology – that create multiple stresses across ecosystems (NDMC, 2023a). However, ecologists describe or define droughts in multiple ways, as captured by Slette et al. (2019) who captured 8 drought categories.

Health Effects of Droughts

Although droughts may differ from region to region, health effects are likely to result and vary, as captured in Figure 9.6.

Warming Waters

Not only will temperatures increase on land but also in waterbodies. Rising water temperatures provide favourable conditions for many climate-sensitive pathogens (see Table 9.3 for examples of climate-sensitive pathogens) resulting in water-borne infections and intoxications (Dupke et al., 2023). For example, water temperatures of 12°C or higher provide favourable conditions for *Vibrio*, while water temperatures above 20°C are conducive for non-cholera *Vibrio* (NCV). As water temperatures rise, so will the number of NCV pathogens in coastal waters.

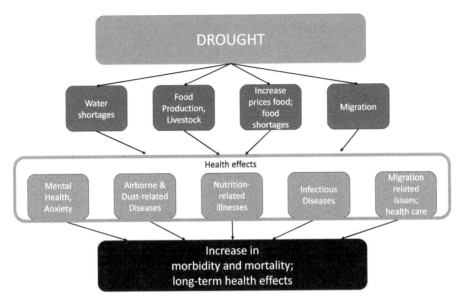

FIGURE 9.6
Summary of health effects from droughts. *Sources*: Asmall et al. (2021); Bellizzi et al. (2020); Sheehan (2022).

Summary of Environment Exposure

The burden caused by global environmental change will not be shared equally with most of the effects on human health expected to occur in the Global South. However, knowing where these burdens and inequalities are will help in providing solutions and reducing the burden.

Planetary Health: Geographic Information and Earth Observations

Planetary health is a concept focused on the interdependencies between human health and the state of the earth's complex natural systems (Figure 9.7). The underlying and ecological drivers combined with proximate causes and mediating factors result in a variety of health outcomes.

The **underlying drivers** are the drivers of the change brought about through human activities. In Chapter 2, these were highlighted (see Table 2.2).

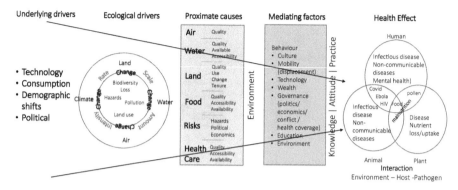

FIGURE 9.7

Planetary health drivers, proximate causes, mediating factors and how they contribute to health outcomes. (Adapted from Myers (2017), Myers et al. (2013) and Baker et al. (2022).)

The **ecological drivers** are the result of the changes brought about by human activities, which are affecting the land, water, air and climate. How these affect health outcomes may be determined by the intensity and rate of the change. The eco-social determinants of health are further affected through **proximate** causes (natural, manmade and other risks in the environment) and **mediating** factors such as behaviour and other societal influences through knowledge, attitude and practice. These act in a similar capacity as the pressure-release model (Wisner et al., 2004) where risk relates to vulnerabilities and hazards. To reduce future health effects not only is it important to understand how current human-related environmental change can affect health and well-being but also what mitigation strategies are needed and how long it will take for effects to be realized.

A Geospatially Enabled Society for Mitigating Health Effects

In conclusion, geographic information, geospatial technologies and spatial data science are vital for health. "GIS provide us with a view of human health that cannot be obtained by other methods" (Cromley, 2019). By coupling what we know about health and disease with geographic information we can evaluate the potential risks and anticipate possible consequences of health outcomes in different geographic locations and in different populations.

Human health and the health of the natural environment need to be considered together in the face of unprecedented environmental challenges we are currently facing, such as those linked to climate change, loss of biodiversity, technology and other anthropogenic changes. Geospatial technologies make it possible to map diseases spatially, monitor trends, assess changes that may be taking place and evaluate eradication efforts. Not only can we monitor the disease itself but gain additional insights when coupled with information about the physical and socio-economic environment. As more and more health systems become digitalized the ability to detect, manage and assess health risks will be enhanced.

As sensors and technologies continue to improve we will further be able to acquire information through non-invasive and diagnostic monitoring about our environment and the distribution of diseases in real-time. For example, thermal imaging of ecosystems and plant species (Still et al., 2019) may be useful for understanding heat absorption, transfer and the role different plants can play in mitigating heat effects. Or the role of multispectral sensors for better understanding the nutrients of agricultural crops (Belgiu et al., 2023) and help combat hidden hunger due to nutrient-poor diets (Gödecke, 2018). Or as disease detection continue to advance from rapid testing to the use of image analysis methods (e.g. for detecting malaria in blood smears (Maturana et al., 2023)), the feasibility to monitor, evaluate and respond to outbreaks in a timely manner will be greatly enhanced. This will further support efforts such as The Malaria Atlas Project (MAP) (Hay and Snow, 2006) which collates information about malaria globally and enable users to visualize geographically the current distribution of malaria and interventions in use. MAP contains a large repository of global, regional and country-level maps capturing the distribution of clinical malaria (*Plasmodium falciparum* and *Plasmodium vivax*) cases and interventions in use that are available as ready-to-use products that are easily accessible. Coupling these risk maps with environmental and socio-economic information can enable public health official to develop targeted response plans.

Geographic Information Science (GIScience), an integrative and technical science is well suited to address a range of **global health** challenges that cross social-ecological systems as captured in Figure 9.8. Diverse information can be integrated using a range of methods (Figure 9.8d) the outputs of which can be disseminated to different stakeholders (Figure 9.8b). Now more than ever we need to understand what changes are taking place spatially and temporally, locally and globally. Many health challenges are complex crossing social-ecological systems (SES) comprised of many components distributed across scales, often compounded by different socio-economic and environmental drivers, many operating non-linearly. By developing a geo-enabled digital society we can tackle the many health challenges outlined in Chapter 1 and achieve the SDG's.

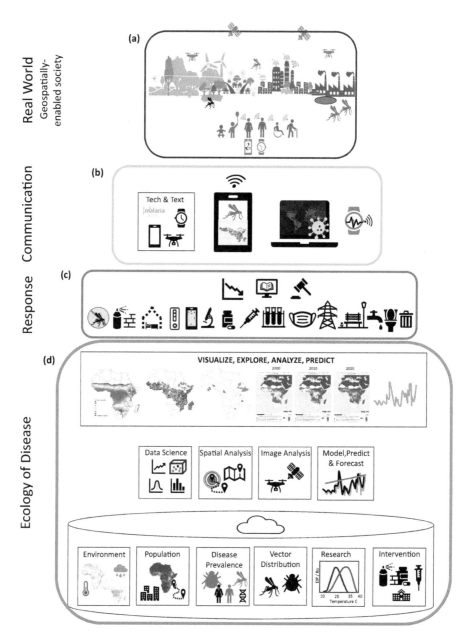

FIGURE 9.8

A (a) geospatially enabled society where (b) health information and health risk information are communicated through different sources; (c) timely interventions are used in response to health risks and needs; (d) an ecosystem of information and geospatial tools are used to understand the ecology of the disease.

References

Abbott, L., P. Newman & J. Benfield (2015) The Influence of Natural Sounds on Attention Restoration, Doctoral dissertation, Pennsylvania State University.

Ferguson, L. A., B. D. Taff, J. I. Blanford, D. Mennitt, A. J. Mowen, M. Levenhagen, C. White, C. A. Monz, C. D. Francis, J. R. Barbe & P. Newman (2024) Understanding park visitors' soundscape perception using subjective and objective measurement. *PeerJ*, 12, e16592.

Abrar, R., S. K. Sarkar, K. T. Nishtha, S. Talukdar, A. Shahfahad, A. R. Rahman, M. T. Islam & A. Mosavi (2022) Assessing the spatial mapping of heat vulnerability under Urban Heat Island (UHI) Effect in the Dhaka metropolitan area. *Sustainability*, 14, 4945.

Aguilar, R., J. Flacke, D. Simon & K. Pfeffer (2023) Stakeholders engagement in noise action planning mediated by OGITO: An Open geo-spatial interactive tool. *Journal of Urban Technology*, 30, 1–24.

Al-Zoughool, M. & D. Krewski (2009) Health effects of radon: A review of the literature. *International Journal of Radiation Biology*, 85, 57–69.

Aletta, F. & J. Kang (2015) Soundscape approach integrating noise mapping techniques: A case study in Brighton, UK. *Noise Mapping*, 2, 1–2.

Ammann, K. (2017) Why Today's Swiss Waterways are Fit for Swimming. https://www.swissinfo.ch/eng/politics/wastewater_bathing-prohibited-in-switzerland-definitely-a-thing-of-the-past/43296836 (last accessed 9 Sep 2023).

Anthonj, C., K. I. H. M. Poague, L. Fleming & S. Stanglow (2024a) Invisible struggles: WASH insecurity and implications of extreme weather among urban homeless in high-income countries-A systematic scoping review. *International Journal of Hygiene and Environmental Health*, 255, 114285.

Anthonj, C., S. N. Stanglow & N. Grunwald (2024b) Co-defining WASH (In)Security challenges among people experiencing homelessness. A qualitative study on the Human Right to Water and Sanitation from Bonn, Germany. *Social Science & Medicine*, 342, 116561.

Asmall, T., A. Abrams, M. Röösli, G. Cissé, K. Carden & M. A. Dalvie (2021) The adverse health effects associated with drought in Africa. *Science of The Total Environment*, 793, 148500.

Baker, R. E., A. S. Mahmud, I. F. Miller, M. Rajeev, F. Rasambainarivo, B. L. Rice, S. Takahashi, A. J. Tatem, C. E. Wagner, L. F. Wang, A. Wesolowski & C. J. E. Metcalf (2022) Infectious disease in an era of global change. *Nature Reviews Microbiology*, 20, 193–205.

Banning-Lover, R. (2014) Using Mobile Money to Buy Water and Solar Power in East Africa. https://www.theguardian.com/global-development-professionals-network/2014/aug/18/mobile-payment-utilitlies-east-africa (last accessed Sep 3 2023).

Barbaro, A. (2022) *Frontiers Report: Noise, Blazes and Mismatches?- Emerging Issues of Environmental Concern.* Rome, Italy: Istituto Superiore Sanita Viale Regina Elena, 299 p.

Barriopedro, D., R. García-Herrera, C. Ordóñez, D. G. Miralles & S. Salcedo-Sanz (2023) Heat waves: Physical understanding and scientific challenges. *Reviews of Geophysics*, 61, e2022RG000780.

Bassil, K. L., C. Vakil, M. Sanborn, D. C. Cole, J. S. Kaur & K. J. Kerr (2007) Cancer health effects of pesticides: Systematic review. *Canadian Family Physician*, 53, 1704–1711.

Bauer, S. E., U. Im, K. Mezuman & C. Y. Gao (2019) Desert dust, industrialization, and agricultural fires: Health impacts of outdoor air pollution in Africa. *Journal of Geophysical Research: Atmospheres*, 124, 4104–4120.

Beermann, S., G. Dobler, M. Faber, C. Frank, B. Habedank, P. Hagedorn, H. Kampen, C. Kuhn, T. Nygren & J. Schmidt-Chanasit (2023) Impact of climate change on vector-and rodent-borne infectious diseases. *Journal of Health Monitoring*, 8, 33.

Belgiu, M., et al. (2023) PRISMA and Sentinel-2 spectral response to the nutrient composition of grains. *Remote Sensing of Environment*, 292, 113567.

Bellizzi, S., C. Lane, M. Elhakim & P. Nabeth (2020) Health consequences of drought in the WHO eastern mediterranean region: Hotspot areas and needed actions. *Environmental Health*, 19, 1–9.

Bello, M. & M. Cardona (2023) Hurricane Idalia Lashes Florida, then Weakens and Turns Fury on Georgia. https://www.reuters.com/world/us/floridas-gulf-coast-braces-major-hurricane-idalia-nears-landfall-2023-08-30/ (last accessed Sep 3 2023).

Benfield, J., B. D. Taff, D. Weinzimmer & P. Newman (2018) Motorized recreation sounds influence nature scene evaluations: The role of attitude moderators. *Frontiers in Psychology*, 9, 495.

Bernard, R. P. (1965) The Zermatt typhoid outbreak in 1963. *Epidemiology & Infection*, 63, 537–563.

Bhowmick, S., S. Pramanik, P. Singh, P. Mondal, D. Chatterjee & J. Nriagu (2018) Arsenic in groundwater of West Bengal, India: A review of human health risks and assessment of possible intervention options. *Science of The Total Environment*, 612, 148–169.

Bild, E., M. Coler, K. Pfeffer & L. Bertolini (2016) Considering sound in planning and designing public spaces: A review of theory and applications and a proposed framework for integrating research and practice. *Journal of Planning Literature*, 31, 419–434.

Blake, A. (2020) Weekly Reported Measles Cases and Code [Dataset]. https://doi.org/10.5061/dryad.1jwstqjrd (last accessed Sep 3 2023).

BOM (2023) Understanding Heatwaves. https://www.bom.gov.au/australia/heatwave/knowledge-centre/understanding.shtml (last accessed Sep 8 2023).

Brown-Luthango, M., E. Reyes & M. Gubevu (2017) Informal settlement upgrading and safety: Experiences from Cape Town, South Africa. *Journal of Housing and the Built Environment*, 32, 471–493.

Brugger, J., K. W. Dunbar, C. Jurt & B. Orlove (2013) Climates of anxiety: Comparing experience of glacier retreat across three mountain regions. *Emotion, Space and Society*, 6, 4–13.

Brusseau, M. L., M. Ramirez-Andreotta, I. L. Pepper & J. Maximillian (2019) Chapter 26 - Environmental Impacts on Human Health and Well-Being. *Environmental and Pollution Science (Third Edition)*, 26, 477–499.

BTS (2023) National Transportation Map. https://maps.dot.gov/BTS/National TransportationNoiseMap/ (last accessed Sep 3 2023).

Butsch, C., L. M. Beckers & E. Nilson (2023) Health impacts of extreme weather events-Cascading risks in a changing climate. *Journal of Health Monitoring*, 8, 33–56.

Cagle, S. (2020) 'Lost Communities': Thousands of Wells in Rural California May Run Dry. https://www.theguardian.com/environment/2020/feb/28/california-water-wells-dry-sgma (last accessed Sep 3 2023).

Cai, M., J. Zou, J. Xie & X. Ma (2015) Road traffic noise mapping in Guangzhou using GIS and GPS. *Applied Acoustics*, 87, 94–102.

Cambaza, E., E. Mongo, E. Anapakala, R. Nhambire, J. Singo & E. Machava (2019) Outbreak of cholera due to cyclone Kenneth in northern Mozambique, 2019. *International Journal of Environmental Research and Public Health*, 16, 2925.

Campbell, I. (2021) *Beating the Heat: A Sustainable Cooling Handbook for Cities*. Nairobi, Kenya: UN Environment Programme.

Carrington, D. & E. Borràs (2024) 'When it's this hot, time stands still': surviving west Africa's blistering heat. https://www.theguardian.com/world/2024/mar/22/west-africa-heatwave (last accessed May 5 2024).

Casanueva, A., A. Burgstall, S. Kotlarski, A. Messeri, M. Morabito, A. D. Flouris, L. Nybo, C. Spirig & C. Schwierz (2019) Overview of existing heat-health warning systems in Europe. *International Journal of Environmental Research and Public Health*, 16, 2657.

Cheng, L., K. von Schuckmann, J. P. Abraham, K. E. Trenberth, M. E. Mann, L. Zanna, M. H. England, J. D. Zika, J. T. Fasullo, Y. Yu, Y. Pan, J. Zhu, E. R. Newsom, B. Bronselaer & X. Lin (2022) Past and future ocean warming. *Nature Reviews Earth & Environment*, 3, 776–794.

Clemmons, N. S., P. A. Gastanaduy, A. P. Fiebelkorn, S. B. Redd & G. S. Wallace (2015) Measles-United States, January 4–April 2, 2015. *Morbidity and Mortality Weekly Report*, 64, 373.

Conti, S., P. Meli, G. Minelli, R. Solimini, V. Toccaceli, M. Vichi, C. Beltrano & L. Perini (2005) Epidemiologic study of mortality during the Summer 2003 heat wave in Italy. *Environmental Research*, 98, 390–399.

Conway, G. R. & J. N. Pretty (2013) *Unwelcome Harvest: Agriculture and Pollution*. London: Routledge.

Conway, D., B. Robinson, P. Mudimu, T. Chitekwe, K. Koranteng & M. Swilling (2019) Exploring hybrid models for universal access to basic solar energy services in informal settlements: Case studies from South Africa and Zimbabwe. *Energy Research & Social Science*, 56, 101202.

Copernicus (2021) Flooding in Europe. https://climate.copernicus.eu/esotc/2021/flooding-july (last accessed May 5 2024).

Copernicus (2023) 2023 Hottest year ever. What's next? https://climate.copernicus.eu/2023-track-be-hottest-year-ever-whats-next (last accessed May 5 2024).

Cromley, E. K. (2019). Using GIS to address epidemiologic research questions. *Current Epidemiology Reports*, 6, 162–173.

Derbyshire, E. (2007) Natural minerogenic dust and human health. *AMBIO: A Journal of the Human Environment*, 36, 73–77.

Deribe, K., S. J. Brooker, R. L. Pullan, A. Hailu, F. Enquselassie, R. Reithinger, M. Newport & G. Davey (2013) Spatial distribution of podoconiosis in relation to environmental factors in Ethiopia: A historical review. *PloS One*, 8, e68330.

Deribe, K., J. Cano, M. L. Trueba, M. J. Newport & G. Davey (2018) Global epidemiology of podoconiosis: A systematic review. *PLoS Neglected Tropical Diseases*, 12, e0006324.

Dockery, D. W., C. A. Pope, Z. Zu, J. D. Spengler, J. H. Ware, M. E. Fay, B. G. Ferris & F. E. Speizer. (1993) An association between air pollution and mortality in six U.S. cities. *The New England Journal of Medicine*, 329, 1753–1759.

Dupke, S., U. Buchholz, J. Fastner, C. Förster, C. Frank, A. Lewin, V. Rickerts & H.-C. Selinka (2023) Impact of climate change on waterborne infections and intoxications. *Journal of Health Monitoring*, 8, 62.

Ebi, K. L. & J. J. Hess (2020) Health risks due to climate change: Inequity in causes and consequences: Study examines health risks due to climate change. *Health Affairs,* 39, 2056–2062.

ESA Sentinel-3 World Fire Atlas. https://s3wfa.esa.int/ (last accessed Sep 3 2023).

Escobar, L. E., S. J. Ryan, A. M. Stewart-Ibarra, J. L. Finkelstein, C. A. King, H. Qiao & M. E. Polhemus (2015) A global map of suitability for coastal vibrio cholerae under current and future climate conditions. *Acta Tropica,* 149, 202–211.

Ferrari, M. J., R. F. Grais, N. Bharti, A. J. Conlan, O. N. Bjornstad, L. J. Wolfson, P. J. Guerin, A. Djibo & B. T. Grenfell (2008) The dynamics of measles in sub-Saharan Africa. *Nature,* 451, 679–684.

Flores Quiroz, N., R. Walls & A. Cicione (2021) Towards understanding fire causes in informal settlements based on inhabitant risk perception. *Fire,* 4, 39.

Fouillet, A., G. Rey, F. Laurent, G. Pavillon, S. Bellec, C. Guihenneuc-Jouyaux, J. Clavel, E. Jougla & D. Hémon (2006) Excess mortality related to the August 2003 heat wave in France. *International Archives of Occupational and Environmental Health,* 80, 16–24.

García-Pando, C. P., M. C. Stanton, P. J. Diggle, S. Trzaska, R. L. Miller, J. P. Perlwitz, J. M. Baldasano, E. Cuevas, P. Ceccato & P. Yaka (2014) Soil dust aerosols and wind as predictors of seasonal meningitis incidence in Niger. *Environmental Health Perspectives,* 122, 679–686.

Garrison, V. H., E. A. Shinn, W. T. Foreman, D. W. Griffin, C. W. Holmes, C. A. Kellogg, M. S. Majewski, L. L. Richardson, K. B. Ritchie & G. W. Smith (2003) African and Asian dust: From desert soils to coral reefs. *BioScience,* 53, 469–480.

Gödecke, T., A. J. Stein & M. Qaim (2018) The global burden of chronic and hidden hunger: Trends and determinants. *Global Food Security,* 17, 21–29.

Godoy, P., M. Alsedà, M. Falguera, T. Puig, P. Bach, M. Monrabà & A. Manonelles (2017) A highly transmissible tuberculosis outbreak: The importance of bars. *Epidemiology & Infection,* 145, 3497–3504.

Goldstone, M. E. (2015) Review of evidence on health aspects of air pollution-REVIHAAP project. *Air Quality and Climate Change,* 49, 35–41.

Gosling, S. N., J. A. Lowe, G. R. McGregor, M. Pelling & B. D. Malamud (2009) Associations between elevated atmospheric temperature and human mortality: A critical review of the literature. *Climatic Change,* 92, 299–341.

Graham, J. P., V. Corella Barud, R. Avitia Diaz & P. Gurian (2005) The in-home environment and household health: A cross-sectional study of informal urban settlements in northern Mexico. *The International Journal of Environmental Research and Public Health,* 2, 394–402.

Guerra, F. M., S. Bolotin, G. Lim, J. Heffernan, S. L. Deeks, Y. Li & N. S. Crowcroft (2017) The basic reproduction number (R_0) of measles: A systematic review. *The Lancet Infectious Diseases,* 17, e420–e428.

Habitat, U. N. (2022) *World Cities Report 2022: Envisaging the Future of Cities.* Nairobi, Kenya: United Nations Human Settlements Programme, pp. 41–44.

Haines, A. & K. Ebi (2019) The imperative for climate action to protect health. *New England Journal of Medicine,* 380, 263–273.

Halonen, J. I., M. Erhola, E. Furman, T. Haahtela, P. Jousilahti, R. Barouki, Å. Bergman, N. E. Billo, R. Fuller & A. Haines (2021) A call for urgent action to safeguard our planet and our health in line with the Helsinki declaration. *Environmental Research,* 193, 110600.

Hammer, M. S., T. K. Swinburn & R. L. Neitzel (2014) Environmental noise pollution in the United States: Developing an effective public health response. *Environmental Health Perspectives*, 122, 115–119.

Hamstead, Z. A. & J. Sauer (2021) Mapping vulnerability to weather extremes: Heat and flood assessment approaches. *Resilient Urban Futures*, 3, 47–66.

Hansson, E., A. Mansourian, M. Farnaghi, M. Petzold & K. Jakobsson (2021) An ecological study of chronic kidney disease in five Mesoamerican countries: Associations with crop and heat. *BMC Public Health*, 21, 840.

Harter, F. (2022) A Common Condition: How Wearing Shoes Could Eliminate One of the World's Most Neglected Tropical Diseases. https://www.theguardian.com/global-development/2022/nov/17/how-wearing-shoes-tropical-diseases-podoconiosis-acc (last accessed 3 Sep 2023).

Hay, S. I. & R. W. Snow (2006) The malaria atlas project: Developing global maps of malaria risk. *PLoS Medicine*, 3, e473.

Hong, J. Y. & J. Y. Jeon (2014) *Soundscape Mapping in Urban Contexts Using GIS Techniques*. Korea: Hanyang University.

Huschke, R. E. (1959) *Glossary of Meteorology*. Boston, MA: American Meteorological Society.

IEA (2020) SDG7: Data and Projections. https://www.iea.org/reports/sdg7-data-and-projections (last accessed Feb 22 2022).

Ifejube, J., C. van Weston, S. Lukiakosa, T. Anish & J. I. Blanford (2024) Analysing the outbreaks of leptospirosis after floods in Kerala, India. *International Journal of Health Geographics*, 23, 11. https://doi.org/10.1186/s12942-024-00372-9.

International, O. (2023) Drought in East Africa: "If the Rains Do Not Come, None of Us will Survive". https://www.oxfam.org/en/drought-east-africa-if-rains-do-not-come-none-us-will-survive (last accessed 3 Sep 2023).

IPCC (2023) Synthesis Report of the IPCC Sixth Assessment Report (AR6). Longer Report. https://www.ipcc.ch/report/ar6/syr/ (last accessed June 18 2023).

Isbell, F., P. Balvanera, A. S. Mori, J. S. He, J. M. Bullock, G. R. Regmi, E. W. Seabloom, S. Ferrier, O. E. Sala & N. R. Guerrero-Ramírez (2023) Expert perspectives on global biodiversity loss and its drivers and impacts on people. *Frontiers in Ecology and the Environment*, 21, 94–103.

Jayasumana, C., P. Paranagama, S. Agampodi, C. Wijewardane, S. Gunatilake & S. Siribaddana (2015) Drinking well water and occupational exposure to Herbicides is associated with chronic kidney disease, in Padavi-Sripura, Sri Lanka. *Environmental Health*, 14, 6.

Jephcote, C., S. N. Clark, A. L. Hansell, N. Jones, Y. Chen, C. Blackmore, K. Eminson, M. Evans, X. Gong, K. Adams, G. Rodgers, B. Fenech & J. Gulliver (2023) Spatial assessment of the attributable burden of disease due to transportation noise in England. *Environment International*, 178, 107966.

Jolly, A. M. & J. L. Wylie (2002) Gonorrhoea and chlamydia core groups and sexual networks in Manitoba. *Sexually Transmitted Infections*, 78, i145–i151.

Jolly, A. M. & J. L. Wylie (2013) Sexual Networks and Sexually Transmitted Infections; "The Strength of Weak (Long Distance) Ties". In *The New Public Health and STD/HIV Prevention: Personal, Public and Health Systems Approaches*, eds. S. O. Aral, K. A. Fenton & J. A. Lipshutz, 77–109. New York: Springer.

Jones, J. (2023) Tropical Storm Hilary Makes Landfall in Mexico. https://www.nytimes.com/article/tropical-storm-hilary-hurricane-california-mexico.html (last accessed Sep 3 2023).

Kaenzig, R. & E. Piguet (2013) Migration and Climate Change in Latin America and the Caribbean. *People on the Move in a Changing Climate: The Regional Impact of Environmental Change on Migration.* New York: Springer, pp. 155–176.

Kampa, M. & E. Castanas (2008) Human health effects of air pollution. *Environmental Pollution,* 151, 362–367.

Kassam, A. (2022) 'They're being cooked': baby swifts die leaving nests as heatwave hits Spain. https://www.theguardian.com/world/2022/jun/16/spain-heatwave-baby-swifts-die-leaving-nest (last accessed Mar 11 2024).

Katz, S. L., A. R. Ninman, J. Temte, L. Pickering & e. al. (2004) Summary and conclusions: Measles elimination meeting, 16–17 March 2000. *The Journal of Infectious Diseases,* 189, S43–S47.

Klovdahl, A. S., E. A. Graviss, A. Yaganehdoost, M. W. Ross, A. Wanger, G. J. Adams & J. M. Musser (2001) Networks and tuberculosis: An undetected community outbreak involving public places. *Social Science & Medicine,* 52, 681–694.

KNAW (2023) Planetary Health. An Emerging Field to be Developed, p. 137. https://www.knaw.nl/nl/publicaties/planetary-health-emerging-field-be-developed (last accessed June 18 2023).

Ko, J. H., S. I. Chang, M. Kim, J. B. Holt & J. C. Seong (2011) Transportation noise and exposed population of an urban area in the Republic of Korea. *Environment International,* 37, 328–334.

Koran, M. (2019) Californians are Turning to Vending Machines for Safer Water. Are they Being swindled? https://www.theguardian.com/us-news/2019/dec/02/california-water-vending-machines-quality (last accessed Sep 3 2023).

Kordas, K., B. Lönnerdal & R. J. Stoltzfus (2007) Interactions between nutrition and environmental exposures: Effects on health outcomes in women and children. *The Journal of Nutrition,* 137, 2794–2797.

Kumar, P., A. Rai, A. Upadhyaya & A. Chakraborty (2022) Analysis of heat stress and heat wave in the four metropolitan cities of India in recent period. *Science of The Total Environment,* 818, 151788.

Lake Iain, R., R. Jones Natalia, M. Agnew, M. Goodess Clare, F. Giorgi, L. Hamaoui-Laguel, A. Semenov Mikhail, F. Solomon, J. Storkey, R. Vautard & M. Epstein Michelle (2017) Climate change and future pollen allergy in Europe. *Environmental Health Perspectives,* 125, 385–391.

Lakhani, N., M. Singh & R. Kamal (2021) Almost Half Million US Households Lack Indoor Plumbing. https://www.theguardian.com/us-news/2021/sep/27/water-almost-half-million-us-households-lack-indoor-plumbing (last accessed 3 Sep 2023).

Lee, S.-W., S. I. Chang & Y.-M. Park (2008) Utilizing noise mapping for environmental impact assessment in a downtown redevelopment area of Seoul, Korea. *Applied Acoustics,* 69, 704–714.

Lee, H., K. Calvin, D. Dasgupta, G. Krinner, A. Mukherji, P. Thorne, C. Trisos, J. Romero, P. Aldunce & K. Barrett (2023) AR6 Synthesis Report: Climate Change 2023. Summary for Policymakers. https://www.ipcc.ch/report/ar6/syr/ (last access 2 Jun 2023).

Liu, J., J. Kang, T. Luo, H. Behm & T. Coppack (2013) Spatiotemporal variability of soundscapes in a multiple functional urban area. *Landscape and Urban Planning,* 115, 1–9.

Logan, J. J., A. M. Jolly & J. I. Blanford (2016) The sociospatial network: Risk and the role of place in the transmission of infectious diseases. *PLoS One,* 11, e0146915.

Loridan, T., L. Coates, D. Argueso, S. E. Perkins-Kirkpatrick & J. McAneney (2016) The excess heat factor as a metric for heat-related fatalitis: Defining heatwave risk categories. *Australian Journal of Emergency Management*, 31, 31–37.

Ma, Y., C. R. Horsburgh, L. F. White & H. E. Jenkins (2018) Quantifying TB transmission: A systematic review of reproduction number and serial interval estimates for tuberculosis. *Epidemiology & Infection*, 146, 1478–1494.

Mahr, K. (2018) How Cape Town was Saved from Running Out of Water. https://www.theguardian.com/world/2018/may/04/back-from-the-brink-how-cape-town-cracked-its-water-crisis (last accessed Sep 3 2023).

Manisalidis, I., E. Stavropoulou, A. Stavropoulos & E. Bezirtzoglou (2020) Environmental and health impacts of air pollution: A review. *Frontiers in Public Health*, 8, 14.

Marghidan, C. P., M. van Aalst, J. I. Blanford, K. Guigma, G. Maure, I. Pinto, J. Arrighi & T. Marrufo (2023) Assessing the Spatio-Temporal Distribution of Extreme Heat Events in Mozambique Using the CHIRTS Temperature Dataset for 1983–2016. *Weather & Climate Extremes, Liverpool, United Kingdom*.

Mark, B. G., J. Bury, J. M. McKenzie, A. French & M. Baraer (2010) Climate change and tropical Andean glacier recession: Evaluating hydrologic changes and livelihood vulnerability in the Cordillera Blanca, Peru. *Annals of the Association of American Geographers*, 100, 794–805.

Maturana, C. R., A. D. de Oliveira, S. Nadal, F. Z. Serrat, E. Sulleiro, et al. (2023) iMAGING: A novel automated system for malaria diagnosis by using artificial intelligence tools and a universal low-cost robotized microscope. *Frontiers in Microbiology*, 14.

McMichael, A. J., R. E. Woodruff & S. Hales (2006) Climate change and human health: Present and future risks. *The Lancet*, 367, 859–869.

Meehl, G. A. & C. Tebaldi (2004) More intense, more frequent, and longer lasting heat waves in the 21st century. *Science*, 305, 994–997.

Mennitt, D., K. Sherrill & K. Fristrup (2014) A geospatial model of ambient sound pressure levels in the contiguous United States. *The Journal of the Acoustical Society of America*, 135, 2746–2764.

Molesworth, A. M., M. C. Thomson, S. J. Connor, M. P. Cresswell, A. P. Morse, P. Shears, C. A. Hart & L. E. Cuevas (2002) Where is the meningitis belt? Defining an area at risk of epidemic meningitis in Africa. *Transactions of The Royal Society of Tropical Medicine and Hygiene*, 96, 242–249.

Mora, C., C. W. W. Counsell, C. R. Bielecki & L. V. Louis (2017) Twenty-seven ways a heat wave can kill you: Deadly heat in the era of climate change. *Circulation: Cardiovascular Quality and Outcomes*, 10, e004233.

Mora, C., T. McKenzie, I. M. Gaw, J. M. Dean, H. von Hammerstein, T. A. Knudson, R. O. Setter, C. Z. Smith, K. M. Webster, J. A. Patz & E. C. Franklin (2022) Over half of known human pathogenic diseases can be aggravated by climate change. *Nature Climate Change*, 12, 869–875.

Morgan, J., S. L. Bornstein, A. M. Karpati, M. Bruce, C. A. Bolin, C. C. Austin, C. W. Woods, J. Lingappa, C. Langkop, B. Davis, D. R. Graham, M. Proctor, D. A. Ashford, M. Bajani, S. L. Bragg, K. Shutt, B. A. Perkins, J. W. Tappero & G. Leptospirosis Working (2002) Outbreak of leptospirosis among triathlon participants and community residents in Springfield, Illinois, 1998. *Clinical Infectious Diseases*, 34, 1593–1599.

Murphy, E. & E. A. King (2010) Strategic environmental noise mapping: Methodological issues concerning the implementation of the EU environmental noise directive and their policy implications. *Environment International*, 36, 290–298.

Myers, S. S. (2017) Planetary health: Protecting human health on a rapidly changing planet. *Lancet*, 390, 2860–2868.

Myers, S. S., L. Gaffikin, C. D. Golden, R. S. Ostfeld, K. H. Redford, T. H. Ricketts, W. R. Turner & S. A. Osofsky (2013) Human health impacts of ecosystem alteration. *Proceedings of the National Academy of Sciences*, 110, 18753–18760.

Nairn, J. R. & R. J. B. Fawcett (2015) The excess heat factor: A metric for heatwave intensity and its use in classifying heatwave severity. *International Journal of Environmental Research and Public Health*, 12, 227–253.

NASA (2022) Worst Drought on Record Parches Horn of Africa. https://earthobservatory. nasa.gov/images/150712/worst-drought-on-record-parches-horn-of-africa (last accessed 3 Sep 2023).

—— (2023) Raging Fires in Nova Scotia. https://earthobservatory.nasa.gov/ images/151407/raging-fires-in-nova-scotia (last accessed 3 Sep 2023).

NDMC (2023a) Types of Drought. https://drought.unl.edu/Education/DroughtIn-depth/TypesofDrought.aspx (last accessed Sep 9 2023).

—— (2023b) U.S. Drought Monitor. https://droughtmonitor.unl.edu/CurrentMap. aspx (last accessed Sep 3 2023).

Niranjan, A. (2023) Health Alerts Issued as Blistering Heat Scorches Southern Europe. https://www.theguardian.com/world/2023/jul/14/health-alerts-blistering-heat-scorches-southern-europe (last accessed May 5 2024).

Nordstrom, D. K. (2002) *Worldwide Occurrences of Arsenic in Ground Water*, pp. 2143–2145. Washington, DC: American Association for the Advancement of Science.

Nrc, U. (2001) Arsenic in Drinking Water: 2001 Update. *National Research Council, Subcommittee to Update the 1999 Arsenic in Drinking*. Washington, DC: National Academies Press.

Oliver, M. & M. Sainato (2023) Millions in US Under Warnings as Record Heat Expected to Continue Next Week. https://www.theguardian.com/world/2023/jul/16/ us-extreme-weather-record-heat (last accessed Aug 8 2023).

Ortiz, L., A. Mustafa, B. Rosenzweig & T. McPhearson (2021) Modeling urban futures: Data-driven scenarios of climate change and vulnerability in cities. *Resilient Urban Futures*, 187, 129–144.

Palinkas, L. A., M. S. Hurlburt, C. Fernandez, J. De Leon, K. Yu, E. Salinas, E. Garcia, J. Johnston, M. M. Rahman & S. J. Silva (2022) Vulnerable, resilient, or both? A qualitative study of adaptation resources and behaviors to heat waves and health outcomes of low-income residents of urban heat islands. *International Journal of Environmental Research and Public Health*, 19, 11090.

Pawankar, R. (2014) Allergic diseases and asthma: A global public health concern and a call to action. *BioMed Central*, 7, 1–3.

Peeples, L. (2019) Rethinking herd immunity. *Nature Medicine*, 25, 1178–1180.

Perkins, S. E. & L. V. Alexander (2013) On the measurement of heat waves. *Journal of climate*, 26, 4500–4517.

Phiri, D., M. Simwanda & V. Nyirenda (2021) Mapping the impacts of cyclone Idai in Mozambique using Sentinel-2 and OBIA approach. *South African Geographical Journal*, 103, 237–258.

Platts-Mills, T. A. E. (2015) The allergy epidemics: 1870–2010. *Journal of Allergy and Clinical Immunology*, 136, 3–13.

Pörtner, H. O., D. C. Roberts, H. Adams, C. Adler, P. Aldunce, E. Ali, R. A. Begum, R. Betts, R. B. Kerr & R. Biesbroek (2022) *Climate Change 2022: Impacts, Adaptation and Vulnerability*. New York: IPCC.

Pritchard, H. D. (2017) Asia's glaciers are a regionally important buffer against drought. *Nature*, 545, 169–174.

—— (2019) Asia's shrinking glaciers protect large populations from drought stress. *Nature*, 569, 649–654.

Puley, G., A. G. Calhoun & L. Toruteva (2022) Extreme Heat: Preparing for the Heat Waves of the Future. https://www.ifrc.org/sites/default/files/2022-10/Extreme-Heat-Report-IFRC-OCHA-2022.pdf (last accessed 3 Sep 2023).

Radl, C., M. Müller, S. Revilla-Fernandez, S. Karner-Zuser, A. de Martin, U. Schauer, F. Karner, G. Stanek, P. Balcke & A. Hallas (2011) Outbreak of leptospirosis among triathlon participants in Langau, Austria, 2010. *Wiener klinische Wochenschrift*, 123, 23–24.

Ratcliffe, R. (2024a) 'Inside an oven': sweltering heat ravages crops and takes lives in south-east Asia. https://www.theguardian.com/environment/article/2024/may/04/inside-an-oven-how-life-in-south-east-asia-is-a-struggle-amid-sweltering-heat (last accessed May 5 2024).

—— (2024b) Schools close and crops wither as 'historic' heatwave hits south-east Asia. https://www.theguardian.com/environment/2024/apr/04/schools-close-and-crops-wither-as-historic-heatwave-hits-south-east-asia (last accessed May 5 2024).

—— (2024c) Wave of exceptionally hot weather scorches south and south-east Asia. https://www.theguardian.com/world/2024/apr/26/asia-heatwaves-philippines-bangladesh-india (last accessed May 5 2024).

Ritchie, H. & M. Roser (2021) Forests and Deforestation. Our World in Data. https://ourworldindata.org/deforestation.

RIVM (2023) Rioolwateronderzoek. https://www.rivm.nl/rioolwateronderzoek (last accessed Sep 3 2023).

Robine, J.-M., S. L. K. Cheung, S. Le Roy, H. Van Oyen, C. Griffiths, J.-P. Michel & F. R. Herrmann (2008) Death toll exceeded 70,000 in Europe during the summer of 2003. *Comptes Rendus Biologies*, 331, 171–178.

Rohr, J. R., A. Sack, S. Bakhoum, C. B. Barrett, D. Lopez-Carr, A. J. Chamberlin, D. J. Civitello, C. Diatta, M. J. Doruska & G. A. De Leo (2023) A planetary health innovation for disease, food and water challenges in Africa. *Nature*, 619, 782–787.

Rojanasakul, M., C. Flavelle, B. Migliozzi & E. Murray (2023) America is Using Up Its Groundwater Like There's No Tomorrow: Overuse is Draining and Damaging Aquifers Nationwide, a New York Times Data Investigation Revealed. https://www.nytimes.com/interactive/2023/08/28/climate/groundwater-drying-climate-change.html (last accessed Sep 3 2023).

Rosado, J. L., D. Ronquillo, K. Kordas, O. Rojas, J. Alatorre, P. Lopez, G. Garcia-Vargas, M. del Carmen Caamaño, M. E. Cebrián & R. J. Stoltzfus (2007) Arsenic exposure and cognitive performance in Mexican schoolchildren. *Environmental Health Perspectives*, 115, 1371–1375.

Rose, S. (2021) 'Our Biggest Challenge? Lack of Imagination': The Scientists Turning the Desert Green. https://www.theguardian.com/environment/2021/mar/20/our-biggest-challenge-lack-of-imagination-the-scientists-turning-the-desert-green (last accessed 9 Sep 2021).

SA, U. H. (2023) Noise Pollution: Mapping the Health Impacts of Transportation Noise in England. *Reducing the Burden of Disease.* https://ukhsa.blog.gov.uk/2023/06/29/noise-pollution-mapping-the-health-impacts-of-transportation-noise-in-england/ (last accessed 3 Sep 2023).

Safford, H. R., K. Shapiro & H. N. Bischel (2022) Wastewater analysis can be a powerful public health tool-if it's done sensibly. *Proceedings of the National Academy of Sciences,* 119, e2119600119.

Sharma, A., G. Andhikaputra & Y.-C. Wang (2022) Heatwaves in South Asia: Characterization, consequences on human health, and adaptation strategies. *Atmosphere,* 13, 734.

Sheehan, M. C. (2022) 2021 climate and health review-uncharted territory: Extreme weather events and morbidity. *International Journal of Health Services,* 52, 189–200.

Shepherd, G. (2017) Sand and Dust Storms: Subduing a Global Phenomenon-Frontiers 2017: Emerging Issues of Environmental Concern. https://api.semanticscholar.org/CorpusID:263480086.

Sims, N. & B. Kasprzyk-Hordern (2020) Future perspectives of wastewater-based epidemiology: Monitoring infectious disease spread and resistance to the community level. *Environment International,* 139, 105689.

Slette, I. J., A. K. Post, M. Awad, T. Even, A. Punzalan, S. Williams, M. D. Smith & A. K. Knapp (2019) How ecologists define drought, and why we should do better. *Global Change Biology,* 25, 3193–3200.

Sznajder, K., T. Eshak, A. Biney, N. Dodoo, M. Wang, T. Toprah, J. I. Blanford, L. Jensen, F. & N-A. Dodoo (2024) Adverse childhood experiences are associated with HIV risk factors in Agbogbloshie, Ghana. *Vulnerable Children and Youth Studies.*

Still, C., R. Powell, D. Aubrecht, Y. Kim, B. Helliker, D. Roberts, A. D. Richardson & M. Goulden (2019) Thermal imaging in plant and ecosystem ecology: Applications and challenges. *Ecosphere,* 10, e02768.

Stoddard, S. T., A. C. Morrison, G. M. Vazquez-Prokopec, V. Paz Soldan, T. J. Kochel, U. Kitron, J. P. Elder & T. W. Scott (2009) The role of human movement in the transmission of vector-borne pathogens. *PLoS Neglected Tropical Diseases,* 3, e481.

Talukder, B., R. Matthew, G. W. vanLoon, M. J. Bunch, K. W. Hipel & J. Orbinski (2021) Melting of Himalayan glaciers and planetary health. *Current Opinion in Environmental Sustainability,* 50, 98–108.

theGFD The Global Flood Database. https://global-flood-database.cloudtostreet.ai/#interactive-map (last accessed Sep 3 2023).

Thompson, V., A. T. Kennedy-Asser, E. Vosper, Y. T. E. Lo, C. Huntingford, O. Andrews, M. Collins, G. C. Hegerl & D. Mitchell (2022) The 2021 western North America heat wave among the most extreme events ever recorded globally. *Science Advances,* 8, eabm6860.

Thomson, M. C. & S. J. Mason (2018) *Climate Information for Public Health Action.* London, UK: Taylor & Francis.

Thomson, M. C., D. Grace, R. S. DeFries, C. J. E. Metcalf, H. Nissan & A. Giannini (2018) *Climate Impacts on Disasters, Infectious Diseases and Nutrition.* London, UK: Taylor & Francis.

Tomaszewski, B., J. I. Blanford, K. Ross, S. Pezanowski & A. M. MacEachren (2011) Supporting geographically-aware webdocument foraging and sensemaking. *Computers, Environment and Urban Systems,* 35, 192–207.

Tomlinson, C. J., L. Chapman, J. E. Thornes & C. J. Baker (2011) Including the urban heat island in spatial heat health risk assessment strategies: A case study for Birmingham, UK. *International Journal of Health Geographics,* 10, 42.

Toreti, A., D. Bavera, J. Acosta Navarro, C. Arias Muñoz, P. Barbosa, A. de Jager, C. Di Ciollo, G. Fioravanti, S. Grimaldi, A. Hrast Essenfelder, W. Maetens, D. Magni, D. Masante, M. Mazzeschi, N. McCormick & P. Salamon (2023) Drought in the Western Mediterranean May 2023. https://publications.jrc.ec.europa.eu/repository/handle/JRC134081 (last accessed 3 Sep 2023).

Tozer, S. J., S. B. Lambert, C. L. Strong, H. E. Field, T. P. Sloots & M. D. Nissen (2014) Potential animal and environmental sources of q fever infection for humans in queensland. *Zoonoses and Public Health*, 61, 105–112.

Tracy, B. L., D. Krewski, J. Chen, J. M. Zielinski, K. P. Brand & D. Meyerhof (2006) Assessment and management of residential radon health risks: A report from the health Canada radon workshop. *Journal of Toxicology and Environmental Health, Part A*, 69, 735–758.

Tripathy, K. P. & A. K. Mishra (2023) How unusual Is the 2022 European compound drought and heatwave event? *Geophysical Research Letters*, 50, e2023GL105453.

Tseng, C.-H., C.-P. Tseng, H.-Y. Chiou, Y.-M. Hsueh, C.-K. Chong & C.-J. Chen (2002) Epidemiologic evidence of diabetogenic effect of arsenic. *Toxicology Letters*, 133, 69–76.

Tye, S. & J. Waslander (2021) *Mainstreaming Climate Adaptation Planning and Action into Health Systems in Fiji, Ghana, and Benin.* Washington, DC: World Resources Institute.

UKMetOffice (2023) What is a Heatwave?. https://www.metoffice.gov.uk/weather/learn-about/weather/types-of-weather/temperature/heatwave (last accessed Sep 8 2023).

Ullah, S., Q. You, D. Chen, D. A. Sachindra, A. AghaKouchak, S. Kang, M. Li, P. Zhai & W. Ullah (2022) Future population exposure to daytime and nighttime heat waves in South Asia. *Earth's Future*, 10, e2021EF002511.

UN (2021) The United Nations World Water Development Report 2021: Valuing Water, 206 p. https://unesdoc.unesco.org/ark:/48223/pf0000375724 (last accessed Aug 5 2023).

UNEP, F. A. O. (2020) *The State of the World's Forests 2020. Forests, Biodiversity and People.* Roma (Itay): FAO.

Van Loon, A. F., S. W. Ploum, J. Parajka, A. K. Fleig, E. Garnier, G. Laaha & H. A. J. Van Lanen (2014) Hydrological drought typology: Temperature-related drought types and associated societal impacts. *Hydrology & Earth System Sciences Discussions*, 11, 3–14.

Wallace-Wells, D. (2021) Ten Million a Year, 43 p. https://www.lrb.co.uk/the-paper/v43/n23/david-wallace-wells/ten-million-a-year (last accessed 25 Sep 2022).

Wang, F., X. Zhao, C. Gerlein-Safdi, Y. Mu, D. Wang & Q. Lu (2017) Global sources, emissions, transport and deposition of dust and sand and their effects on the climate and environment: A review. *Frontiers of Environmental Science & Engineering*, 11, 1–9.

Wasserman, G. A., X. Liu, F. Parvez, H. Ahsan, P. Factor-Litvak, A. van Geen, V. Slavkovich, N. J. Lolacono, Z. Cheng & I. Hussain (2004) Water arsenic exposure and children's intellectual function in Araihazar, Bangladesh. Environmental Health Perspectives, 112, 1329–1333.

Watts, N., M. Amann, N. Arnell, S. Ayeb-Karlsson, J. Beagley, K. Belesova, M. Boykoff, P. Byass, W. Cai & D. Campbell-Lendrum (2021) The 2020 report of the lancet countdown on health and climate change: Responding to converging crises. *The Lancet*, 397, 129–170.

West, S. E., P. Büker, M. Ashmore, G. Njoroge, N. Welden, C. Muhoza, P. Osano, J. Makau, P. Njoroge & W. Apondo (2020) Particulate matter pollution in an informal settlement in Nairobi: Using citizen science to make the invisible visible. *Applied Geography*, 114, 102133.

WHO, Air Quality and Health. https://www.who.int/teams/environment-climate-change-and-health/air-quality-and-health/health-impacts/types-of-pollutants (last accessed 3 Sep 2023).

— (2000). *Air Quality Guidelines for Europe.* Geneva, Switzerland: World Health Organization. Regional Office for Europe.

— 2010. *WHO Guidelines for Indoor Air Quality: Selected Pollutants.* Geneva, Switzerland: World Health Organization. Regional Office for Europe.

— (2015) *Connecting Global Priorities: Biodiversity and Human Health: A State of Knowledge Review.* Geneva, Switzerland: World Health Organization (WHO).

— (2016) Infographic of How the Environment Impacts Our Health. Geneva, Switzerland: World Health Organization. Regional Office for Europe.

— 2021. *WHO Global Air Quality Guidelines: Particulate Matter (PM2. 5 and PM10), Ozone, Nitrogen Dioxide, Sulfur Dioxide and Carbon Monoxide.* Geneva, Switzerland: World Health Organization.

— (2023) Climate Change and Health. https://www.who.int/news-room/fact-sheets/detail/climate-change-and-health (last accessed May 5 2024).

WHO (2007) Global Surveillance, Prevention and Control of Chronic Respiratory Diseases: A Comprehensive Approach. In *Global Surveillance, Prevention and Control of Chronic Respiratory Diseases: A Comprehensive Approach.* Geneva, Switzerland: World Health Organization (WHO), pp. 7–146.

Wilhite, D. A. & M. H. Glantz (1985) Understanding: The drought phenomenon: The role of definitions. *Water International*, 10, 111–120.

Wisner, B., P. Blaikie, T. Cannon & I. Davis (2003) The disaster pressure and release model. *At Risk: Natural Hazards, People's Vulnerability, and Disasters*, London, Routledge, pp 496.

Wynwood, S. J., G. C. Graham, S. L. Weier, T. A. Collet, D. B. McKay & S. B. Craig (2014) Leptospirosis from water sources. *Pathogens and Global Health*, 108, 334–338.

Yevjevich, V., W. A. Hall & J. D. Salas (1978) Drought Research Needs, Water Resources. In *Drought Research Needs, Water Resources Publications.* Fort Collins, CO.

Yu, H., M. Chin, T. Yuan, H. Bian, L. A. Remer, J. M. Prospero, A. Omar, D. Winker, Y. Yang & Y. Zhang (2015) The fertilizing role of African dust in the Amazon rainforest: A first multiyear assessment based on data from cloud-aerosol lidar and infrared pathfinder satellite observations. *Geophysical Research Letters*, 42, 1984–1991.

Zachariah, M. et al., (2023) *Environmental Research:* Climate, 2, 045005.

Zipprich, J., K. Winter, J. Hacker, D. Xia, J. Watt & K. Harriman (2015) Measles outbreak-California, December 2014–February 2015. *Morbidity and Mortality Weekly Report*, 64, 153.

Zuberbier, T., J. Lötvall, S. Simoens, S. V. Subramanian & M. K. Church (2014) Economic burden of inadequate management of allergic diseases in the European Union: A GA2LEN review. *Allergy*, 69, 1275–1279.

Zürcher, K., M. Ballif, M. Zwahlen, H. L. Rieder, M. Egger & L. Fenner (2016) Tuberculosis mortality and living conditions in Bern, Switzerland, 1856–1950. *PLoS One*, 11, e0149195.

Index